METHODS IN MOLECULAR BIOLOGY

Series Editor
John M. Walker
School of Life and Medical Sciences
University of Hertfordshire
Hatfield, Hertfordshire, UK

For further volumes:
http://www.springer.com/series/7651

For over 35 years, biological scientists have come to rely on the research protocols and methodologies in the critically acclaimed *Methods in Molecular Biology* series. The series was the first to introduce the step-by-step protocols approach that has become the standard in all biomedical protocol publishing. Each protocol is provided in readily-reproducible step-by step fashion, opening with an introductory overview, a list of the materials and reagents needed to complete the experiment, and followed by a detailed procedure that is supported with a helpful notes section offering tips and tricks of the trade as well as troubleshooting advice. These hallmark features were introduced by series editor Dr. John Walker and constitute the key ingredient in each and every volume of the *Methods in Molecular Biology* series. Tested and trusted, comprehensive and reliable, all protocols from the series are indexed in PubMed.

Cell Cycle Control

Methods and Protocols

Edited by

Anna Castro

Université de Montpellier, Centre de Recherche en Biologie cellulaire de Montpellier (CRBM), CNRS UMR 5237, Montpellier Cedex 5, France; Programme équipes Labellisées Ligue Contre le Cancer, Paris, France

Benjamin Lacroix

Université de Montpellier, Centre de Recherche en Biologie cellulaire de Montpellier (CRBM), CNRS UMR 5237, Montpellier Cedex 5, France

Humana Press

Editors
Anna Castro
Université de Montpellier, Centre de Recherche en Biologie cellulaire de Montpellier (CRBM)
CNRS UMR 5237
Montpellier Cedex 5, France

Programme équipes Labellisées
Ligue Contre le Cancer
Paris, France

Benjamin Lacroix
Université de Montpellier, Centre de Recherche en Biologie cellulaire de Montpellier (CRBM)
CNRS UMR 5237
Montpellier Cedex 5, France

ISSN 1064-3745 ISSN 1940-6029 (electronic)
Methods in Molecular Biology
ISBN 978-1-0716-3556-8 ISBN 978-1-0716-3557-5 (eBook)
https://doi.org/10.1007/978-1-0716-3557-5

This Humana imprint is published by the registered company Springer Science+Business Media, LLC, part of Springer Nature.
The registered company address is: 1 New York Plaza, New York, NY 10004, U.S.A.

Paper in this product is recyclable.

Preface

Cell division is a crucial process by which one cell divides into two genetically identical daughter cells. During this process, the genetic information has to be appropriately duplicated and segregated to ensure the correct transmission of the genetic information to the next generation. Because cell viability and therefore the sustainability of every form of life depends on the correct duplication and partition of the DNA, the cell cycle is a highly regulated process.

Cell proliferation is under the control of fine-tuned molecular cascades that ensure correct chromatin duplication even under damaged DNA. Cycling cells are undergoing crucial sequences of molecular and physical events including chromosomes duplication and condensation, nuclear envelope reorganization, and assembly and disassembly of the mitotic structures. Likewise, a tight regulation of protein phosphorylation/dephosphorylation during mitotic division is required to timely and spatially orchestrate all these processes. In this book, we put together several chapters dedicated to investigating these processes and assessing how cells respond when these complicated pathways are simplified by using synthetic biology and in vitro reconstitutions.

Besides the regular molecular mechanisms required to progress into the cell cycle, cells ensure a safe division by additionally sensing and responding to environmental conditions. Physical constraints and metabolic stress impact on cell cycle regulation. This is a new essential and evolving research area for which we also decided to collect some protocol chapters that will help the readers to address this field.

In the last part of the book, we report some new protocols in different model systems and cellular types to visualize cellular architecture during cell division, and we emphasize innovative single cell microscopy techniques to highlight the heterogeneity of the cell population with respect to cell cycle progression. Our book finally provides a computer-assisted protocol dedicated to the analysis and quantification of the cell cycle.

We would like to thank all the authors who contributed to this protocol book by sharing their protocols supported by key advice for their development and practical application. We are also indebted to the Editor-in-Chief of Methods in Molecular Biology, Dr. John Walker, for his great support and guidance.

Montpellier, France

Anna Castro
Benjamin Lacroix

Contents

Contributors

ANTOINE AZE • *Genome Surveillance and Stability Laboratory, Institute of Human Genetics, UMR9002, CNRS-University of Montpellier, Montpellier, France*

LUCIE BARBIER • *Center for Interdisciplinary Research in Biology (CIRB), Collège de France, CNRS, INSERM, Université PSL, Paris, France*

ANNE-MARIE BERISHA • *Univ Rennes, CNRS, INSERM, IGDR (Institut de Génétique et Dé veloppement de Rennes)-UMR6290-U1305, Rennes, France*

JAN BRUGUÉS • *Max Planck Institute of Molecular Cell Biology and Genetics, Dresden, Germany; Max Planck Institute for the Physics of Complex Systems, Dresden, Germany; Center for Systems Biology Dresden, Dresden, Germany; Physics of Life, TU Dresden, Dresden, Germany*

ROSE BULTEAU • *Université Paris-Saclay, Univ Evry, CNRS, LAMBE, Paris, France; Center for Interdisciplinary Research in Biology (CIRB), Collège de France, CNRS, INSERM, Université PSL, Paris, France*

SÉBASTIEN CACCHIA • *Université de Montpellier, Centre de Recherche en Biologie cellulaire de Montpellier (CRBM), CNRS UMR 5237, Montpellier Cedex 5, France*

CLÉMENT CAMPILLO • *Université Paris-Saclay, Univ Evry, CNRS, LAMBE, Paris, France; Institut Universitaire de France (IUF), Paris, France*

STEFANIA CASTAGNETTI • *Sorbonne Universités, CNRS, Laboratoire de Biologie du Dé veloppement de Villefranche-sur-mer (LBDV), Villefranche-sur-mer, France*

ANNA CASTRO • *Université de Montpellier, Centre de Recherche en Biologie cellulaire de Montpellier (CRBM), CNRS UMR 5237, Montpellier Cedex 5, France; Programme é quipes Labellisées Ligue Contre le Cancer, Paris, France*

JANET CHENEVERT • *Sorbonne Universités, CNRS, Laboratoire de Biologie du Développement de Villefranche-sur-mer (LBDV), Villefranche-sur-mer, France*

XIANRUI CHENG • *Department of Biological Sciences, University of Southern California, Los Angeles, CA, USA*

DAMIEN COUDREUSE • *Institute of Genetics and Development of Rennes, CNRS UMR 6290 and University of Rennes, Rennes, France; Institute of Biochemistry and Cellular Genetics, CNRS UMR 5095 and University of Bordeaux, Bordeaux, France*

JULIEN DUMONT • *CNRS, Institut Jacques Monod, Université Paris Cité, Paris, France*

GREGORY EOT-HOULLIER • *Univ Rennes, CNRS, INSERM, IGDR (Institut de Génétique et Développement de Rennes)-UMR6290-U1305, Rennes, France*

JAMES E. FERRELL JR. • *Department of Chemical and Systems Biology, Stanford University School of Medicine, Stanford, CA, USA; Department of Biochemistry, Stanford University School of Medicine, Stanford, CA, USA*

YANNICK GACHET • *MCD, Centre de Biologie Intégrative, Université de Toulouse, CNRS, UPS, Toulouse Cedex, France*

GILLES GADEA • *INSERM U1194, IRCM, Institut de Recherche en Cancérologie de Montpellier, Institut régional du Cancer de Montpellier, Université de Montpellier, Montpellier, France*

ROMAIN GIBEAUX • *Univ Rennes, CNRS, IGDR (Institut de Génétique et Développement de Rennes) – UMR 6290, Rennes, France*

Régis Giet • *Univ Rennes, CNRS, INSERM, IGDR (Institut de Génétique et Dé veloppement de Rennes)-UMR6290-U1305, Rennes, France*

Stefan Golfier • *Max Planck Institute of Molecular Cell Biology and Genetics, Dresden, Germany; Max Planck Institute for the Physics of Complex Systems, Dresden, Germany; Center for Systems Biology Dresden, Dresden, Germany; B CUBE, Center for Molecular Bioengineering, TU Dresden, Dresden, Germany*

Gabriel Guilloux • *Univ Rennes, CNRS, IGDR (Institut de Génétique et Développement de Rennes) – UMR 6290, Rennes, France*

Rebecca Heald • *Department of Molecular and Cell Biology, University of California, Berkeley, CA, USA*

William Y. C. Huang • *Department of Chemical and Systems Biology, Stanford University School of Medicine, Stanford, CA, USA*

James R. A. Hutchins • *Genome Surveillance and Stability Laboratory, Institute of Human Genetics, UMR9002, CNRS-University of Montpellier, Montpellier, France*

Safia El Jailani • *Université Paris Cité, CNRS, Institut Jacques Monod, Paris, France*

Akanksha Jain • *Institute of Genetics and Development of Rennes, CNRS UMR 6290 and University of Rennes, Rennes, France; Institute of Biochemistry and Cellular Genetics, CNRS UMR 5095 and University of Bordeaux, Bordeaux, France*

Katarzyna M. Kedziora • *Department of Cell Biology, Center for Biologic Imaging (CBI), University of Pittsburgh, Pittsburgh, PA, USA*

Maiko Kitaoka • *Department of Molecular and Cell Biology, University of California, Berkeley, CA, USA; Whitehead Institute of Biomedical Research and Howard Hughes Medical Institute, Cambridge, MA, USA*

Dorothee C Kommer • *College of Health, Medicine and Life Science, Brunel University London, London, UK*

Elsa Labrune • *Hospices Civils de Lyon, Service de Médecine de la Reproduction, Bron, France; Faculté de Médecine, Université Claude Bernard Lyon 1, Lyon, France; INSERM U1208, Stem Cells and Brain Institute, Bron, France*

Benjamin Lacroix • *Université de Montpellier, Centre de Recherche en Biologie cellulaire de Montpellier (CRBM), CNRS UMR 5237, Montpellier Cedex 5, France*

Guillaume Lamour • *Université Paris-Saclay, Univ Evry, CNRS, LAMBE, Paris, France*

Daniel L. Levy • *Department of Molecular Biology, University of Wyoming, Laramie, WY, USA*

Tong Li • *MCD, Centre de Biologie Intégrative, Université de Toulouse, CNRS, UPS, Toulouse Cedex, France; Wellcome Sanger Institute, Cambridge, UK*

Laetitia K Linares • *INSERM U1194, IRCM, Institut de Recherche en Cancérologie de Montpellier, Institut régional du Cancer de Montpellier, Université de Montpellier, Montpellier, France*

Thierry Lorca • *Université de Montpellier, Centre de Recherche en Biologie cellulaire de Montpellier (CRBM), CNRS UMR 5237, Montpellier Cedex 5, France*

Domenico Maiorano • *Genome Surveillance and Stability Laboratory, Institute of Human Genetics, UMR9002, CNRS-University of Montpellier, Montpellier, France*

Marcos Malumbres • *Cell Division and Cancer group, Spanish National Cancer Research Center (CNIO), Madrid, Spain; Cancer Cell Cycle group, Vall d'Hebron Institute of Oncology (VHIO), Barcelona, Spain; ICREA, Barcelona, Spain*

Alex McDougall • *Sorbonne Universités, CNRS, Laboratoire de Biologie du Développement de Villefranche-sur-mer (LBDV), Villefranche-sur-mer, France*

Nicolas Minc • *CNRS, Institut Jacques Monod, Université Paris Cité, Paris, France; Equipe Labellisée LIGUE Contre le Cancer, Paris, France*

BATOOL OSSAREH-NAZARI • *Université Paris Cité, CNRS, Institut Jacques Monod, Paris, France; Programme Equipe Labellisée Ligue Contre le Cancer, Paris, France*
LIONEL PINTARD • *Université Paris Cité, CNRS, Institut Jacques Monod, Paris, France; Programme Equipe Labellisée Ligue Contre le Cancer, Paris, France*
TRISTAN PIOLOT • *Center for Interdisciplinary Research in Biology (CIRB), Collège de France, CNRS, INSERM, Université PSL, Paris, France*
THOMAS QUAIL • *Max Planck Institute of Molecular Cell Biology and Genetics, Dresden, Germany; Max Planck Institute for the Physics of Complex Systems, Dresden, Germany; Center for Systems Biology Dresden, Dresden, Germany; EMBL Heidelberg, Cell Biology and Biophysics Unit, Heidelberg, Germany*
BLANCHE RIQUIER-MORCANT • *INSERM U1194, IRCM, Institut de Recherche en Cancé rologie de Montpellier, Institut régional du Cancer de Montpellier, Université de Montpellier, Montpellier, France*
ROMAIN RISCAL • *INSERM U1194, IRCM, Institut de Recherche en Cancérologie de Montpellier, Institut régional du Cancer de Montpellier, Université de Montpellier, Montpellier, France*
MORGANE L. V. ROBERT • *Université de Montpellier, Centre de Recherche en Biologie cellulaire de Montpellier (CRBM), CNRS UMR 5237, Montpellier Cedex 5, France*
JÉRÉMY SALLÉ • *CNRS, Institut Jacques Monod, Université Paris Cité, Paris, France; Equipe Labellisée LIGUE Contre le Cancer, Paris, France*
WAYNE STALLAERT • *Department of Computational and Systems Biology, UPMC Hillman Cancer Center, University of Pittsburgh, Pittsburgh, PA, USA*
KONSTANTINOS STAMATIOU • *College of Health, Medicine and Life Science, Brunel University London, London, UK*
MARIE-EMILIE TERRET • *Center for Interdisciplinary Research in Biology (CIRB), Collège de France, CNRS, INSERM, Université PSL, Paris, France*
SANDRA A. TOUATI • *Université Paris Cité, CNRS, Institut Jacques Monod, Paris, France*
SYLVIE TOURNIER • *MCD, Centre de Biologie Intégrative, Université de Toulouse, CNRS, UPS, Toulouse Cedex, France*
PAOLA VAGNARELLI • *College of Health, Medicine and Life Science, Brunel University London, London, UK*
DIANA VARA-CIRUELOS • *Cell Division and Cancer group, Spanish National Cancer Research Center (CNIO), Madrid, Spain*
GRISELDA VELEZ-AGUILERA • *Université Paris Cité, CNRS, Institut Jacques Monod, Paris, France; Programme Equipe Labellisée Ligue Contre le Cancer, Paris, France*
KATJA WASSMANN • *Université Paris Cité, CNRS, Institut Jacques Monod, Paris, France*
PEI-YUN JENNY WU • *Institute of Genetics and Development of Rennes, CNRS UMR 6290 and University of Rennes, Rennes, France; Institute of Biochemistry and Cellular Genetics, CNRS UMR 5095 and University of Bordeaux, Bordeaux, France*
JING XIE • *CNRS, Institut Jacques Monod, Université Paris Cité, Paris, France; Equipe Labellisée LIGUE Contre le Cancer, Paris, France*
HEE WON YANG • *Department of Pathology and Cell Biology, Columbia University Irving Medical Center, New York, NY, USA; Herbert Irving Comprehensive Cancer Center, Columbia University Irving Medical Center, New York, NY, USA*

Chapter 1

Single-Molecule Approaches to Study DNA Condensation

Stefan Golfier, Thomas Quail, and Jan Brugués

Abstract

Proteins drive genome compartmentalization across different length scales. While the identities of these proteins have been well-studied, the physical mechanisms that drive genome organization have remained largely elusive. Studying these mechanisms is challenging owing to a lack of methodologies to parametrize physical models in cellular contexts. Furthermore, because of the complex, entangled, and dense nature of chromatin, conventional live imaging approaches often lack the spatial resolution to dissect these principles. In this chapter, we will describe how to image the interactions of λ-DNA with proteins under purified and cytoplasmic conditions. First, we will outline how to prepare biotinylated DNA, functionalize coverslips with biotin-conjugated poly-ethylene glycol (PEG), and assemble DNA microchannels compatible for the imaging of protein–DNA interactions using total internal fluorescence microscopy. Then we will describe experimental methods to image protein–DNA interactions in vitro and DNA loop extrusion using *Xenopus laevis* egg extracts.

Key words Single-molecule biophysics, Genome organization, Loop extrusion, TIRF microscopy, Quantitative imaging, Lysate-based approaches, In vitro biochemistry

1 Introduction

The collective behaviour of chromatin-associated proteins and their interaction with DNA drives the organisation of the genome across nuclear length scales [1]. Single-molecule techniques provide an alternative to live imaging of protein dynamics in intact interphase nuclei or mitotic chromosomes. Methods such as single-molecule Förster resonance energy transfer (smFRET) that examine the interactions of purified recombinant proteins with short pieces of nucleic acid (<100 bp) have provided insights into the molecular mechanisms driving a range of biological processes [2], including DNA damage repair [3–5], DNA replication [6], transcription [7], and chromatin remodeling [8]. These experiments generate detailed information related to protein binding dynamics and how these quantities connect to the biochemistry of individual proteins [9]. Combining these assays with structural studies using

Anna Castro and Benjamin Lacroix (eds.), *Cell Cycle Control: Methods and Protocols*, Methods in Molecular Biology, vol. 2740, https://doi.org/10.1007/978-1-0716-3557-5_1,

techniques such as cryo-electron microscopy and cryo-electron tomography can disentangle structure–function relationships of specific proteins [10]. However, proteins collectively work together to perform functions in space and time in the cell nucleus, a feature that is difficult to capture with these assays. Another caveat is that discoveries from in vitro studies do not necessarily reflect the full scope of complexity in the cell nucleus and mechanisms relevant in physiological contexts.

To study how proteins interact with longer pieces of DNA, many groups have adopted optical tweezers-based approaches [11]. In these assays, manipulating the positions of optically trapped beads with long pieces of DNA attached to them gives quantitative information related to the forces at play. These experiments have laid the groundwork for much of our dynamic understanding of protein–DNA interactions [12, 13], and combining these force measurements with confocal fluorescence microscopy has yielded insights into biological mechanisms [14, 15]. Establishing these assays, however, is time-consuming, costly, and requires specialized expertise.

To address these problems, methodologies have been developed to visualize the interactions of proteins with long pieces of DNA (on the order of 50 kbp) attached to functionalized coverslips. These DNA carpet or DNA curtain assays have yielded direct imaging evidence for processes ranging from genome integrity to DNA loop extrusion [16–19]. While most DNA carpet-/curtain-based assays examine interactions between DNA and purified proteins, more physiological lysate-based approaches have also been adopted, enabling the visualization of DNA replication [20] and parental histone recycling[21]. Here, we describe how to image the interactions of λ-DNA with proteins under purified and cytoplasmic conditions.

This method has been widely adopted in the single-molecule biophysics field to provide dynamic measurements, direct imaging evidence, and mechanistic insight for diverse processes in the nucleus, including DNA loop extrusion, protein condensation, histone recycling, and DNA replication. This assay retains qualitatively similar features to combined optical tweezer/confocal microscopy measurements, though there are a number of advantages worth discussing. First, establishing this assay is comparatively easy, cheap, and does not require specialized optics knowledge. Second, these assays offer an advantage on statistics, as they allow for imaging between 10 and 15 full-length DNA molecules in parallel in one single field of view, in contrast to optical tweezer measurements that are, in general, performed one single DNA molecule at a time. Third, sampling many different conditions—such as protein concentrations or buffer conditions for in vitro experiments—can be performed without much technical difficulty owing to the design of the microchannels. And lastly, microfluidics workflows enable flexibility for designing novel sample preparation work flows for new experiments. The main advantage, on the other hand, for combined optical tweezer/confocal microscopy approaches is the sensitivity of the force measurements and the ability to move a single molecule of DNA between channels. Scanning for conditions using the carpet assay and measuring forces and step sizes using optical tweezers combines the strengths of both approaches.

Using these assays to image the interactions of labeled proteins with increasingly complex DNA templates, including positioned nucleosomes, mononucleosomes, and sequence-specific motifs, represents promising avenues for future studies. Combining these assays with proteins that are compatible with FRET to determine conformational states of proteins in different contexts will yield significant insights. A very attractive, far-in-the-distance goal is of course combining these measurements with structural biology approaches such as cryo-EM/ET to unravel protein structure–function relationships. Determining novel physical mechanisms of genome organization requires direct imaging evidence of proteins with chromatinized templates—and this assay represents a useful technique to achieve this.

2 Materials

2.1 DNA Biotinylation

1. λ-phage DNA from bacteriophage λ, 48,502 bp in storage buffer (10 mM Tris-HCl, pH 8.0, 1 mM EDTA) (NEB, N3011S).
2. DNA pol I large fragment Klenow polymerase with 5′ -> 3′ polymerase (NEB, M0210S).
3. 10× polymerase buffer NEB2 (supplied with Klenow): 500 mM Tris-HCl (pH 7.2 at 25 °C), 100 mM $MgSO_4$, 1 mM DTT (NEB, B7002S).
4. dNTPs: dTTP, dCTP, biotin-dATP, and biotin-dUTP (Jena Bioscience).
5. Ultrapure DNase-free water.
6. Heat block and freezer.
7. Pure ethanol and 3M sodium acetate stock solution for DNA extraction. Nanodrop to measure DNA concentration.

2.2 Cover Slip PEGylation

1. Safety goggles.
2. Diamond pen.
3. Ethanol.
4. Methanol.
5. Acetone.
6. Acetic acid.
7. Ultrapure DNase-free water.

8. (3-Aminopropyl)-trimethoxysilane (Sigma, 281778).
9. Potassium hydroxide.
10. Sodium bicarbonate.
11. Sonicator (Branson 2510).
12. Teflon tweezers.
13. Benchtop centrifuge and high-precision scale.
14. Teflon slide holder (similar to https://proscitech.com.au/products/wash-n-dry-coverslip-rack-pp).
15. Rectangular glass container (i.e., IKEA "Koloni") to contain the Teflon slide holder during the washes.
16. Pressurized nitrogen.
17. 5 kDa methoxy-PEG-NHS (RAPP Polymere, 125000-35).
18. 5 kDa biotin-PEG-NHS (RAPP Polymere, 135000-25-35).

2.3 Flow Channel Assembly and DNA Visualization

1. Filter paper (Whatmann, 1001-090).
2. Plastic petri dish 90 × 14 mm.
3. Valap slide sealant (Vaseline, lanoline, paraffin).
4. Channel washing buffer: 40 mM Tris-HCl pH 8.0, 40 mM NaCl, 0.4 mM EDTA.
5. Streptavidin (S4762, Sigma Aldrich).
6. SYTOX Orange (S11368, Thermo Fisher).
7. Microscope slides (Fisher, 1156-2203).
8. Mucasol.
9. Ethanol.
10. Pressurized nitrogen.
11. Double-sided tape (Tesa, 4970-25).
12. Acetone.
13. Teflon slide holder (similar to https://proscitech.com.au/products/wash-n-dry-coverslip-rack-pp).

2.4 Histone Immunodepletion of Cytoplasmic Xenopus Egg Extract

1. High binding affinity beads rProtein A Sepharose (GE HealthCare, Cytiva).
2. 130 μg of the mouse monoclonal anti-H4K12ac antibody for H3–H4 depletion of 50 μL cytoplasmic extract.
3. 130 μg random mouse IgG antibodies for the mock depletion (IgG from mouse polyclonal unconjugated, Jackson Immuno Research).
4. Antibody coupling buffer: 10 mM K-HEPES pH 8.0, 150 mM NaCl.
5. *Xenopus* buffer: 10 mM HEPES pH 7.7, 100 mM KCl, 150 mM Sucrose, 1 mM $MgCl_2$.

6. Titanium tweezers with sharp tips.
7. Metal Eppendorf rack and Styrofoam box to incubate on ice.

3 Methods

In this section, we describe methods to image the interactions of purified and labelled DNA-binding proteins or lysates with single molecules of DNA. As a dsDNA template, we use linearized λ-DNA (48.5 kbp in length). To attach biotin molecules to the ends of λ-DNA, we combine Klenow polymerase I fragment and biotinylated nucleotides to fill the 12-bp 5′ overhangs (Fig. 1a). We then immobilize λ-DNA to PEGylated glass coverslips using streptavidin and surface-bound PEG-biotin (Fig. 1b). On the glass surface, PEG-biotin is combined with 500 kDA PEG-CONH molecules to generate a chemically inert carpet that prevents random sticking of the DNA or protein to the glass surface. Flow channels are generated using double-sided tape to attach functionalized coverslips to cleaned glass slides, optionally with drilled holes, which facilitate rapid buffer exchange. We then use total internal reflection fluorescence microscopy (TIRF) to image dozens of individual single molecules of DNA in parallel per field of view (Fig. 2b), with various DNA extensions with high signal-to-noise and spatiotemporal resolution.

3.1 DNA Biotinylation

1. 10 μg λ-DNA (20 μL from the commercial 0.5 mg/mL stock) are added to 22 μL ddH_2O and 5 μL 10× NEB2 polymerase buffer into a 1.5 mL Eppendorf tube.

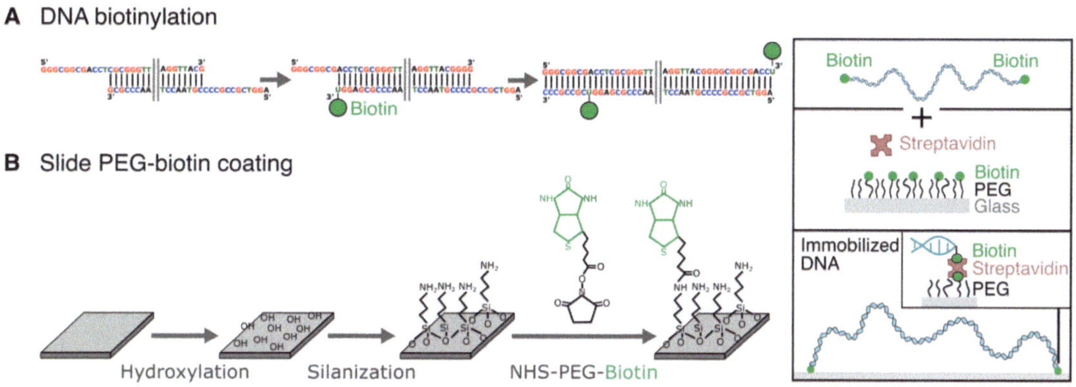

Fig. 1 DNA biotinylation and slide PEGylation. (**a**) Schematic representation of the 12 base pair 5′ overhang at the ends of the λ-phage DNA (left). Middle and right panels depict the incorporation of biotin and other dNTPs by Klenow polymerase I activity. (**b**) PEGylation of a glass coverslip. After hydroxylation by KOH or piranha treatment the glass surface is silanized using APTES, followed by coating with PEG-biotin. While inert CONH-PEG molecules (not shown) passivate the slide against unwanted sticking of proteins on the surface, about 5% of the PEG molecules contain a terminal biotin group (green), which allows the attachment of biotin DNA molecules through a streptavidin linker as shown in right panel

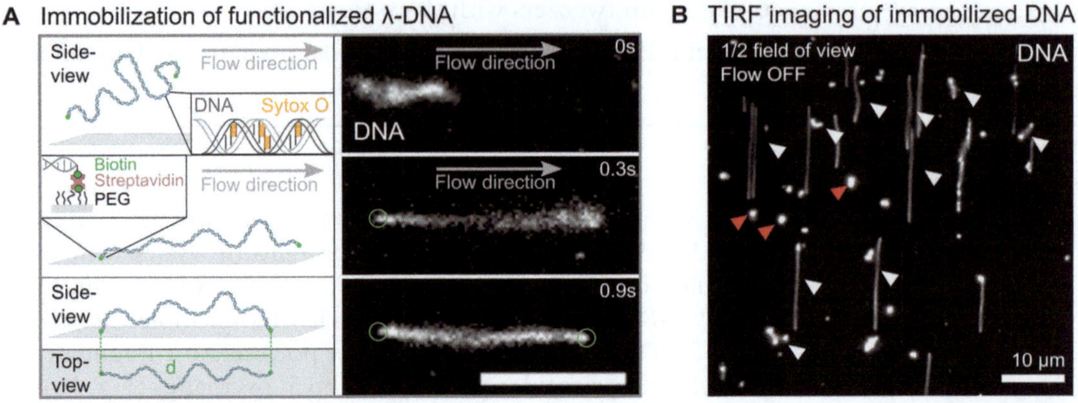

Fig. 2 TIRF imaging of individual λ-DNA molecules. (**a**) DNA immobilization. A suspension of biotinylated λ-DNA molecules is introduced into a custom-built microfluidic channel on a PEG-biotin–streptavidin decorated glass cover slip. Schematic drawings (left) and TIRF microscopy images (right) illustrate the steps of DNA immobilization. (Top) Free-floating DNA molecules visualized by SYTOX Orange, a very bright intercalating dye, are brought into close proximity to the cover slip surface through hydrodynamic flow. (Middle) Upon contact between one biotin-labeled DNA end and the streptavidin-decorated surface, DNA molecules become bound with one end to the cover slip (green circle). (Bottom) Stretching the immobilized DNA molecule by a continuous flow of buffer, the unbound end of the DNA molecule is brought into contact with the surface until a bond is established between the biotinylated DNA and the streptavidin-coated cover slip. Scale bar is 5 μm. (**b**) Typical field of view during TIRF microscopy at 150× magnification, showing several double-bound (white arrows) and single-bound (red arrow) DNA molecules

2. The tube is inserted into a heat block and heated at 65 °C for 5 min to break apart the sticky ends of the λ-DNA.
3. 0.5 μL of each of the 100 mM nucleotide stock (dCTP, biotin-dUTP, dTTP, and biotin-dATP) is added to the reaction, resulting in a 100× molar excess of dNTP to λ-DNA. Including both biotinylated dATP and biotinylated dUTP results in two biotins per DNA end, which we have found to improve binding stability of the DNA molecule to the surface; yet in principle, one biotin group at each end of the DNA molecule is sufficient for stable DNA immobilization.
4. 1 μL (5 U) of Klenow enzyme is added to the reaction which is then carefully mixed by pipetting up and down with a cut pipette tip to minimize shearing the λ-DNA molecules.
5. The reaction is incubated overnight at room temperature, yet the reaction reaches a plateau already after 1 h at 37 °C. (Note that the Klenow enzyme functions as a polymerase in the 5′ to 3′ direction, yet as an exonuclease in the 3′ to 5′ direction. Consequently, when functionalizing DNA templates other than λ-DNA, one should ensure that the overhangs are always on the 3′ end.)
6. To purify the biotinylated DNA, perform an ethanol precipitation (see next section).

3.2 Ethanol Precipitation

1. Add 1/10 volume of 3 M sodium acetate pH 5.5 to the reaction mixture.
2. Add 2 volumes of ice-cold pure ethanol (100 μL).
3. Mix by inverting the tube a couple of times (do not pipette up and down as precipitated DNA can get stuck at the tip).
4. Leave at −20 °C for 30 min or longer. In the meantime, begin cooling the centrifuge to 4 °C.
5. Spin at 13,000 rpm for 20 min at 4 °C.
6. Remove carefully the ethanol without touching the DNA pellet.
7. Wash with 1 mL 70% ethanol and spin again at 13,000 rpm for 5 min at room temperature.
8. Remove ethanol and let the pellet dry (5–10 min). Do not over-dry the pellet!
9. Functionalized DNA molecules are resuspended in TE buffer (10 mM Tris and 1 mM EDTA, pH 8) or nuclease-free water to a final concentration of roughly 250 ng/μL.
10. Measure DNA concentration using NanoDrop.
11. Functionalized DNA molecules can be stored in TE buffer—or in water—at 4 °C for up to a year or at −20 °C for longer periods of time, yet repeated freeze–thaw cycles should be avoided as these can damage the long DNA molecules.

3.3 Slide PEGylation

To allow specific binding of the biotinylated ends of the DNA molecules to the surface of a glass coverslip and to prevent nonspecific sticking of the DNA molecule to this surface, cover slips are covalently functionalized with PEG-CONH and PEG-biotin as shown in Fig. 2. This protocol does not use Piranha solution and uses an extended KOH etching step instead, as we found it sufficient for investigating DNA organisation in cell lysate and purified protein solutions. For studying single molecule – DNA interaction dynamics that require an extremely clean surface for sufficient signal-to-noise ratio, Piranha cleaning approaches might be required. In the following we describe the cleaning and preparation of PEG-biotin/PEG-CONH functionalized cover slips for DNA immobilisation.

1. To later discriminate the functionalized side of the slide from the non-functionalized side, scratch the letter "R" into the top right corner of each coverslip using a diamond pen. While scratching, place the coverslips on a piece of Kimwipe and wear safety goggles as the coverslip might break.
2. Place the coverslips into the Teflon slide holder, leaving one empty slot between each slide (in our case 16 slides per slide holder). This ensures that there is enough room to prevent the

slides from sticking together. Place the Teflon holder inside a glass container.

3. Coverslips are then cleaned by subsequent sonication in acetone for 5 min followed by 5 rinses with Milli-Q water and another 40 min sonication step in freshly prepared 5 M KOH.
4. Cover slips are then rinsed three times with Milli-Q water and three times with methanol and carefully dried with nitrogen.
5. For silanization, the cleaned coverslips are incubated in a mixture of 250 mL methanol, 12.5 mL acetic acid, and 5 mL (3-aminopropyl)-trimethoxy silane for 20 min at room temperature, interrupted by a brief 1 min sonication after 10 min.
6. Coverslips are then rinsed once with methanol, once with water, then once again with methanol, and dried again with nitrogen.
7. Using a high-precision scale, combine 98 mg of 5 kDa methoxy-PEG-NHS and 2 mg of the 5 kDa biotin-PEG-NHS in 450 μL of freshly made 0.1 M sodium bicarbonate buffer (pH = 8.5).
8. Remove air bubbles from the solution by briefly (1 min) spinning the solution at 13,000 RPM using a table-top centrifuge.
9. To PEGylate the slides, add a drop of 25 μL PEG mixture to one side a coverslip and then add another coverslip on top, forming a sandwich. Make sure that the sides marked with the scratched "R"—are facing towards each other.
10. To prevent drying of the slides, make a humidity chamber. For example, you can fill the bottom of an empty pipette tip box with distilled water and place the coverslips over the holes of the pipette tips and close the box.
11. Cover slips are then placed in the humidity chamber and incubated in a dark level place overnight.
12. The next day, cover slips are carefully disassembled by completely immersing them in 0.1 M sodium bicarbonate buffer (pH = 8.5) and carefully peeling them apart using Teflon tweezers. Avoid sheering the slides, as this might damage the PEG-carpet.
13. Slides are subsequently rinsed with Milli-Q water and dried with nitrogen.
14. Functionalized coverslips can be stored inside nitrogen-gas-filled 50 mL falcon tubes at −20 °C for several months.

3.4 Flow-Chamber Assembly and DNA Visualization

To allow for buffer exchange and hydrodynamic stretching of DNA, microfluidic channels are created by placing thin strips of double-sided tape over a cleaned glass slide with drilled holes for buffer exchange and placing a PEGylated coverslip on top, with the functionalized side facing down towards the glass slide. The channels are then sealed with Valap with the drilled holes in the microscope slide serving as inlets and outlets. Drilling through glass slides is delicate and requires an upright drill and ideally a dental drill bit of 1 mm diameter; thus, these slides are cleaned after each experiment and can, in principle, be used indefinitely.

1. Inlets and outlets are drilled through the microscope slide using a dental drill bit and an upright drill with a mount to hold the slide in place. Use oil to lubricate the drill or immerse the slide completely in water, which will cool the drill bit and prevent the scattering of glass shards. Apply subtle pressure and always wear safety goggles. For an alternative channel assembly approach without the need to drill holes into glass (*see* **Note 4**).
2. The drilled microscope slides are then cleaned by placing them into a Teflon rack inside a glass container followed by each 15 min sonication in 1:20 Mucasol, ethanol, and ddH_2O. Rinse the slides and glass container thoroughly with ddH_2O between each step. (This procedure yielded sufficiently clean slides that later served to close the DNA channels. However, proteins and small molecules will stick to the slide, which is not a concern for imaging, which will be performed on the immobilized DNA on the PEGylated coverslips. Yet if protein concentration is of concern, these slides can be passivated by PEGylation similar to the coverslips by following the aforementioned protocol with the modification of only using the 5 kDa methoxy-PEG-NHS (no biotin) to prevent DNA from sticking to the other side of the channel.)
3. Double-sided tape is cut into thin (approximately 3 mm wide) strips and placed on the cleaned glass slide generating channels of about 2 mm in width.
4. A functionalized PEG-biotinylated coverslip is placed on top of the double-sided tape with the functionalized side facing down.
5. The open channel ends are sealed with Valap and subsequently flushed with 10 μL of 0.1 mg/mL streptavidin (S4762, Sigma Aldrich) by carefully pipetting through the holes, followed by a brief (2 min) incubation.
6. Unbound streptavidin is removed from the channel by washing each channel with 100 μL channel washing buffer, using the drilled holes as channel inlets and outlets and a piece of thick filter paper as a sink for the buffer on the outlet.
7. Functionalized DNA is diluted 1:1000 (to about 0.25 ng/μL or 1–10 pM) in channel washing buffer and 20 μL are introduced per channel.

8. Following 45–60 s of incubation, each channel is then washed with 200 μL of channel washing buffer to remove unbound DNA molecules. The speed of buffer flow determines the final elongation of the immobilized DNA molecules (*see* Fig. 2). Do not generate flows that are too fast, but aim for a flow rate of about 200 μL per minute, so pipetting takes about 1min per channel. At this flow rate, most DNA molecules will be immobilized with both ends attached to the coverslip. Flowing too quickly will create many examples of stretched-out DNA pieces with not enough slack to facilitate DNA compaction, whereas flowing too slowly yields too few DNA molecules attached to the coverslip at both ends.
9. Finally, SYTOX Orange DNA dye is diluted to a final concentration between 5 and 50 nM in the desired buffer (usually in channel washing buffer).
10. Verify DNA fidelity using TIRF microscopy. If the number of double-bound DNA molecules or general DNA concentration is too low, the previous steps can be repeated until the desired number of double-bound DNA molecules per field of view (usually between 10 and 30, *see* Fig. 2) is reached or read further about troubleshooting steps in **Notes 1–2.**
11. Until the experiment is ready to proceed to imaging, channels should be kept in a dark petri dish under a moist atmosphere using wet filter paper to prevent evaporation. If you observe dye-induced DNA breaking at the beginning of the microscopy session, address this problem using the steps outlined in **Note 3.**

3.5 Reconstituting Protein–DNA Interactions in Native Xenopus laevis *Egg Extract*

Preparing DNA-coated slides as described earlier allows for the study of protein–DNA interactions using purified proteins or cell-lysate-based approaches. By examining protein–DNA interactions using purified components, we can interrogate specific proteins or enzymes in isolation, whereas by using cell extracts we can address the dynamics of DNA compaction in physiological contexts (Fig. 3).

In this assay, examining protein–DNA interactions using labeled recombinant proteins is straightforward, requiring the experimenter to find the appropriate concentration (typical range of a few nM) and the buffer condition within which the protein is stable [16, 17, 22]. Using *Xenopus laevis* egg extracts, however, is more difficult and often requires protein immunodepletions to interrogate phenotypes. That being said, such approaches combining frog egg extracts and single-molecule approaches are powerful at dissecting diverse biological mechanisms, including nucleosomal assembly [21, 23], DNA replication [24], chromatin compaction and remodeling [25–27], as well as DNA loop extrusion [28]. Additional advantages of frog egg extract are as follows:

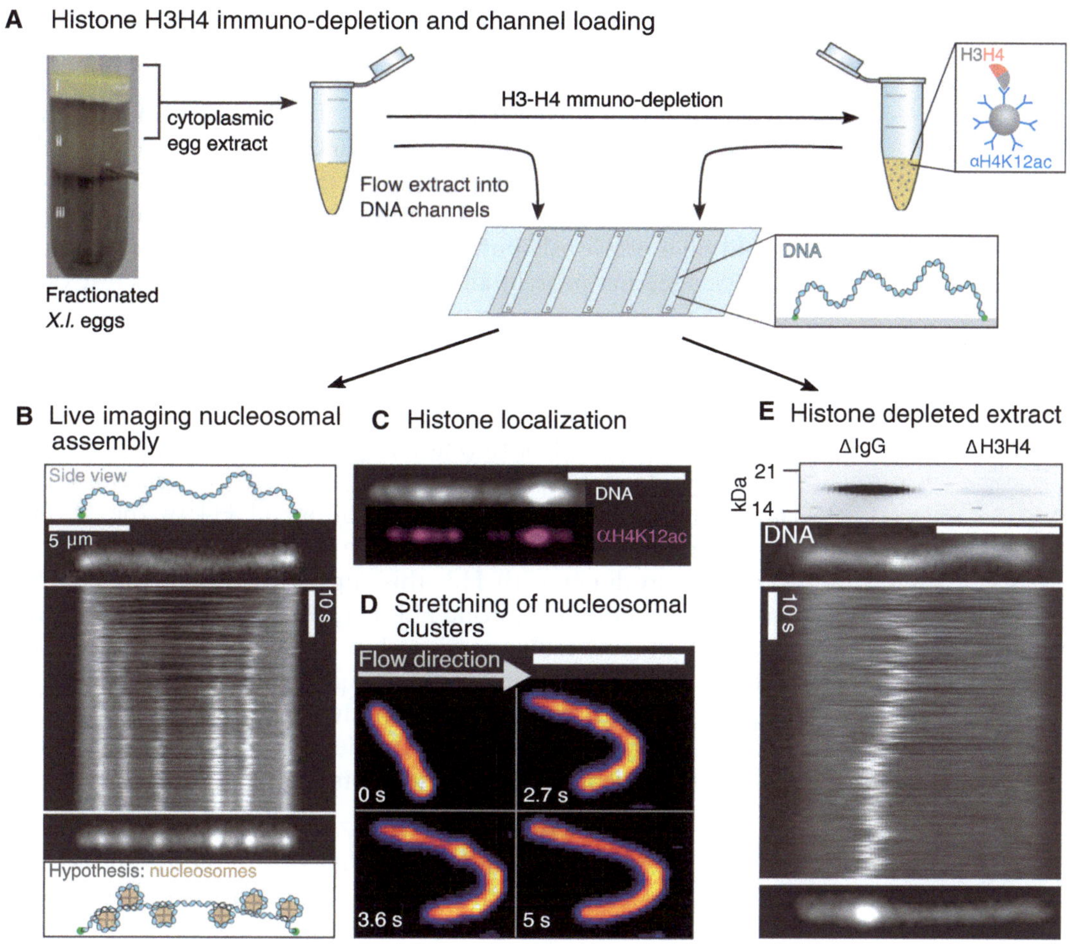

Fig. 3 Reconstitution of nucleosomal assembly and DNA loop extrusion in *Xenopus* egg extract. (**a**) Schematic representation of the workflow for introducing cytoplasmic *Xenopus* egg extract or immunodepleted (here histone H3 and H4 depleted) extract into the λ-DNA-coated channels. The former can be used to study nucleosomal assembly, whereas the histone H3–H4 depletion will prevent the formation of nucleosomes and allow the reconstitution of native DNA loop extrusion. (**b**) Nucleosomal assembly in undepleted cytoplasmic egg extract. (Top) Schematic drawing and microscopy image of an immobilized λ-DNA molecule in buffer prior to extract application. (Middle) Kymograph showing the compaction of slack in an immobilized λ-DNA molecule into numerous dense clusters upon application of egg extract. (Bottom) TIRF microscopy image of the final DNA configuration with schematics depicting hypothetical nucleosomal assembly, with each nucleosome in the schematics most likely representing a cluster of nucleosomes in the microscopy image. (**c**) (Top) Localization of histone H4 on a λ-DNA molecule that was previously exposed to egg extract, using a fluorescently labeled H4K12ac mouse monoclonal antibody. Scale-bar is 5 μm. (**d**) Stretching of possible nucleosomal clusters on λ-DNA, displaying stepwise unzipping of DNA clusters at constant hydrodynamical forces. A different colormap is used to achieve better visual contrast between clusters and nonclustered DNA. Scale-bar is 5 μm. (**e**) Reconstitution of native DNA loop extrusion on λ-DNA in histone-depleted egg extract. (top) Western blot of histone 3, showing 95% depletion of histone H3–H4 dimers from the cytoplasmic extract. TIRF microscopy images and Kymographs depicting DNA loop extrusion upon introduction of histone H3–H4 depleted extract into the DNA channels. Scale-bar is 5 μm

(1) the phase of the cell cycle is easily controlled, and (2) it is an open system and easily biochemically manipulable.

In this section, we will describe how to image DNA loop extrusion using *X. laevis* egg extract. To do this, we found that it is necessary to deplete the extract of H3–H4 heterodimers, which we will describe as a representative example of protein immunodepletion in metaphase-arrested *X. laevis* egg extract. This method is based on a previously published protocol [29]. To the best of our knowledge, any protein and protein complexes of interest can be depleted from the extract using a similar approach. Also, several proteins can be depleted in one reaction by coupling the appropriate amounts of antibodies onto the resin in one single reaction.

To deplete H3–H4 heterodimers from extract, we coupled mouse monoclonal antibodies targeting H4K12Ac (a generous gift from Hiroshi Kimura, Tokyo Institute of Technology) to high-binding affinity resin. Since almost all H4 in the *X. laevis* egg extract carry the acetylation mark on lysine 12 and exist in a stable heterodimer with H3, this approach allowed the complete immunodepletion of histone H3–H4, which inhibited nucleosome assembly as reported in previous literature [29]. It is important to provide a negative control for each experiment by performing mock depletions which replace the antibody against the protein of interest with random mouse IgG antibodies. The quality of depletions is assessed by quantitative Western blots.

1. To immunodeplete soluble H3–H4 heterodimers from an extract reaction of 50 μL, 130 μg of the mouse monoclonal anti-H4K12ac antibody is coupled to 12.5 μL rProtein A Sepharose (GE HealthCare) bead slurry in antibody coupling buffer inside a 0.6 mL Eppendorf tube.
2. Incubate overnight on a rotator at 4 °C at 20 rpm. (A faster way is to incubate the beads with the antibodies on a rotator at room temperature for one hour, which has given comparable results.)
3. Right before the final step in the extract preparation, the antibody-coupled beads are washed several times with *Xenopus* buffer.
4. Beads are then separated from the buffer by poking a small hole in the bottom of the Eppendorf tube using titanium tweezers with sharp tips. The hole must be small enough to retain the beads yet enable buffer flow-through.
5. Place the 0.6 mL Eppendorf tube containing the beads and buffer inside a 1.5 mL empty Eppendorf tube and briefly (30 s) centrifugate at 5000 rpm, separating buffer and beads, retaining the AB-coupled beads in the 0.6 mL tube.
6. Repeat the previous washing procedure two more times to completely remove unbound AB.

7. The beads are then resuspended in 50 μL fresh CSF extract and incubated for 1.5 h on ice, occasionally flicking the tubes in order to prevent the beads settling to the bottom.
8. Extract is recovered using the same method as extracting the beads from buffer (poking a small hole into the bottom of the Eppendorf tube) yet now retaining the flow-through which is the depleted cytoplasmic extract.
9. The depleted extract can be kept functional ideally for several hours, if stored on ice, yet we strongly suggest to immediately proceed with the experiment.
10. Before the end of the experiment, store a sample of at least 5 μL from each condition in the −20 °C freezer for later quantitative Western blot to assess the quality of the depletion.

3.6 Image Acquisition Using TIRF Microscopy

We visualized DNA molecules and protein–DNA interactions using total internal reflection fluorescence (TIRF) microscopy. TIRF microscopy uses an inclined excitation laser beam that allows for internal reflection at the glass–buffer interface, creating an evanescent light field that penetrates a few tens of nanometers into the sample and allows the study of surface-bound molecules with very low background signal. As the depth of excitation into the sample can be tuned with the angle of the excitation beam, this system is perfectly suited to visualize fluorescently labeled DNA molecules in proximity to the glass surface. Although we have tested other forms of microscopy, TIRF has proven to be most robust and give a good compromise in terms of signal-to-noise, spatial, and temporal resolution.

1. Before loading the protein of interest or cytosolic extract into the channel, the channel should be filled with the appropriate buffer containing an appropriate concentration of DNA dye. Note that the signal of many DNA intercalating dyes can change dramatically with salt or protein concentration, for more information (*see* **Note 5**). For our study, we used 5–20 nM SYTOX Orange in Xenopus buffer. For further analysis, consider that the DNA dye changes the contour length of the molecule in a concentration-dependent manner, (*see* **Note 6**).
2. Transfer the slide to an inverted fluorescence microscope with a TIRF unit. We used a commercial fluorescence microscope (Nikon Eclipse Ti-E) with the Nikon TIRF illuminator, a Nikon 100×/1.49NA oil SR Apo TIRF objective, an Andor iXon3 EMCCD camera and an appropriate filter set, as well as a Gemini beam-splitter unit for simultaneous dual-color imaging (optional).
3. SYTOX Orange-labeled DNA should be excited using a 561 nm laser with the lowest workable power to minimize

DNA damage. A typical frame rate we used was 10 fps at an EM gain of 200, yet the optimal frame rate depends on the DNA signal and temporal resolution needed for the experiment. Many protein–DNA interaction dynamics happen very fast and demand higher frame rates as in our example.

4. Initialize image acquisition using appropriate software (we use NIS-Elements, Nikon Microscopes) and adjust the TIRF angle such that it yields a decent signal-to-noise ratio for the entire DNA molecule. As molecules are fluctuating due to thermal motion, the amplitude of these fluctuations can lift parts of the DNA molecule a few μm above the slide, leading to dark spots that are too far away from the TIRF evanescent field to be excited. Consequently, the optimal TIRF angle that avoids this issue might be slightly above the critical TIRF angle. We found it best to work in a highly inclined and laminated optical sheet (HILO) configuration, which creates a few μm-thick light sheet parallel to the slide surface.
5. Verify DNA integrity and choose a channel with a high number of DNA molecules bound at both ends. For us, a typical field of view contained 5–20 individual DNA molecules tethered on both ends.
6. Introduce 1–2 μL of the (histone-depleted) extract into the selected channel by slowly (!!!) pipetting through one of the drilled openings of the channel. Usually, this volume of extract only fills about one-quarter to one-half of the channel and continues to diffuse further into the channel. Pay careful attention to not introduce air bubbles into the channel, as a large bubble can render the channel unusable.
7. As many DNA–protein interactions are rapid processes, the best chance to observe them is to follow the edge of the diffusing extract. The signal of DNA intercalating dyes decays rapidly inside the extract, which is likely due to the high concentration of proteins outcompeting the DNA dye.

4 Notes

In this section, we provide a collection of notes that address common challenges that we encountered setting up and optimizing our system. We encourage the reader to set up the system step-by-step and perform spot-checks using the microscope after each step, verifying that the procedure has worked. For the use of purified proteins, buffer conditions and protein concentrations are crucial and generally have to be determined individually for each protein and experimental condition.

1. Improving DNA Attachment and Integrity

When preparing channels, an important step is to check for successful DNA attachment and proper integrity of the double-bound molecules. If too few molecules are seen per field of view, the introduction of streptavidin and biotinylated DNA can be repeated. If still too few DNA molecules are observed, the surface functionalization may be compromised and we recommend preparing new channels on another functionalized coverslip.

If most DNA molecules are only bound on one end, we found that an additional washing step where the channel washing buffer is introduced at a flow rate of about 200–500 μl/min with frequent stops to be quite helpful. By periodically altering the flow rate, surface-bound DNA molecules are retracted and extended, increasing the chances of the unbound end of the DNA to bind to the surface.

2. DNA Sticking to the Slide or High Background Due to Protein Accumulation

Insufficient blocking of the glass surface is the most common reason underlying this common imaging problem. This can be addressed by improving the surface cleaning approach—through a piranha-based passivation strategy—or by further blocking the PEG carpet using other inert molecules such as BSA or short-stranded PEG. This can be achieved either by introducing a BSA-blocking buffer directly before the experiment (option 1) or adding an additional PEGylation step during the slide preparation (option 2).

Blocking option 1: blocking with BSA

- Prepare fresh blocking buffer (20 mM Tris pH 7.5, 50 mM NaCl, 2 mM EDTA, 0.2 mg/mL BSA) and filter it using a 0.2 μm Luer lock filter.
- Fill each channel with blocking buffer and incubate for at least 15 min at room temperature in a moist and dark environment.
- Wash each channel three times with appropriate buffer before the experiment.

Blocking option 2: blocking with short-stranded PEG

- Directly after the main PEGylation step, dissolve 7 μL of 250 mM MS4-PEG in 63 μL sodium bicarbonate buffer.
- Put 30 μL of this buffer on the PEGylated surface of the glass coverslip and make a sandwich with another PEGylated slide similar to the PEGylation procedure described earlier. Incubate in dark and humid environment for 30 min.

- Take slides apart under sodium bicarbonate buffer as described earlier, wash slides with ddH_2O, dry with nitrogen, and proceed to channel assembly.

Another less obvious reason for nonspecific sticking could be photon-induced radicals, which can damage DNA molecules but also create covalent bonds with the surface. Reducing the laser power or employing an oxygen-scavenging system can reduce DNA sticking (see next section). Finally, we noticed that, when introducing proteins or cytoplasmic extract into the channel, a slow flow rate really helped to prevent sticking of the molecules to the surface.

3. Dye-Induced DNA Breaking

Every fluorescent dye has a certain probability to convert to a triplet state, in which the excitation energy is not converted to fluorescence but can form radicals that damage the DNA molecule. This obviously depends on the concentration of dye and laser power used. Decreasing the laser power can be effective in reducing the DNA breaking rate yet may sacrifice signal. For that reason, a proper TIRF angle, good slide passivation to minimize background, and high NA objectives are crucial to achieve a good compromise between minimal laser power and workable signal-to-noise ratio.

Another possibility that dramatically increases DNA lifetime is the use of an oxygen-scavenging system that prevents radical-induced DNA breakage. The most commonly used is based on glucose oxidase and catalase, yet it produces gluconic acid that will reduce the pH of the system on time scales of minutes, if no appropriate buffer is used. In principle, this could be circumvented by using pyranose oxidase instead of glucose oxidase, yet we found this system less effective. Second, it is important to keep in mind that, as the name implies, this system reduces the concentration of oxygen in the buffer, which might be detrimental for living systems.

GODCAT Oxygen-Scavenging System

- Mix 1 μL glucose oxidase (10 mg/mL stock = 100 U/mL) or pyranose oxidase with 2 μL glucose (2 M stock) and 1 μL catalase (2 mg/mL stock).
- Optionally include 10 μL Trolox (10 mM stock = 0.25 mg/mL), which reduces blinking.
- Fill up to 100 μL with appropriate buffer containing the DNA dye, fill channels, and proceed to experiment.

4. Alternative Channel Preparation
An alternative technique that circumvents the need to drill inlet and outlet holes into glass is also possible. First, channels are created using double-sided tape as described earlier, then the PEGylated coverslip is placed onto the channels, making sure to leave the channel ends open until after the last wash. All buffer exchanges are then performed by adding buffer to the channels by slowly pipetting it into one channel opening and sucking it through with a piece of filter paper at the other opening.

5. DNA Disappears After Proteins or Extract Have Been Introduced
In cytoplasmic extract and under high protein concentrations, this is to be expected as DNA binding proteins can outcompete the DNA dye or induce fluorophore quenching. Due to the low residence time of the dye on the DNA, it is important to add the DNA dye to the final buffer that is introduced into the channels, such that it is constantly present throughout the experiment.

Adding DNA dye directly to the cytoplasmic extract however did not increase DNA signal and only increased background fluorescence in our case. However, diluting the extract or introducing it slowly and only partially (halfway up the channel) allowed a small window at the front of the diffusing extract to study DNA in presence of the cytoplasm.

6. Dye-Induced DNA Lengthening
One has to bear in mind that DNA intercalating dyes change the contour length of the DNA molecule. By intercalating between the bases in dsDNA, cyanine dyes increase the distance between adjacent base pairs, thus lengthening the DNA molecule (increasing its absolute contour length) [30]. This process is highly nonlinear in dye concentration and ionic strength of the surrounding medium, the tension on the DNA molecule, and the chemical composition of the dye molecule itself. In our study, we employed SYTOX Orange and SYTOX Green, which are small mono-intercalators that introduce one dye molecule in between two adjacent bases. This process increases the DNA contour length by 0.34 nm per intercalated dye molecule (see Fig. 4). At the nM concentrations we employed, thousands of dye molecules are bound to each DNA molecule at every time point, resulting in an increase in DNA contour length of several micrometers. Figure 4 visualizes this effect by exposing an immobilized λ-DNA molecule to an increasing concentration of SYTOX Orange. Despite its constant end-to-end distance, one and the same DNA molecule undergoes substantial lengthening due to intercalating dye molecules. Not accounting for this effect can have broad consequences on downstream analysis, as, for example, the tension on the DNA molecule not only depends on the end-to-end

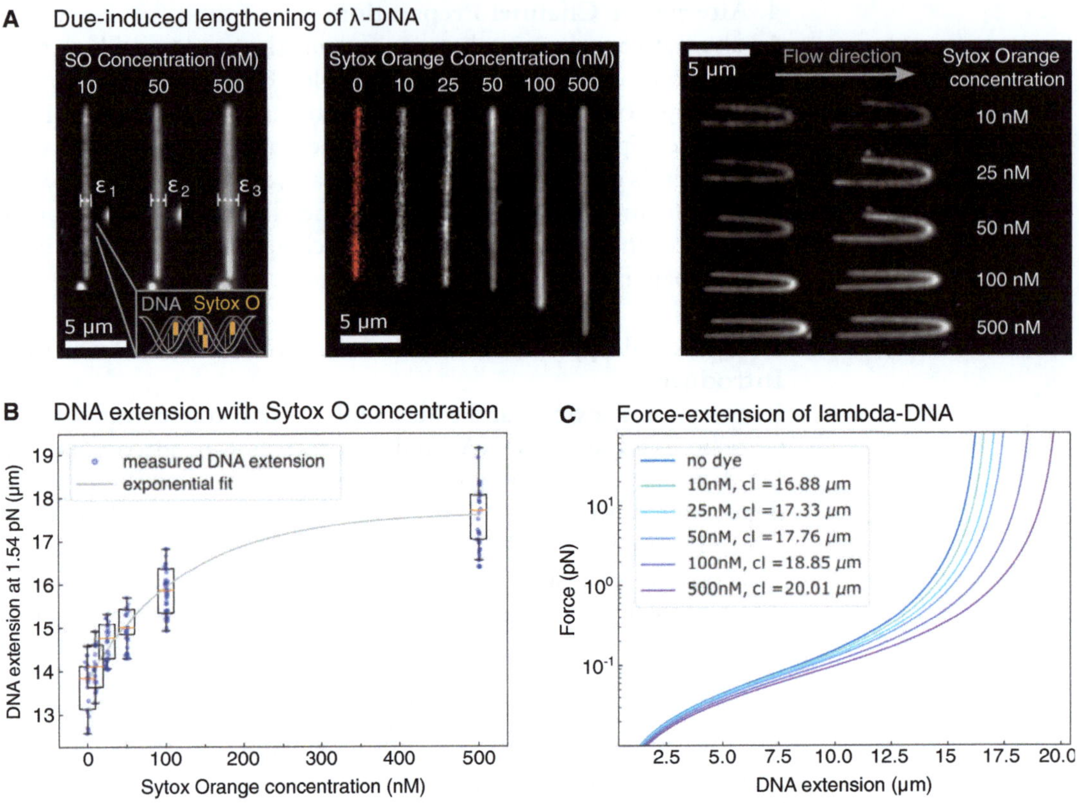

Fig. 4 Quantification of dye-induced lengthening of λ-DNA. (**a**) (Left) Time averages of the same DNA molecule reveal increasing transversal fluctuations for increasing dye concentrations, indicative of growing amounts of slack due to dye-induced lengthening of the DNA molecule. Inset depicts schematic dye intercalation between adjacent DNA bases. (Middle) Examples of maximally extended, immobilized λ-DNA molecules that were exposed to increasing dye concentrations prior and during the immobilization. Cy5 backbone labeled λ-DNA (red) serves as control in the absence of SYTOX Orange. (Right) Hydrodynamic stretching of λ-DNA molecules with constant flow rate and a buffer containing increasing SYTOX Orange concentrations. (**b**) Quantification of dye concentration-dependent λ-DNA length at constant hydrodynamic stretching. (**c**) Absolute contour length (cl) of λ-DNA for several dye concentrations, calculated using the wormlike chain model for DNA

distance, but also the dye concentration employed. Unfortunately, alternative DNA labeling techniques, such as covalent fluorescent backbone labeling, have proven unsuitable for our experiments due to their inferior brightness and compromised DNA integrity.

References

1. Misteli T (2020) The self-organizing genome: principles of genome architecture and function. Cell 183:28–45. https://doi.org/10.1016/j.cell.2020.09.014
2. Joo C, Ha T (2012) Single-molecule FRET with total internal reflection microscopy. Cold Spring Harb Protocols
3. Sefer A (2022) Structural dynamics of DNA strand break sensing by PARP-1 at a single-molecule level. Nature Commun
4. Reid DA, Keegan S, Leo-Macias A et al (2015) Organization and dynamics of the nonhomologous end-joining machinery during DNA

double-strand break repair. Proc Natl Acad Sci 112:E2575–E2584
5. Carney SM, Moreno AT, Piatt SC et al (2020) XLF acts as a flexible connector during non-homologous end joining. elife 9:e61920
6. Santoso Y, Joyce CM, Potapova O et al (2010) Conformational transitions in DNA polymerase I revealed by single-molecule FRET. Proc Natl Acad Sci 107:715–720
7. Duchi D, Bauer DL, Fernandez L et al (2016) RNA polymerase pausing during initial transcription. Mol Cell 63:939–950
8. Sabantsev A, Levendosky RF, Zhuang X et al (2019) Direct observation of coordinated DNA movements on the nucleosome during chromatin remodelling. Nat Commun 10: 1720
9. Feng XA, Poyton MF, Ha T (2021) Multicolor single-molecule FRET for DNA and RNA processes. Curr Opin Struct Biol 70:26–33
10. Abid Ali F, Renault L, Gannon J et al (2016) Cryo-EM structures of the eukaryotic replicative helicase bound to a translocation substrate. Nat Commun 7:10708
11. Leicher R, Liu S (2022) Probing the interaction between chromatin and chromatin-associated complexes with optical tweezers. In: Optical tweezers: methods and protocols. Springer, Cham, pp 313–327
12. Mihardja S, Spakowitz AJ, Zhang Y, Bustamante C (2006) Effect of force on mononucleosomal dynamics. Proc Natl Acad Sci 103: 15871–15876
13. Lisica A, Engel C, Jahnel M et al (2016) Mechanisms of backtrack recovery by RNA polymerases I and II. Proc Natl Acad Sci 113: 2946–2951
14. Keenen MM, Brown D, Brennan LD et al (2021) HP1 proteins compact DNA into mechanically and positionally stable phase separated domains. elife 10:e64563
15. Renger R, Morin JA, Lemaitre R et al (2022) Co-condensation of proteins with single-and double-stranded DNA. Proc Natl Acad Sci 119:e2107871119
16. Ganji M, Shaltiel IA, Bisht S et al (2018) Real-time imaging of DNA loop extrusion by condensin. Science 360:102–105
17. Davidson IF, Bauer B, Goetz D et al (2019) DNA loop extrusion by human cohesin. Science 366:1338–1345
18. Sternberg SH, Redding S, Jinek M et al (2014) DNA interrogation by the CRISPR RNA-guided endonuclease Cas9. Nature 507: 62–67
19. Xue C, Salunkhe SJ, Tomimatsu N et al (2022) Bloom helicase mediates formation of large single–stranded DNA loops during DNA end processing. Nat Commun 13:2248
20. Vrtis KB, Dewar JM, Chistol G et al (2021) Single-strand DNA breaks cause replisome disassembly. Mol Cell 81:1309–1318
21. Gruszka DT, Xie S, Kimura H, Yardimci H (2020) Single-molecule imaging reveals control of parental histone recycling by free histones during DNA replication. Sci Adv 6: eabc0330
22. Quail T, Golfier S, Elsner M et al (2021) Force generation by protein–DNA co-condensation. Nat Phys 17:1007–1012
23. Yan J, Maresca TJ, Skoko D et al (2007) Micromanipulation studies of chromatin fibers in Xenopus egg extracts reveal ATP-dependent chromatin assembly dynamics. Mol Biol Cell 18:464–474
24. Yardimci H, Loveland AB, Habuchi S et al (2010) Uncoupling of sister replisomes during eukaryotic DNA replication. Mol Cell 40:834–840
25. Hirano T, Mitchison TJ (1994) A heterodimeric coiled-coil protein required for mitotic chromosome condensation in vitro. Cell 79: 449–458
26. Losada A, Hirano M, Hirano T (1998) Identification of Xenopus SMC protein complexes required for sister chromatid cohesion. Genes Dev 12:1986–1997
27. Shintomi K, Inoue F, Watanabe H et al (2017) Mitotic chromosome assembly despite nucleosome depletion in Xenopus egg extracts. Science 356:1284–1287
28. Golfier S, Quail T, Kimura H, Brugués J (2020) Cohesin and condensin extrude DNA loops in a cell cycle-dependent manner. elife 9: e53885
29. Zierhut C, Jenness C, Kimura H, Funabiki H (2014) Nucleosomal regulation of chromatin composition and nuclear assembly revealed by histone depletion. Nat Struct Mol Biol 21: 617–625
30. Ganji M, Kim SH, Van Der Torre J et al (2016) Intercalation-based single-molecule fluorescence assay to study DNA supercoil dynamics. Nano Lett 16:4699–4707

Chapter 2

Studying Translesion DNA Synthesis Using *Xenopus* In Vitro Systems

Antoine Aze, James R. A. Hutchins, and Domenico Maiorano

Abstract

Cell-free extracts derived from *Xenopus* eggs have been widely used to decipher molecular pathways involved in several cellular processes including DNA synthesis, the DNA damage response, and genome integrity maintenance. We set out assays using *Xenopus* cell-free extracts to study translesion DNA synthesis (TLS), a branch of the DNA damage tolerance pathway that allows replication of damaged DNA. Using this system, we were able to recapitulate TLS activities that occur naturally in vivo during early embryogenesis. This chapter describes protocols to detect chromatin-bound TLS factors by western blotting and immunofluorescence microscopy upon induction of DNA damage by UV irradiation, monitor TLS-dependent mutagenesis, and perform proteomic screening.

Key words *Xenopus laevis*, Cell-free extract, DNA replication, DNA damage tolerance, Mutagenesis, Chromatin, Proteomics

1 Introduction

Cells are continually exposed to endogenous and exogenous sources of DNA damage that threaten the integrity of the genome. In response to these cues, of external and internal nature, cells activate the DNA damage response, a multifaceted biochemical process that allows to detect, sense, and repair DNA lesions and as such preserve and maintain genome integrity [1]. Comprehension and characterization of the protein networks involved in this process are of particular interest for understanding how cell malignancy is driven, since DNA damage can generate mutations. Further, a better understanding of this process can be instrumental for the development of novel and optimized cancer therapies. As part of the DNA damage response, cells have evolved DNA damage tolerance (DDT) pathways, including Translesion DNA synthesis

Authors Antoine Aze and James R. A. Hutchins have equally contributed to this chapter.

Anna Castro and Benjamin Lacroix (eds.), *Cell Cycle Control: Methods and Protocols*, Methods in Molecular Biology, vol. 2740, https://doi.org/10.1007/978-1-0716-3557-5_2,

(TLS), template switch (TS), and repriming [2]. These processes are particularly important when DNA lesions persist during DNA synthesis (S-phase of the cell cycle), since they can favor recovery of DNA replication forks stalled by DNA damage. Within this context, TLS and TS are initiated by the transfer of one ubiquitin moiety onto PCNA, the processive factor of replicative polymerases, by the ubiquitin ligases Rad6 (E2) and Rad18 (E3). Hence, when replication forks encounter DNA lesions, PCNA is monoubiquitinated, allowing recruitment of TLS polymerases (pols) of the Y-family, such as Pol eta (η), Pol kappa (κ), Pol iota (ι), and Rev1. These polymerases have the property of replicating through damaged DNA templates, thanks to catalytic sites more open than that of replicative polymerases, however at the expense of increased mutagenesis, since they are error prone. By so doing, TLS pols favor proliferation under DNA damage conditions, but they may insert mutations, which can drive tumorigenesis [3] and be involved in chemotherapy resistance [2]. The use of cell-free extracts derived from *Xenopus laevis* eggs has been a key stepping stone towards major discoveries in genome maintenance mechanisms (*see* [4] for a review). *Xenopus* eggs contain a large stockpile of mRNAs and proteins needed to complete cell division cycles during early embryogenesis until the activation of zygotic transcription. Addition of demembranated sperm nuclei into egg extract recapitulates in vitro several key nuclear processes, including chromatin decondensation and assembly, formation of intact nuclei, and synchronous semiconservative DNA synthesis in a transcription-free context, where protein translation is unnecessary [5]. This in vitro system is also able to recapitulate DNA damage signalling generated upon addition of DNA damaging agents [6].

We have been using *Xenopus* egg extracts to understand the molecular basis of checkpoint silencing and radioresistance of early embryos. We found that early *Xenopus* embryos are characterized by DNA damage-tolerant replisomes, as a consequence of constitutive $PCNA^{mUb}$ and recruitment of at least one TLS pol [7]. In line with this observation, we have more recently observed TLS-driven mutagenesis in very early embryos of *X. laevis* and have shown that this process is also conserved in *Drosophila melanogaster*, making TLS a novel source of genetic diversity [8]. Hence, the *Xenopus* egg has proven to be a pertinent biological system to investigate DDT mechanisms.

We set out in vitro conditions using *Xenopus* egg extracts to mimic very early embryogenesis, in which TLS is constitutive in the absence of DNA damage. These conditions allow biochemical and proteomic investigations intended to further our knowledge about TLS activation and regulation. In this chapter, we describe techniques using *Xenopus* egg extracts to analyze constitutive, as well as UV-C-induced TLS in vitro. In particular, we detail methods to detect recruitment to chromatin of endogenous TLS factors (such as PCNA^{mUb} and TLS pols) upon UV irradiation of sperm nuclei.

2 Materials

2.1 UV Irradiation of Demembranated Sperm DNA and Replication Reaction

1. Creatine kinase: stock solution 10 mg/mL dissolved in 10 mM HEPES-KOH, pH 7.6, with 20% (v/v) glycerol final. Aliquots are stored at −20 °C.
2. Sodium creatine phosphate dibasic tetrahydrate: 1 M solution in sterile H_2O, aliquoted and stored at −20 °C.
3. ATP: Adenosine 5′ triphosphate disodium salt hydrate: prepare a 200 mM stock solution in 20 mM HEPES-NaOH, pH 7.0. Store aliquots at −20 °C.
4. Cycloheximide: prepare a 10 mg/mL stock solution in sterile H_2O. Store aliquots of 50 μL at −20 °C.
5. Demembranated sperm nuclei, prepared as described [9], at a concentration of 60,000 nuclei/μL, are stored as frozen aliquots at −80 °C.
6. *Xenopus* low speed egg extract, prepared as described [9], is stored as 50 μL aliquots at −80 °C.
7. Energy mix 20×: combine 200 μg/mL creatine kinase, 200 mM creatine phosphate, 20 mM ATP, 20 mM $MgCl_2$, 2 mM EGTA. Aliquot in 20 μL and store at −20 °C.
8. dCTP-Cy3.
9. Eppendorf 1.5 mL tubes (or similar).
10. Stratalinker UV 1800 crosslinker with UV-C bulbs (Stratagene).
11. Parafilm "M" (Amcor).
12. Stainless-steel tube holder of a dry block incubator.
13. Cut tips (200 μL).
14. UV radiometer (Vilber, VLX-3W).

2.2 Purification of Nuclei and Immunofluorescence Staining of Endogenous PCNA^{mUb}

1. Glass microscope slides; coverslips, round, 13 mm diameter.
2. 24-well microtitre plates, flat bottom.
3. XB buffer: 10 mM HEPES-KOH pH 7.7, 100 mM KCl, 0.1 mM $CaCl_2$, 1 mM $MgCl_2$, 50 mM sucrose, 1 mM DTT.
4. XB/0.1% Triton: XB (above) plus 0.1% (w/v) Triton X-100. (Alternatively, NP-40 can be used.)
5. XB/0.5 M sucrose: XB (above) plus 0.5 M sucrose, protease inhibitors (10 μg/mL each leupeptin, aprotinin, pepstatin).
6. 16% (w/v) formaldehyde solution, methanol-free.

7. 4% formaldehyde solution: dilute a 16% stock solution of formaldehyde in PBS 1×.
8. Wash buffer: PBS 1×, 0.1% Tween-20.
9. Blocking solution: 3% BSA, PBS 1×, 0.1% Tween-20.
10. Wet chamber (a 15 cm cell culture plate containing wet tissue paper).
11. Ubiquityl-PCNA (Lys164; D5C7P) rabbit monoclonal antibody (Cell Signaling, 13439).
12. Secondary antibody anti-rabbit Alexa 488 (Thermo Fisher Scientific, A-11034).
13. Hoechst 33342: stock solution 10 mg/mL in sterile water, stored as aliquots at −20 °C.
14. ProLong Gold antifade mountant (Thermo Fisher Scientific, P36930).
15. Megafuge ST Centrifuge (Thermo Fisher Scientific) equipped with a TX-100 swinging bucket rotor.

2.3 Analysis of TLS in Early Versus Late Embryogenesis Using Egg Extracts

In addition to materials described in Subheading 2.2:

1. Plasmid purification kit; QIAquick gel extraction kit (QIAGEN).
2. Horizontal agarose gel electrophoresis apparatus.
3. Agarose.
4. TAE buffer (40 mM Tris, 20 mM acetic acid, 1 mM EDTA).
5. GelRed nucleic acid gel stain (Biotium, 41003).
6. STOP MIX (40 mM EDTA-NaOH, pH 8.0; 1% SDS).
7. Proteinase K (10 mg/mL in water).
8. Phenol/chloroform/isoamyl alcohol (25:24:1).
9. Chloroform/isoamyl alcohol (24:1).
10. Sodium acetate (NaAc), 3 M, pH 5.3.
11. Absolute ethanol.
12. 70% ethanol.
13. *Dpn*I restriction enzyme.
14. *E. coli* electrocompetent bacteria (*see* **Note 1**).

2.4 Sample Processing for MS Analysis

1. Solubilization solution: 20% (w/v) SDS.
2. S-Trap™ micro column (ProtiFi, NY, USA).
3. TEAB buffer: 1 M triethylammonium bicarbonate, pH 8.5.
4. Acids: acetic acid, formic acid (FA), phosphoric acid (12%), trifluoracetic acid.
5. Solvents: acetonitrile (ACN), methanol.

6. Other reagents: dithiothreitol (DTT), iodoacetamide (IAA).
7. Trypsin Gold Mass Spectrometry Grade (Promega, V5280): prepare at 1 μg/μL in 50 mM acetic acid, and store as aliquots at −20 °C.

2.5 Mass Spectrometry Apparatus

1. Nano-flow (RSLC U3000, Thermo Fisher Scientific) coupled to a mass spectrometer equipped with a nanoelectrospray source (Q Exactive HFX, Thermo Fisher Scientific).
2. Capillary column (Acclaim PepMap 100, reverse phase C18, 0.075 mm × 250 mm, NanoViper, Thermo Fisher Scientific).

2.6 Software for Mass Spectrometry Data Analysis

1. MaxQuant (https://www.maxquant.org/).
2. Perseus (https://www.maxquant.org/perseus/)
3. Spreadsheet: e.g. Microsoft Excel (https://www.microsoft.com/excel) or LibreOffice Calc (https://www.libreoffice.org/discover/calc/).

2.7 Protein Reference Sequences

Sequences can be from the UniProt database and can be downloaded using the following links:

1. *X. laevis* reference proteome:
   https://www.uniprot.org/proteomes?query=Xenopus%20laevis
2. Human keratins:
   https://www.uniprot.org/uniprotkb?query=(taxonomy_id:9606)%20AND%20(protein_name:%22keratin,%20type%22)%20AND%20(reviewed:true)
3. Porcine trypsin:
   https://rest.uniprot.org/uniprotkb/P00761.fasta

3 Methods

3.1 Nuclear Assembly and DNA Replication of UV-Damaged Sperm Nuclei in Low Speed Extract

The procedure comprises assembly of replicating nuclei from damaged demembranated sperm chromatin, purification of replicating nuclei, and immunostaining of fixed nuclei.

3.1.1 Assembly of UV-Irradiated Nuclei in Egg Extracts

UV irradiation of demembranated sperm nuclei

1. Place the stainless-steel block upside down, in contact with ice to cool it down. Stick a Parafilm sheet on it with a drop of water and flatten the surface using a tissue paper. Remove the protecting layer from the Parafilm.

2. Using a yellow tip, apply 5 μL of sperm nuclei in a line directly on the Parafilm.
3. Set the Stratalinker to 300 J/m^2. Using a radiometer, verify that the radiation delivered is the one displayed on the settings. Then transfer the stainless-steel block into the chamber of the Stratalinker. Start the radiation.
4. Recover the sperm using a yellow tip and transfer into a new tube. Keep the UV-damaged sperm on ice until use.

Nuclear assembly reaction

5. Supplement two freshly thawed 50 μL aliquots of low-speed extract with 2.5 μL 20× energy mix, 1 μL cycloheximide, and 1 μL dCTP-Cy3 to monitor DNA synthesis (*see* **Note 2**). At this stage, two additional 50 μL reactions that *do not* contain dCTP-Cy3 can be set up to detect TLS factors binding to chromatin by western blot (see Subheading 3.2).
6. Transfer the supplemented extract into the tube holder of a dry block incubator set to 22 °C.
7. Add 1 μL of undamaged (control) or UV-damaged demembranated sperm nuclei.
8. Incubate the reaction for 60–90 min and at least until dCTP-Cy3 incorporation is detectable in UV-irradiated nuclei (*see* **Note 2**).
9. Timepoints can be sampled as 10 μL aliquots of the reaction and processed as described subsequently.

A DNA replication reaction of damaged template in egg extract is a prerequisite to monitor the recruitment of TLS factors onto chromatin, which occurs when replication fork progression is challenged by DNA lesions. Methods to monitor DNA replication of sperm nuclei in egg extract have been widely covered in the literature; we refer readers to the following: [10, 11].

3.1.2 Purification and Fixation of Replicating Nuclei

1. Cool the centrifuge to 4 °C.
2. Add one coverslip per well of a 24-well microtitre plate. Cover the coverslips with 2 mL of ice-cold XB buffer/0.5 M sucrose (hereafter called sucrose cushion). Keep the plate at 4 °C.
3. For each timepoint, take 10 μL of the mixture and transfer into an Eppendorf tube containing 100 μL of ice-cold XB/0.1% Triton buffer and leave for 5 min on ice.
4. Carefully overlay the diluted mixture over the sucrose cushion using a yellow tip whose extremity is cut with a scalpel blade to avoid disturbance of the sucrose cushion layer.
5. Spin the plate at 100 *g* for 10 min.

6. Slowly discard the top layer with a cut 1 mL pipette tip. Then discard the sucrose cushion.
7. Fix the nuclei attached to the coverslips with 4% formaldehyde solution for 10 min.
8. Wash the coverslips with PBS 1×/0.1% Tween-20 for 5 min.
9. Block the coverslips in a solution of 3% BSA, PBS 1×/0.1% Tween-20 for 1 h.

3.1.3 Immunostaining of Fixed Nuclei with PCNA^{mUb}

So far it has been impossible to detect PCNA^{mUb} in isolated nuclei of any cell type. We have recently developed a protocol that allows detection of PCNA^{mUb} in single mammalian cells [12], that we have also adapted to detect nuclei assembled in in vitro *Xenopus* egg extracts.

1. Transfer coverslips prepared in **step 9** of Subheading 3.1.2 into a wet chamber and lay them over a Parafilm sheet, nuclei side up.
2. Add one 35 μL drop of anti-ubiquityl-PCNA (Lys164) rabbit monoclonal antibody, diluted 1/100 in PBS 1×/0.1% Tween-20.
3. Incubate the coverslips in the wet chamber for 2 h at room temperature.
4. Transfer coverslips into a 24-well microtitre plate and wash them three times in 1 mL PBS 1×/0.1% Tween-20 for 5 min each.
5. Repeat **step 9**.
6. Onto each coverslip, apply a 35 μL drop of diluted secondary antibody (goat anti-rabbit Alexa 488, 1/250 in PBS 1×/0.1% Tween-20).
7. Transfer coverslips into a 24-well microtitre plate and wash them twice, as in **step 12**.
8. To stain nuclei with Hoechst, incubate coverslips for 5 min in PBS 1×/0.1% Tween-20 supplemented with 2 μg/mL Hoechst 33342.
9. Wash coverslips one last time with PBS 1×.
10. Add one drop (~12 μL) of ProLong Gold mountant onto a glass slide.
11. Remove excess liquid from coverslips by touching the edges onto an absorbent tissue.
12. Place coverslips nuclei side down onto the drop of ProLong Gold mountant.
13. Leave overnight at room temperature in the dark.
14. Observe slides by fluorescence microscopy with a 63× objective lens and ApoTome facility (Fig. 1a).

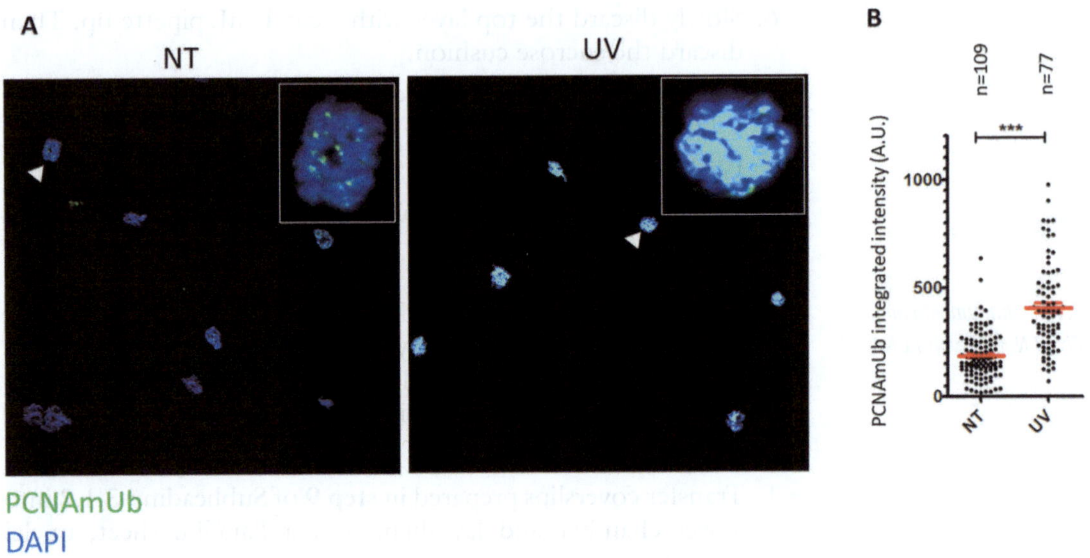

Fig. 1 Detection of PCNA^{mUb} in non-treated (NT) and UV-C-treated demembranated sperm nuclei. (**a**) Immunofluorescence microscopy images. (**b**) Quantification following image analysis with CellProfiler

3.1.4 Image Acquisition and Data Analysis

We use a ZEISS Axio Imager fluorescence microscope in ApoTome mode for image acquisition. ZEN software (ZEISS) is used for image format conversion. CellProfiler, Microsoft Excel, and GraphPad Prism software are used for data analysis, statistical calculations, and graphing. The ability of the anti-PCNA^{mUb} antibody to detect endogenous PCNA^{mUb}-containing nuclear foci was validated by [12].

1. Use ZEN software to convert ".czi" images to "raw converted. czi" format.
2. Set the appropriate pipeline on CellProfiler to identify nuclei as the primary object.
3. Adapt the parameters of the pipeline to discriminate the PCNA^{mUb} signal from the background. The size and shape of the foci are quantified from the nuclei.
4. Export results as an Excel spreadsheet file. The mean integrity intensity values of each focus per nucleus are used to generate a scatter plot in GraphPad Prism (Fig. 1b). Count at least 100 nuclei per condition.

3.2 Analysis of TLS in Early Versus Late Embryogenesis Using Egg Extracts

Previous work showed that egg extracts can recapitulate the embryonic developmental transition resulting in the activation of the DNA damage checkpoint that restrains mitosis in the presence of unreplicated or damaged DNA [13–15]. This can be achieved by altering the amount of sperm nuclei added into the egg extract, so as to increase the nuclear-to-cytoplasmic ratio (N/C ratio) to a value close to that observed at late stages of embryogenesis (mid-blastula transition). In this section, we describe a procedure to analyze DNA synthesis and TLS at low versus high N/C ratios.

1. UV-C irradiate sperm chromatin as described earlier (Subheading 3.1.1). To increase the signal of chromatin-bound PCNA^{mUb} and TLS pols, sperm chromatin can be irradiated up to 800 J/m^2
2. Supplement freshly thawed low speed egg extract with energy mix and cycloheximide as described earlier (Subheading 3.1.1). Set up four reactions as follows: two reactions at low N/C ratio and two reactions at high N/C ratio containing sperm chromatin UV-irradiated or not. In order to recover enough material for Western blot analysis, the volume of the low N/C ratio reaction must be at least five times higher that of the high N/C ratio reaction. Typically, if setting up a 50 μL reaction at high N/C ratio, at least a 250 μL reaction at low N/C ratio must be used.
3. Add the same volume of sperm chromatin, UV-C irradiated or not, so to reach a final concentration of either 200 nuclei/μL (low N/C) or 2000 nuclei/μL (high N/C) in the egg extract. Sperm nuclei can be diluted with XB buffer to obtain the low N/C ratio condition.
4. The kinetics of DNA synthesis in the different conditions can be monitored in parallel by setting up similar reactions in the presence of 1 μL dCTP-Cy3 (*see* **Note 2**), except that in this case two reactions of identical egg extract volume can be used as there will be enough labelled DNA to analyze.
5. Transfer the reactions into the tube holder of a dry block incubator set at 22 °C.
6. To monitor DNA synthesis, proceed as described in Subheading 3.1.1.
7. To analyze chromatin binding of TLS factors (namely, PCNA^{mUb} and TLS pols), stop the reaction after 50–60 min incubation by diluting 10 times with ice-cold XB buffer containing proteases inhibitors (+PIs). Transfer the samples into ice-cold 2 mL Eppendorf tubes.
8. Centrifuge samples at 1500 *g*, 5 min at 4 °C in a microcentrifuge with a swing-out rotor to isolate the nuclei.
9. Gently decant the supernatant and wash the nuclear pellet once with 0.5 mL of ice-cold XB buffer +PIs.
10. Centrifuge as in **step 8**.

11. Resuspend the nuclear pellet in 50 μL of ice-cold XB buffer + PIs containing 0.1% of either Triton X-100 or NP-40.
12. Detergent extract the nuclei by incubation on ice for 5 min.
13. Centrifuge the samples at 6000 *g*, 5 min at 4 °C in a microcentrifuge.
14. Gently decant the supernatant.
15. Resuspend the chromatin pellet in 40 μL of 2× Laemmli buffer.
16. Analyze samples by SDS-PAGE followed by Western blotting. The material is enough to load two separate lanes per sample. Typical results are as shown in Kermi et al. [7].

3.3 Using In Vitro Egg Extracts to Generate TLS-Dependent Mutagenesis on Plasmid DNA Templates

TLS-dependent mutagenesis, as measured in early *Xenopus* embryos obtained by in vitro fertilization of *Xenopus* eggs [8], can equally be reproduced using egg extracts and is described in this section. To this end, it is crucial to prepare undamaged plasmid DNA template bearing a reporter gene; we use a plasmid containing the bacterial *lacZ* gene [8].

3.3.1 Plasmid DNA Replication and Purification

1. Prepare plasmid DNA more than 90% supercoiled, as previously described [16].
2. Check the quality of the DNA by analyzing an aliquot by agarose gel electrophoresis in TAE without intercalating agents. Stain the gel after electrophoresis using GelRed.
3. Supplement freshly thawed low speed egg extract with energy mix and cycloheximide as previously described above (Subheading 3.1.1), and plasmid DNA at a concentration of 6 ng/μL of egg extract.
4. Incubate at 22 °C for 2–3 h.
5. Stop reactions by diluting in 10 volumes of STOP MIX containing 0.2 mg/mL of Proteinase K.
6. Incubate at 37 °C for 1 h.
7. Purify the DNA from the proteins by extraction with an equal volume of phenol/chloroform/isoamyl alcohol (25:24:1).
8. Centrifuge at 10,000 *g* for 8 min at room temperature.
9. Recover the aqueous solution (supernatant) into a fresh Eppendorf tube.
10. Extract the aqueous solution with an equal volume of chloroform/isoamyl alcohol.
11. Centrifuge as in **step 8**.
12. Recover the aqueous solution into a fresh Eppendorf tube.

13. Recover the DNA by precipitation by adding 1/10th volume of NaAc 3 M, pH 5.3, and 2.5 volumes of −20 °C absolute ethanol.
14. Incubate the samples overnight at −20 °C.
15. The next day, recover the precipitated DNA by centrifugation at maximum speed in a microcentrifuge for 30 min at 4 °C.
16. Discard the supernatant.
17. Wash the pellet with 1 mL of 70% ethanol.
18. Centrifuge for 15 min as in **step 15**.
19. Discard the supernatant.
20. Air-dry the pellet for 1 h at room temperature.
21. Dissolve the pellet in 82 μL of dH_2O and incubate at 37 °C for 1 h.

3.3.2 Digestion of Unreplicated DNA

1. Set up a restriction enzyme digestion reaction with *Dpn*I as follows:

DNA	82 μL
10× buffer	10 μL
*Dpn*I	8 μL (160 U)
Final volume:	100 μL

2. Incubate the reaction at 37 °C for 2 h.
3. Purify plasmid DNA using the QIAquick gel extraction kit, following the manufacturer's instructions.
4. Elute the DNA with 30 μL of dH_2O.
5. Transform 30 μL of electrocompetent bacteria cells with 3 μL of plasmid DNA recovered in **step 4**. If using a *lacZ*-based reporter gene, colonies bearing the mutant plasmid can be scored using blue/white selection on X-Gal/IPTG indicator plates.

Typical results are described in Lo Furno et al. [11].

3.4 Proteomic Analysis of Xenopus Egg Extracts

Insights into processes such as TLS can be gleaned by proteomic analysis of components isolated using targeted approaches such as immunoprecipitation [17] or affinity purification [18], performed with starting material such as low- or high-speed egg extracts [9], nucleoplasmic extracts (NPE) [19], or chromatin fractions [20–22]. Here we employ the gel-free suspension trapping (S-Trap) method [23] that allows efficient column-based solubilization and trapping of proteins, digestion, and peptide elution (*see* **Note 3**). Starting from a complex mixture, this approach enables identification of over 2000 proteins and their relative quantification with a dynamic range of over five orders of magnitude (Fig. 2). In the

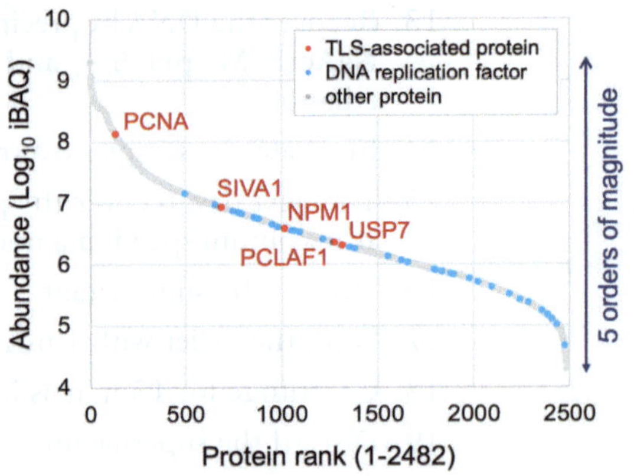

Fig. 2 Quantitative profiling of proteins identified by proteomic analysis of low speed *Xenopus* egg extract using the S-Trap method shows proteins can be detected across five orders of magnitude of abundance. In the analysis shown, numerous DNA replication factors as well as TLS-associated proteins are identified

example shown, we were able to identify numerous factors involved in the DNA replication and TLS pathways (*see* **Note 4**).

1. Protein solubilization: starting with up to 100 μg of protein, add concentrated SDS solution to give a final concentration of 5% (w/v) SDS.
2. Protein reduction: add 1 M DTT solution (freshly prepared), to give a final DTT concentration of 20 mM. Incubate at 95 °C for 10 min.
3. Protein alkylation: add 1/10th volume of 0.5 M IAA (freshly prepared), to give a final IAA concentration of 40 mM. Incubate at room temperature for 30 min in the dark.
4. Protein acidification: add 1/10th volume phosphoric acid solution (12%) to the sample until the pH ≤ 1.0.
5. Protein denaturation: add 9 volumes of 90% (v/v) methanol in 100 mM TEAB. The sample may develop a translucent appearance, which is normal.
6. Column loading: transfer the sample, including any insoluble material, onto the S-Trap micro column. Centrifuge at 4000 *g* for 30 s to trap proteins.
7. Column washing: add 150 μL 90% (v/v) methanol in 100 mM TEAB, centrifuge at 4000 *g* for 30 s, discard the flow-through. Repeat three times.
8. Protein digestion: starting with 1 μg of Trypsin Gold (1 μg/μL in acetic acid), dilute in 50 mM TEAB to a final volume of 20 μL. Add this to the stoppered S-Trap column and incubate at 47 °C for 1 h without mixing.

9. Peptide elution: add 40 μL of elution buffer (50% (v/v) ACN, 0.2% (v/v) formic acid in 50 mM TEAB). Centrifuge at 4000 *g*, for 1 min, collecting the flow-through. Repeat this twice.
10. Desiccate the eluate using a vacuum concentrator to dryness.
11. Redissolve the dried peptides in 0.05% (v/v) trifluoracetic acid, 2% (v/v) ACN.
12. Inject the sample for on-line analysis using a nano-flow HPLC coupled to a mass spectrometer equipped with a nanoelectrospray source. For peptide separation, use a capillary column (0.075 mm × 250 mm); as a gradient, we suggest 2–40% buffer B over 120 min (buffer A = 0.1% (v/v) FA; buffer B = 0.1% (v/v) FA, 80% ACN), at a flow rate of 300 nL/min.
13. To identify proteins in the sample, search raw mass spectrum data using the Andromeda search algorithm [24] within the MaxQuant program [25] against the *X. laevis* reference proteome, plus a "contaminant" sequence database comprising, for example, human keratins and porcine trypsin. As quantification options, include iBAQ [26] and label-free quantification (LFQ) generated by the MaxLFQ algorithm [27] (*see* **Note 5**).
14. Identify and filter out "contaminant" entries, for example, porcine trypsin and human keratins. This can be performed using the Perseus statistical and graphing software [28] or using the filter function of spreadsheet software.
15. A *Xenopus* protein whose entry name is "uncharacterised" can often be identified either via its gene name, or by performing a BLAST search against the human proteome (*see* **Note 6**).
16. Protein entries can be annotated with terms including gene ontology (GO), using tools from the UniProt website. Methods for these approaches have been described previously [29].

4 Notes

1. We have used the MBM7070 *E. coli* indicator strain that harbors an amber mutation in the *lacZ* gene.
2. Addition of dCTP-Cy3 into egg extract is optional. Replication of UV-C-treated sperm nuclei may be delayed compared to the control. It is therefore critical to ensure that damaged template has started DNA synthesis before sampling. When dCTP-Cy3 is supplemented to extract, ongoing incorporation of dCTP-Cy3 into DNA can easily be monitored by epifluorescence microscopy by putting a drop (2–5 μL) of extract mixture in fix solution (45 mM PIPES pH 7.2, 45 mM NaCl, 240 mM KCl, 10% formalin, 50% glycerol, 2 μg/mL Hoechst 33342).

Moreover, dCTP-Cy3 incorporation will be detectable in fixed nuclei together with the $PCNA^{mUb}$ signal after the immunostaining procedure.

3. Detailed protocols and troubleshooting information for S-Trap columns are available from https://protifi.com/pages/protocols.
4. The TLS-associated proteins identified were proliferating cell nuclear antigen (gene symbol PCNA), the apoptosis regulatory protein Siva (SIVA1), the multifunctional protein nucleophosmin (NPM1), PCNA-clamp-associated factor (PCLAF), and ubiquitin carboxyl-terminal hydrolase 7 (USP7). Curiously, neither TLS Pols nor Rad18 could be identified using this procedure, although they could be detected by Western blot (Aze and Maiorano, unpublished). At this stage, it is not clear whether this is a problem of relative abundance or sensitivity of the detection procedure.
5. The iBAQ (intensity-based absolute quantification) algorithm is suited to compare and rank the abundance of proteins within one sample. LFQ is suited to compare the relative abundance of proteins between samples of similar overall composition.
6. The *X. laevis* genome was sequenced relatively recently [30], and despite rapid advances in annotation of corresponding proteome sequences, some 5% of *X. laevis* UniProt protein entries are named "uncharacterized protein." In many cases, a gene name is present; this can be used to retrieve relevant information about the human orthologue. The following query from UniProt enables the corresponding human Swiss-Prot entry (where present) to be retrieved (example for the gene siva1): https://www.uniprot.org/uniprotkb?query=taxonomy_id%3A9606%20AND%20reviewed%3Atrue%20AND%20gene%3A**siva1**

   Where a protein entry has neither a recognizable protein nor gene name, a BLAST search of the protein accession code against the human proteome can provide clues as to orthology. This can be performed here: https://www.uniprot.org/blast

Acknowledgments

We thank Tom Egger, Nathalie Malirat, Montpellier Ressources Imagérie (MRI), and Serge Urbach (Montpellier Proteomics Platform, BioCampus Montpellier) for help and advice. Research in the laboratory of D.M. is supported by grants from Fondation ARC Pour la Recherche sur le Cancer (to A.A.) and MSD Avenir (GNOSTIC, to D.M.).

References

1. Blackford AN, Jackson SP (2017) ATM, ATR, and DNA-PK: the trinity at the heart of the DNA damage response. Mol Cell 66:801–817. https://doi.org/10.1016/j.molcel.2017.05.015
2. Cybulla E, Vindigni A (2023) Leveraging the replication stress response to optimize cancer therapy. Nat Rev Cancer 23:6–24. https://doi.org/10.1038/s41568-022-00518-6
3. Yang Y, Gao Y, Zlatanou A et al (2018) Diverse roles of RAD18 and Y-family DNA polymerases in tumorigenesis. Cell Cycle 17:833–843. https://doi.org/10.1080/15384101.2018.1456296
4. Hoogenboom WS, Klein Douwel D, Knipscheer P (2017) Xenopus egg extract: a powerful tool to study genome maintenance mechanisms. Dev Biol 428:300–309. https://doi.org/10.1016/j.ydbio.2017.03.033
5. Hutchison CJ, Cox R, Drepaul RS et al (1987) Periodic DNA synthesis in cell-free extracts of Xenopus eggs. EMBO J 6:2003–2010. https://doi.org/10.1002/j.1460-2075.1987.tb02464.x
6. Garner E, Costanzo V (2009) Studying the DNA damage response using in vitro model systems. DNA Repair (Amst) 8:1025–1037. https://doi.org/10.1016/j.dnarep.2009.04.015
7. Kermi C, Prieto S, van der Laan S et al (2015) RAD18 is a maternal limiting factor silencing the UV-dependent DNA damage checkpoint in Xenopus embryos. Dev Cell 34:364–372. https://doi.org/10.1016/j.devcel.2015.06.002
8. Lo Furno E, Busseau I, Aze A et al (2022) Translesion DNA synthesis-driven mutagenesis in very early embryogenesis of fast cleaving embryos. Nucleic Acids Res 50:885–898. https://doi.org/10.1093/nar/gkab1223
9. Lemaitre J-M, Vassetzky Y, Méchali M (2001) Analysis of chromatin assembly, chromatin domains, and DNA replication using xenopus systems. In: Mapping protein/DNA interactions by cross-linking. Institut national de la santé et de la recherche médicale, Paris
10. Menut S, Lemaitre J-M, Hair A, Mechali M (1999) DNA replication and chromatin assembly using Xenopus egg extracts. In: Advances in molecular biology: a comparative methods approach to the study of oocytes and embryos, pp 196–226
11. Lo Furno E, Recolin B, van der Laan S et al (2021) Studying the DNA damage response in embryonic systems. Methods Enzymol 661: 95–120. https://doi.org/10.1016/bs.mie.2021.08.016
12. Egger T, Aze A, Maiorano D (2023) Detection of endogenous translesion DNA synthesis in single mammalian cells. Cell Rep Methods, In Press
13. Newport J, Dasso M (1989) On the coupling between DNA replication and mitosis. J Cell Sci Suppl 12:149–160. https://doi.org/10.1242/jcs.1989.supplement_12.13
14. Dasso M, Newport JW (1990) Completion of DNA replication is monitored by a feedback system that controls the initiation of mitosis in vitro: studies in Xenopus. Cell 61:811–823. https://doi.org/10.1016/0092-8674(90)90191-g
15. Conn CW, Lewellyn AL, Maller JL (2004) The DNA damage checkpoint in embryonic cell cycles is dependent on the DNA-to-cytoplasmic ratio. Dev Cell 7:275–281. https://doi.org/10.1016/j.devcel.2004.07.003
16. Carbone A, Fioretti FM, Fucci L et al (2012) High efficiency method to obtain supercoiled DNA with a commercial plasmid purification kit. Acta Biochim Pol 59:275–278
17. ten Have S, Boulon S, Ahmad Y, Lamond AI (2011) Mass spectrometry-based immuno-precipitation proteomics – the user's guide. Proteomics 11:1153–1159. https://doi.org/10.1002/pmic.201000548
18. Dunham WH, Mullin M, Gingras A-C (2012) Affinity-purification coupled to mass spectrometry: basic principles and strategies. Proteomics 12:1576–1590. https://doi.org/10.1002/pmic.201100523
19. Lebofsky R, Takahashi T, Walter JC (2009) DNA replication in nucleus-free Xenopus egg extracts. Methods Mol Biol 521:229–252. https://doi.org/10.1007/978-1-60327-815-7_13
20. Nishiyama A, Frappier L, Méchali M (2011) MCM-BP regulates unloading of the MCM2-7 helicase in late S phase. Genes Dev 25:165–175. https://doi.org/10.1101/gad.614411
21. Recolin B, Van der Laan S, Maiorano D (2012) Role of replication protein A as sensor in activation of the S-phase checkpoint in Xenopus egg extracts. Nucleic Acids Res 40:3431–3442. https://doi.org/10.1093/nar/gkr1241
22. Räschle M, Smeenk G, Hansen RK et al (2015) DNA repair. Proteomics reveals dynamic assembly of repair complexes during bypass of DNA cross-links. Science 348:1253671. https://doi.org/10.1126/science.1253671

23. Zougman A, Selby PJ, Banks RE (2014) Suspension trapping (STrap) sample preparation method for bottom-up proteomics analysis. Proteomics 14:1006–1000. https://doi.org/10.1002/pmic.201300553
24. Cox J, Neuhauser N, Michalski A et al (2011) Andromeda: a peptide search engine integrated into the MaxQuant environment. J Proteome Res 10:1794–1805. https://doi.org/10.1021/pr101065j
25. Cox J, Mann M (2008) MaxQuant enables high peptide identification rates, individualized p.p.b.-range mass accuracies and proteome-wide protein quantification. Nat Biotechnol 26:1367–1372. https://doi.org/10.1038/nbt.1511
26. Schwanhäusser B, Busse D, Li N et al (2011) Global quantification of mammalian gene expression control. Nature 473:337–342. https://doi.org/10.1038/nature10098
27. Cox J, Hein MY, Luber CA et al (2014) Accurate proteome-wide label-free quantification by delayed normalization and maximal peptide ratio extraction, termed MaxLFQ. Mol Cell Proteomics 13:2513–2526. https://doi.org/10.1074/mcp.M113.031591
28. Tyanova S, Temu T, Sinitcyn P et al (2016) The Perseus computational platform for comprehensive analysis of (prote)omics data. Nat Methods 13:731–740. https://doi.org/10.1038/nmeth.3901
29. Hutchins JRA (2014) What's that gene (or protein)? Online resources for exploring functions of genes, transcripts, and proteins. Mol Biol Cell 25:1187–1201. https://doi.org/10.1091/mbc.E13-10-0602
30. Session AM, Uno Y, Kwon T et al (2016) Genome evolution in the allotetraploid frog Xenopus laevis. Nature 538:336–343. https://doi.org/10.1038/nature19840

Chapter 3

Cell Cycle–Specific Protein Phosphatase 1 (PP1) Substrates Identification Using Genetically Modified Cell Lines

Dorothee C Kommer, Konstantinos Stamatiou, and Paola Vagnarelli

Abstract

The identification of protein phosphatase 1 (PP1) holoenzyme substrates has proven to be a challenging task. PP1 can form different holoenzyme complexes with a variety of regulatory subunits, and many of those are cell cycle regulated. Although several methods have been used to identify PP1 substrates, their cell cycle specificity is still an unmet need. Here, we present a new strategy to investigate PP1 substrates throughout the cell cycle using clustered regularly interspersed short palindromic repeats (CRISPR)-Cas9 genome editing and generate cell lines with endogenously tagged PP1 regulatory subunit (regulatory interactor of protein phosphatase one, RIPPO). RIPPOs are tagged with the auxin-inducible degron (AID) or ascorbate peroxidase 2 (APEX2) modules, and PP1 substrate identification is conducted by SILAC proteomic-based approaches. Proteins in close proximity to RIPPOs are first identified through mass spectrometry (MS) analyses using the APEX2 system; then a list of differentially phosphorylated proteins upon RIPPOs rapid degradation (achieved via the AID system) is compiled via SILAC phospho–mass spectrometry. The "in silico" overlap between the two proteomes will be enriched for PP1 putative substrates. Several methods including fluorescence resonance energy transfer (FRET), proximity ligation assays (PLA), and in vitro assays can be used as substrate validations approaches.

Key words Protein phosphatase 1 (PP1), CRISPR-Cas9, APEX2 biotinylation system, AID degradation system, Phosphoproteomics, Mass spectrometry, Synchronization

1 Introduction

The tightly controlled interplay of phosphorylation and dephosphorylation events is crucial for several cellular processes including cell proliferation, signal transduction, and cell cycle progression [1–3]. Phosphorylation is performed by hundreds of kinases, whereas only a limited repertoire of phosphatases is involved in the opposing dephosphorylation events [4, 5]. For cells to progress correctly through the cell cycle, the phosphorylation–dephosphorylation interplay is tightly controlled by cyclin-dependent kinase (CDK)/

Authors Dorothee C Kommer and Konstantinos Stamatiou have equally contributed to this chapter.

Anna Castro and Benjamin Lacroix (eds.), *Cell Cycle Control: Methods and Protocols*, Methods in Molecular Biology, vol. 2740, https://doi.org/10.1007/978-1-0716-3557-5_3,

cyclin complexes and mitotic phosphatases [2, 3, 6–8]. The net protein phosphorylation increases throughout the cell cycle from the G_1 phase to peak at the beginning of mitosis [6, 9, 10]. When cells exit mitosis, dephosphorylation of mitotic proteins increases, and with decreasing kinase activity, the net protein phosphorylation status decreases accordingly [11]. Phosphorylation/dephosphorylation events coupled to protein degradation and the formation of different CDK/cyclin complexes at each stage of the cell cycle ensure the unidirectionality of the process [3, 7]. The active PP1 holoenzyme is formed by the PP1 catalytic subunit (PP1c) and one or more regulatory subunits termed RIPPO (regulatory interactors of protein phosphatase one), which determine PP1 localization and sometimes substrate specificity [12–15]. The binding of most RIPPOs to PP1c is mediated by small linear motifs (SLiMs), with the RVxF motif being the most commonly used [16–18]. PP1 can form many highly dynamic, and therefore transient, holoenzyme complexes with over 200 different RIPPOs that counteract phosphorylation events from hundreds of kinases [14, 19]. The identification of PP1 holoenzyme substrates is not only challenged by the fast kinetic of the dephosphorylation reactions [14], but also because phosphatase complexes oppose up to 400 phosphorylation events performed by as many substrate-specific kinases [4, 13].

PP1 substrate identification has mainly been driven by serendipity or educated guesses. Gerty Cori and Arda Green discovered PP1 and its first substrate, phosphorylase a, by accident while working on glycogen phosphorylase [20]. The research of PP1 in muscle has been a hot research topic for years, and over decades, further substrates like myosin were discovered [21–26].

Some of PP1 substrates were easier discovered than others due to their stronger binding affinities to the PP1 holoenzyme complex. These could be identified by mass spectrometry (MS) analyses of pull-down experiments as for the case of histone H3, a substrate of the PP1/Repo-Man complex [27]. Examples of confirmed substrates include serine 573 of the regulatory subunit B56δ/PPP2R5D of protein phosphatase 2A (PP2A) and serine 22 of lamin A (dephosphorylation target of PP1/Repo-Man complex) [28, 29], serine 51 of eukaryotic initiation factor 2α (eIF2α) (dephosphorylation target of PP1/R15) [30], serine 367 and 455 of IRSp53, and the depolymerization factors cofilin and destrin (dephosphorylation target of PP1/Phactr complex and PP1/Phactr4 complex) [31, 32], just to quote a few.

Multiple technical approaches have been developed, improved, and combined to facilitate the identification of PP1 substrates. Each one has its own advantages and limitations, but we are still far from having a robust and efficient method to identify PP1 substrates at each stage of the cell cycle.

In vivo, artificially induced aberrant protein expression, like depletion or overexpression of a PP1-RIPPO complex, is a simple way to investigate cell phenotypes and compare with those in normal cells to narrow down substrates for a specific holoenzyme complex. However, it already requires a specific research question, and the observed phenotypes need to be thoroughly investigated to discriminate between direct or downstream effects.

Mass spectrometry is widely used to study protein–protein interactions. Affinity purification–mass spectrometry (AP-MS) can be used to identify these interactions by adding an affinity tag (i.e. FLAG, GFP, Strep, His_6-tag) N- or C-terminally to a protein of interest (POI) which is part of the protein complex [33, 34]. Considering the transient nature of phosphatase complexes/substrates binding, it is better to choose a single-step purification over a multi-step one as it allows for weaker and more dynamic complex–substrate to be captured [35]. A few false-positive interactions can be identified using MS approaches with proximate proteins being detected and mistaken as substrates [35]. In addition, unspecific binding within the cell lysate must be considered as cell compartments are destroyed during the procedure, and proteins that would never get into contact in an intact cell are artificially brought together [33]; moreover, using this approach, many transiently binding substrates can be missed due to experimental handling during co-purification steps like sample lysis and washing, as these processes can be harsh for complex–substrate binding [36]. MS methods which fuse PP1 to a specified RIPPO to facilitate substrate trapping and subsequent substrate identification have also been developed [28]. Limitations of these MS approaches are that they take advantage of protein overexpression in cells resulting in protein levels which differ from the physiological ones and could potentially lead to changes in protein stoichiometry [37, 38]. To improve in the quest of PP1 substrates, we developed an approach based on tagging the RIPPO of interest endogenously by genome editing using CRISPR Cas9; in this way, the RIPPO is expressed under physiological conditions [39].

In-cell approaches have also been developed using proximity-dependent biotinylation (PDB) called proximity-dependent biotin identification (BioID) combined with MS analysis [40]. These involve a fusion construct with BirA, an *Escherichia coli* biotin ligase [40]. Proximate proteins to the phosphatase of interest are biotinylated, isolated by a biotin affinity capture step, and then analyzed by MS [40]. However, BirA is a very large tag which could potentially lead to incorrect protein localization and protein interaction impairment [40]. More limiting, from a cell cycle point of view, are the slow biotinylation kinetics (18–24 h), leading to difficulties in capturing dynamic phosphatase–substrate complexes [40, 41]. This method was improved by engineering a smaller biotin label [42] that brings the advantage of better labelling efficiencies and specificity, thus providing the opportunity for capturing proximate proteins and short-lived complex–substrate formations by shorter labelling pulses [42, 43]. However, some limitations remain, such as the distinction between interacting proteins and neighboring proteins. A novel approach termed *split–luciferase-based protein–protein interaction assay,* which features one part of the luciferase sensor fused to PP1 and the complementary part of the luciferase sensor fused to its substrate, can be used to determine PP1 holoenzyme interaction partners or to observe disruption of this interaction [44–46]. A luciferase signal is emitted or disrupted when two luciferase sensors are complemented or separated [44, 45]. This approach was recently employed to show that robenidine, used as an antibiotic in animal feed, can be used as a cytoprotective therapeutic without disrupting PP1/PP1R15A-dependent dephosphorylation of the eIF2α [46].

Immunofluorescence and fluorescence microscopy approaches have proven very insightful for studying the co-localization of PP1 holoenzyme complex with a substrate. For these microscopy approaches, proteins of the PP1 holoenzyme complex are overexpressed in the cell. Caveats intrinsic to the methods are protein localization [47, 48], protein aggregation [49], cell organelle morphology [50, 51], or even protein–protein interactions being artifactual [37, 38]. Substrate signal enrichment or depletion, substrate recruitment, and PP1 holoenzyme complex localization can also be studied by microscopy using RNA interference (RNAi) experiments; for example, RIPPO depletion gave insights into dephosphorylation targets and changes in the phosphorylation levels of substrates [27, 52], PP1 localization [53–55], and abnormal formation of specific PP1 holoenzyme complexes leading to changes in their activity [56].

A popular microscopy-based approach is fluorescence resonance energy transfer (FRET), which is used to study protein–protein interactions and their spatial occurrence [15]. This technique takes advantage of two fluorophores with different excitation and emission wavelengths that, once in close proximity, transfer energy from the excited fluorophore to the non-excited fluorophore, which subsequently emits light [57–59]. Fluorescence lifetime imaging *microscopy* (FLIM) and FRET were successfully combined to study the lifetime of PP1 holoenzyme complexes [60] and substrate binding of a PP1 holoenzyme complex [27]. There are inherent limitations of these approaches, including (I) the choice of FRET pairs due to emission/excitation profile overlap, (II) challenges in capturing images can occur due to adjusting capture settings to investigate protein dynamics which can lead to inadequate measurement precision, and (III) false positives in signal recovery of the fluorophore instead of giving a real insight into the protein dynamics itself [61].

Besides the previously mentioned wet-lab approaches, PP1 holoenzyme substrate identification can also be explored using a bioinformatic approach, as a variety of software and web pages like GPS [62], DISPHOS [63], NetPhos [64], NetPhosYeast [65], and GANN-Phos [66] can predict phosphorylation sites which can be subsequently linked to specific molecular pathways and possible phosphatases involved. Furthermore, databases like EKPD [67], HuPho [68], and DEPOD [69] harbor in-depth information about phosphatases (and their RIPPOs) and kinases alike, as well as pathways, structures, de-phosphorylation sites, and previously discovered substrates. These bioinformatic tools can provide valuable information to facilitate the identification of potential PP1 holoenzyme substrates, but the results are only predictive and still need to be verified and investigated using an experimental approach. For example, DEPOD predicts which protein is dephosphorylated by a phosphatase complex, but it cannot predict which exact residue is dephosphorylated [70]. However, computational approaches need a robust training set for users to achieve reliable results; as the repertoire of confirmed substrates is still limited, this also affects the robustness of bioinformatics prediction [71, 72]. A breakthrough deep learning-based model has been developed which predicts probable phosphatase dephosphorylation sites. The transfer learning-based approach can learn from the pattern of smaller data sets with confirmed dephosphorylation sites of phosphatases [73]. Extensive phosphoproteomics studies have been recently undertaken to gain insight into global phosphorylation changes throughout the cell cycle [74, 75]. It was shown that PP1 depletion led to a delay in dephosphorylation events during mitosis and increase in phosphorylation status of a subset of proteins in early mitosis, highlighting around 300 new possible PP1 substrates [74]. The analyses of mitotic exit timing also revealed that PP1 presents dephosphorylation preferences for phospho-threonine (TP) when upstream of a basic amino acid sequence over a phospho-serine (SP) [74, 76]. The preference of dephosphorylation of phospho-threonine during mitotic exit in budding yeast has been linked to the yeast phosphatase $PP2A^{Cdc55}$ [75, 77] and was also shown for the human counterpart $PP2A^{B55}$ [74, 78, 79].

In human cells as well, global TP sites are rapidly lost during mitotic exit compared to SP sites and the amino acids surrounding the phosphosite are also extremely important for the timing of de-phosphorylation with TPG motives being more efficiently dephosphorylated [80]. More studies will be required to build a robust list of PP1 substrates and understand their specificity.

Here, we present an alternative perspective for the identification of PP1 substrates. Our experimental approach includes gene editing of RIPPOs of interest, thus overcoming the problems associated with protein overexpression. We used the human HCT116 cell line due to its high rate of homology-directed repair and stable karyotype [81]. Protein tags are added to the gene of interest (GOI) together with a fluorophore tag. The fluorescent tag allows localization/co-localization of the tagged RIPPO to be followed using microscopy approaches and can be also used for immunoprecipitation experiments and FRET experiments. The other protein tag is either used for protein degradation [82] or biotinylation of proximate proteins [83] for MS analysis and co-precipitation experiments. Protein degradation is achieved by the auxin-inducible degron (AID) technology which drives ubiquitination of the tagged protein after the addition of indole-3-acetic acid (IAA), a plant hormone, and marks it for degradation by the proteasome [82]. The cell line also expresses the *Oryza sative*-derived F-box protein (OsTir1) [84]. OsTir1 forms with SCF a functional E3 ubiquitin ligase which subsequently polyubiquitylate the tagged RIPPO [84]. Protein degradation occurs with a fast kinetic (30′–2 h for most of the proteins analyzed), thus allowing easy synchronization and interrogation of events in specific windows of the cell cycle without compromising others, therefore overcoming the limitations of RNAi methods.

This strategy, in combination with the protein tag ascorbate peroxidase 2 (APEX2) [83], allows to study PP1 holoenzyme complex–substrate binding dynamics in different cell cycle phases. The in-cell biotinylation approach avoids false-positive results and substrate dissociation caused by the breakage of cell compartments [33] and harsh purification, thus allowing the capture of highly dynamic and transient complex–substrate formations [85].

In summary, this method fulfills the so far unmet need of being able to identify PP1 holoenzyme complexes and substrates, including short-lived complex–substrate formations, throughout all the stages of the cell cycle.

2 Materials

All experiment components should be prepared fresh (unless otherwise specified). All buffers used for subsequent MS analyses are filtered with a 0.22 μm filter. All buffers are made in dH_2O if not otherwise specified (*see* **Note 1**).

We used HCT116 cell lines expressing OsTIR1 under the control of the constitutive cytomegalovirus (CMV) (CMV-OsTIR1 HCT116) or conditional TET-ON promoter (TET-ON-OsTIR1 HCT116) generated by Masato T. Kanemaki's lab [82].

2.1 Cell Culture Reagents

1. HCT116 culture medium: McCoy's 5A (Modified) GlutaMax™ Medium supplemented with 1% penicillin/streptomycin and 10% foetal bovine serum.
2. HCT116 TET-ON SILAC medium: McCoy's 5A Media for SILAC supplemented with 1% penicillin/streptomycin and 10% dialysed foetal bovine serum. 0.1 mg/mL light arginine [$C_6H_{14}N_4O_2$ HCl], 0.1 mg/mL light lysine [$H_2N(CH_2)4CH(NH_2)CO_2H{\cdot}HCl$], 0.1 mg/mL heavy labelled arginine [H215N13C(15NH)15NH (13CH_2)313CH(15NH_2)13$CO_2H{\cdot}HCl$], 0.1 mg/mL heavy labelled lysine [H215N(13CH_2)413CH (15NH_2)13$CO_2H{\cdot}HCl$], 0.1 mg/mL L-leucine, and 2 mM L-glutamine are added.
3. Indole-3-acetic acid (IAA): dissolve in DMSO to achieve 500 mM stock concentration.
4. Doxycycline: dissolve in dH_2O to achieve a 2 mg/mL stock concentration and then filter with a 0.22 μm filter.
5. DNA/siRNA transfection reagent such as jetPRIME®: versatile DNA/siRNA transfection reagent.
6. Hygromycin B.
7. Geneticin: G418.
8. 37% Paraformaldehyde (PFA): dissolve 0.9 g PFA in 1.25 mL distilled H_2O by bringing it to the boil and adding potassium hydroxide (KOH) until solution appears clear. Dilute the 37% PFA in PBS to create a 4% PFA solution and store at 37 °C until usage.
9. Biotin–phenol: dissolve biotin–phenol powder in DMSO to achieve a stock concentration of 500 mM.
10. H_2O_2 30% w/w solution.
11. 4× Laemmli buffer: 250 mM Tris-HCl (pH 6.8), 8% SDS, 40% glycerol, bromophenol blue, 20% β-mercaptoethanol. Dilute 4× Laemmli buffer 1:2 or 1:4 to achieve final concentration of 2× Laemmli buffer or 1× Laemmli buffer.
12. Thymidine: dissolve thymidine powder in sterile H_2O to achieve a stock concentration of 200 mM.
13. RO-3306: dissolve thymidine powder in sterile H_2O to achieve a stock concentration of 10 mM.
14. Nocodazole: dissolve nocodazole powder in DMSO to achieve a stock concentration of 1 mg/mL.
15. MG132: dissolve MG132 powder in sterile H_2O to achieve a stock concentration of 20 mM.
16. Spermidine.

2.2 Reagents

2.2.1 APEX2 Biotinylation

1. Stop buffer: 10 mM sodium ascorbate solution, 5 mM Trolox solution, 10 mM sodium azide in PBS.
2. IGEPAL solution: 50 mM Tris-HCl (pH 7.5), 0.5% IGEPAL, 200 mM NaCl.
3. Lysis buffer 10 mL: 6 mL IGEPAL solution, 10 mM sodium ascorbate solution, 5 mM Trolox solution, 10 mM sodium azide.

2.2.2 Phospho-proteomics

1. Phosphoproteomics lysis buffer: 10 mM Tris-HCl (pH 7.4), 2 mM $MgCl_2$, 2 mM EGTA, 0.1% Triton X-100, 590 mM NaCl.
2. Hypotonic buffer: 10 mM HEPES-NaOH (pH 7.9), 1.5 mM $MgCl_2$, 10 mM KCl (*see* **Note 2**).
3. Polylysine solution: Dilute 0.1% of polylysine in PBS to 0.01% final concentration (v/v).

2.3 Websites

1. www.ensembl.org/index.html
2. http://crispor.tefor.net

2.4 Primers

1. Primers with base-pair sequences to amplify antibiotic resistance cassettes: neomycin_FW 5′- CGTTGGCTACCCGTGATATT-3′, neomycin_REV 5′- GCCCAGTCATAGCCGAA TAG-3′, hygromycin_FW 5′- GCTGTGTAGAAGTACTC GCC-3′, hygromycin_REV 5′-AGACGCTGTCGAACTTT TCG-3′.
2. Forward and reverse primers with base-pair sequences specific to the POI.

2.5 Cell Lines

1. HCT116 TET-ON.
2. HCT116 CMV.

2.6 Cloning Reagents

1. pGEM®-T Easy Vector Systems (Promega, Cat# A1360).
2. BamHI restriction enzyme and CutSmart™ Buffer.
3. Calf intestinal alkaline phosphatase (CIP).
4. T4 ligase and T4 ligase buffer.
5. BbsI restriction enzyme and CutSmart™ Buffer.
6. pX330-U6-Chimeric-BB-CBh-hSpCas9 (Addgene Plasmid #42230).

3 Methods

HCT116 cell lines are cultured at 37 °C (5% CO_2 in a humidified incubator) if not stated otherwise. Perform incubation conditions at 37 °C (5% CO_2 in a humidified incubator) if not stated otherwise.

3.1 Generation of Recipient Plasmid and CRISPR Gene Editing Donor Plasmids

1. For C-terminal tagging of the POI, identify the genomic DNA sequence around the STOP codon using ENSMBL (www.ensembl.org/index.html) (*see* **Note 3**).
2. Design specific primers to amplify approximately 500 bp upstream and downstream of the STOP codon (Fig. 1(1)).
3. Isolate genomic DNA from cultured HCT116 TET-ON/CMV cell line.
4. Amplify the sequence of interest using a standard PCR protocol. Verify the PCR product size with standard agarose gel electrophoresis.
5. Clone the PCR product into a pGEM®-T Easy Vector (Promega, Cat# A1360) and subsequently Sanger sequence to confirm the genomic DNA sequence (Fig. 1(1)). This step is essential due to allelic polymorphisms that can affect the targeting efficiency. Screen the region around the STOP codon for suitable guide RNA binding sequences using the CRISPOR website (http://crispor.tefor.net) (*see* **Note 4**) (Fig. 1(4)).
6. Introduce silent mutations within the chosen protospacer adjacent motif (PAM) sequences to prevent future Cas9 re-cutting at the same location once the genome was edited (Fig. 1(2)). To generate the recipient plasmid containing the homology arm sequence of POI and to insert the AID or APEX2 tag, change the STOP codon to a BamHI restriction enzyme digest sequence (Fig. 1(2)). Synthesize the modified PCR product in a plasmid vector of choice (*see* **Note 5**).
7. To insert the AID or APEX2 tag into the plasmid with the synthesised PCR product (recipient plasmid), digest the plasmids using 10 units of BamHI (*see* **Note 6**) (Fig. 1(3)). Incubate digest for 1 h at 37 °C.
8. Add 1 unit of CIP directly to the digestion mix of the recipient plasmid to prevent plasmid self-ligation. Incubate for 1 h at 37 °C. Verify the digestion with standard agarose gel electrophoresis. Isolate and purify the desired bands from the agarose gel using a standard gel purification kit.
9. Ligate the AID or APEX2 tag plasmid fragments into the linearised recipient plasmid using 400 units of T4 ligase overnight at 16 °C (*see* **Note** 7) (Fig. 1(3)).
10. Verify the successful generation of CRISPR gene editing donor plasmids by Sanger sequencing.

3.2 Generation of the guideRNA

1. Synthesize the guideRNA sequences (sense and antisense) best suited to the last exon of your POI with BbsI overhangs (Fig. 1 (4)).
2. To insert the guideRNA sequence into the plasmid carrying the Cas9 gene, digest the plasmid pX330-U6-Chimeric-BB-CBh-

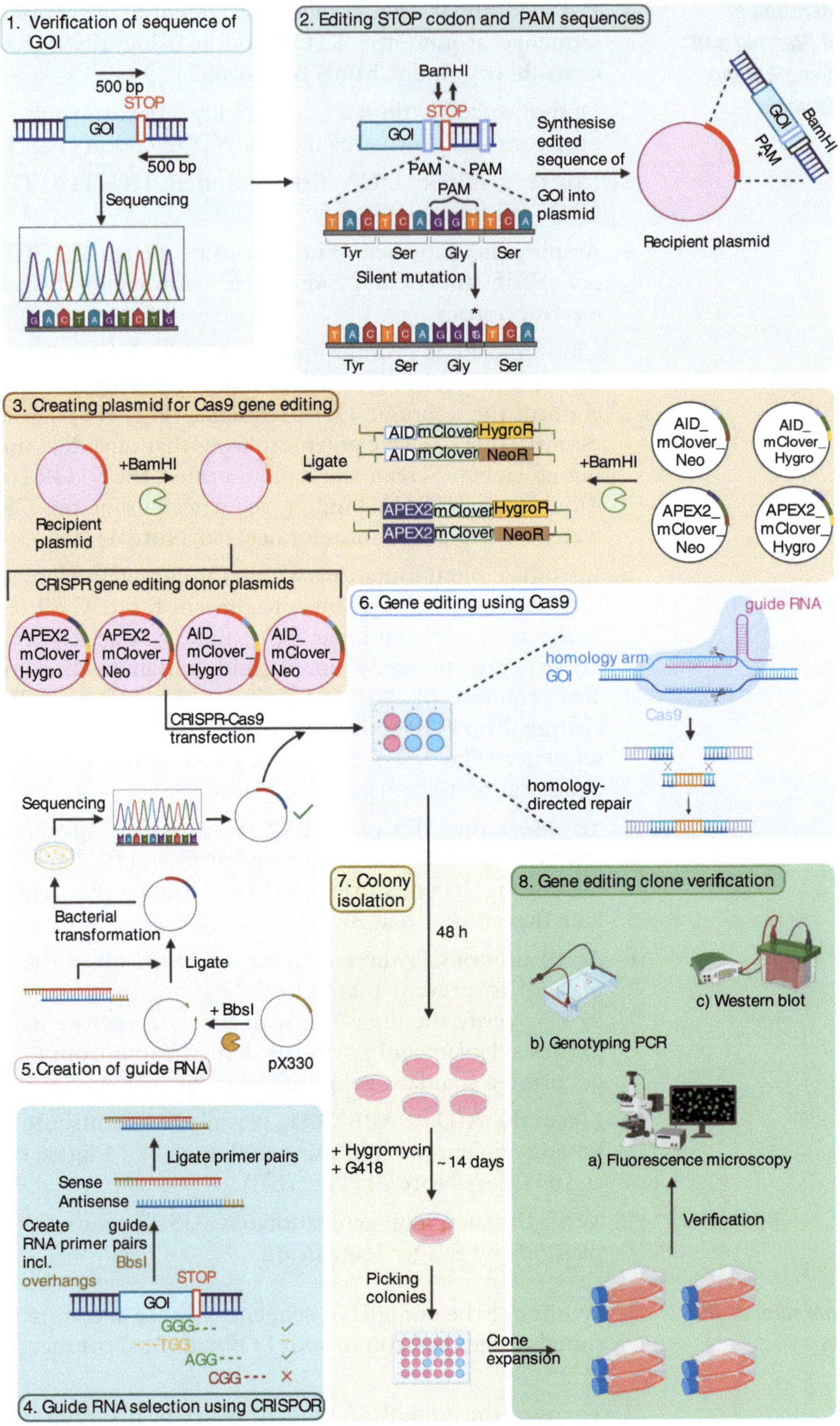

Fig. 1 Schematic workflow illustrating the gene editing procedure at the C-terminus of the GOI in HCT116 TET-ON cell line using CRIPSR Cas9. (**1**) Schematic representation of genomic sequence verification of

hSpCas9 (Addgene Plasmid #42230) with 10 units of BbsI for 1 h at 37 °C (Fig. 1(5)). Isolate and purify the linearized plasmid by agarose gel electrophoresis.

3. To anneal the corresponding sense and antisense guideRNA sequences, dilute 10 μM of sense and antisense guideRNA in 10× T4 ligase buffer. Place the ligation mix in a PCR machine and anneal using the following program: 30 min at 37 °C, 5 min at 95 °C, and ramp to 25 °C (with 5 °C decrease per minute) (Fig. 1(4)).
4. Ligate the linearized pX330-U6-Chimeric-BB-CBh-hSpCas9 with the annealed guideRNA using 400 units of T4 ligase overnight at 16 °C. The generated plasmid is the plasmid construct pX330-U6-Chimeric-BB-CBh-hSpCas9-gRNA (Fig. 1(5)).
5. Verify the ligation by Sanger sequencing.

3.3 Generation of Endogenously Tagged HCT116 TET-ON/CMV Cell Line and Validation

1. Seed 2 * 10^5 cells of HCT116 TET-ON/CMV per well in a 6-well plate in culture medium. This should be done at least 24 h before transfection.
2. Transfect cells using a DNA/siRNA transfection reagent with 2 μg of total plasmid DNA: pX330-U6-Chimeric-BB-CBh-hSpCas9-guideRNA, CRISPR gene editing donor plasmids

Fig. 1 (continued) ~500 base pairs around the STOP codon (red square) of the GOI. (**2**) Schematic representation of editing the STOP codon sequence to a BamHI restriction site and introducing silent mutations (GGT (glycine) → GGG* (glycine)) in the PAM sequences to prevent continuous Cas9 cleavage of genomic DNA once edited (the genomic DNA sequence is only for illustration purposes; the actual sequence can differ). Synthesize the edited GOI sequence into a plasmid vector (*see* **Note 5**) to obtain the recipient plasmid (pink plasmid). (**3**) Schematic representation of the restriction digest (BamHI) of the donor plasmids harboring the protein tags: AID (light grey cassette) or APEX 2 (dark blue cassette), mClover2 (green cassette), neomycin (neo, brown cassette), or hygromycin (hygro, yellow cassette). The recipient plasmid is digested with BamHI. The digested plasmid fragments containing AID or APEX2 protein tags are ligated into the digested recipient plasmid to create the different CRISPR gene editing donor plasmids. (**4**) Schematic representation of designing suitable guideRNA sequences for the GOI using CRISPOR (http:/crispor.tefor.net/) (sequences themselves and the location of the PAM sequences may differ). Sequences are synthesized as primers with BbsI restriction enzyme overhangs and annealed. (**5**) Schematic representation of creating guideRNA plasmid by ligating the annealed RNA sequence into a BbsI digested pX330 vector. Confirmed guideRNA plasmids are used for CRISPR transfection. (**6**) Schematic representation of the CRISPR transfection. The transfection mix consists of either one or two CRISPR gene editing donor plasmids with different resistance cassettes and one guideRNA plasmid. Also shown is a cartoon representation of the gene editing process using CRISPR Cas9. GuideRNA (pink) guides the Cas9 (grey) to the homology arm of the GOI (blue). The protein tags (orange) are inserted C-terminally of the GOI by homology-directed repair using the CRISPR gene editing donor plasmids as template. (**7**) Schematic representation of clone isolation of successfully gene edited cells. (**8**) Successful gene editing is confirmed by (I) fluorescence microscopy, (II) genotyping PCR, and (III) Western blot. (Illustration created with BioRender.com)

Hygromycin B resistance, and Neomycin resistance (1:1:1 ratio) (Fig. 1(6)). Incubate the transfected cells for 48 h.

3. Distribute the transfected cells into five 10 cm cell culture dishes in culture medium containing 100 μg/mL hygromycin B and 700 μg/mL G418 (*see* **Note 8**) (Fig. 1(7)).
4. Replace the medium every 3–4 days with fresh medium containing the antibiotics hygromycin B and G418 (*see* **Note 9**).
5. After ~14 days, transfer the colonies from each 10 cm dish into 24-well plates for expansion (*see* **Notes 9–11**).
6. For the verification of POI localization, seed some cells from each colony on coverslips and screen by microscopy for the correct localization of the POI throughout the cell cycle (Fig. 1 (8)). Continue to expand positive clones to perform genotyping PCR.
7. Isolate genomic DNA from the expanded clones.
8. Design primers which bind further upstream and downstream the POI homology arms used to generate the CRISPR gene editing donor plasmids. Design primers for the antibiotic resistance genes neomycin and hygromycin (e.g. neomycin_FW 5′-CGTTGGCTACCCGTGATATT-3′, neomycin_REV 5′- GC CCAGTCATAGCCGAATAG-3′, hygromycin_FW 5′- GCTG TGTAGAAGTACTCGCC-3′, hygromycin_REV 5′-AGACGC TGTCGAACTTTTCG -3′). Using primers outside of the homology arms, we can confirm if the GOI was edited when compared to not gene-edited HCT116 Tet-ON (Fig. 1(8)).
9. To verify the efficiency of the AID system, seed cells from the selected clones on coverslips in culture medium and treat them with 1–2 μg/mL doxycycline for 24 h to activate the expression of OsTIR1 prior to induce protein degradation (*see* **Note 12**). To degrade the protein, add 500 μM IAA to the medium (*see* **Notes 13** and **14**). Fix the coverslips 1, 2, 3, and 4 h after the addition of IAA with 4% paraformaldehyde (PFA) for 5 min. Immunostain the coverslips with an antibody against the POI. This is to validate the protein degradation by comparing the loss of fluorescent signal and the loss of antibody signal.
10. To verify the APEX2 biotinylation system, seed cells from the selected clones in 6-well plates in culture medium and treat for 30 min with 1 mM, 2.5 mM, or 5 mM biotin–phenol diluted in culture medium.
11. Following biotinylation, discard the culture medium containing biotin–phenol and add 2 mM H_2O_2 solution (*see* **Note 14**) to each well and incubate for 1 min at room temperature. Then, remove the H_2O_2 solution and wash the cells in each well three times with stop buffer (*see* **Note 14**).

12. Proceed to direct lysis adding 30 μL 1× Laemmli buffer per 500,000 cells. Use the total cell lysate for Western blotting. Immunoblot with HRP-conjugated streptavidin (*see* **Note 15**).
13. In addition, seed cells of the positive clones on coverslips in culture medium and treat with biotin–phenol as described earlier. Following the same as in Subheading 3.3, **step 11**. Fix the coverslips with 4% PFA and immunostain with Cy3-conjugated streptavidin. Analyze the Cy3 signal by fluorescence microscope (*see* **Note 16**).

3.4 Subcloning of the Endogenously Tagged HCT116 TET-ON Cell Line (See Note 17)

1. Culture positive clones to 100% confluency in culture medium. Collect the conditioned medium centrifuge at 500× *g* for 5 min and filter with 0.22 μm filter. Add to the filtered conditioned medium 20% FBS and make up with fresh culture medium to a final volume of 21 mL.
2. Trypsinize and count the cells.
3. Add 35 cells to the medium prepared in **step 1** (*see* **Note 18**)
4. Distribute this mix into each well of a 96-well plate and let grow for 2 weeks.
5. Expand clones and check the POI degradation efficiency as described in Subheading 3.3, **step 10**.

3.5 Sample Preparation for Mass Spectrometry Analysis (APEX2 Interactome): Finding Interactors at G1-/S-Phase

1. Grow HCT116-APEX2 (and HCT116 TET-ON unedited cell line as control) in SILAC culture medium for at least six passages (*see* **Note 19**).
2. Treat the cell lines with 2 mM thymidine diluted in SILAC culture medium containing the specific heavy and light amino acids for 18 h (Fig. 2(1)).
3. Dilute biotin–phenol in pre-warmed SILAC culture medium containing the specific heavy and light amino acids (37 °C) to a final concentration of 2.5 mM and follow the same procedure as in Subheading 3.3, **step 12**.
4. Remove the stop buffer, trypsinize the cells, resuspend the cells in stop buffer, count the cells, and mix them in a 1:1 (heavy/light) ratio. Centrifuge the cells at 500× *g* for 5 min and discard the supernatant (Fig. 2(1)).
5. Resuspend the cell pellet with 2 mL IGEPAL lysis buffer and incubate on ice for 30 min.
6. Equilibrate glutathione agarose beads with IGEPAL lysis buffer. Centrifuge the beads at 500× *g* for 5 min at 4 °C. Repeat this three times and leave them on ice until needed.
7. After the 30 min incubation, sonicate the cells (*see* **Note 20**).
8. Transfer the lysate to the equilibrated glutathione beads. After mixing, incubate for 1 h at 4 °C with rotation.

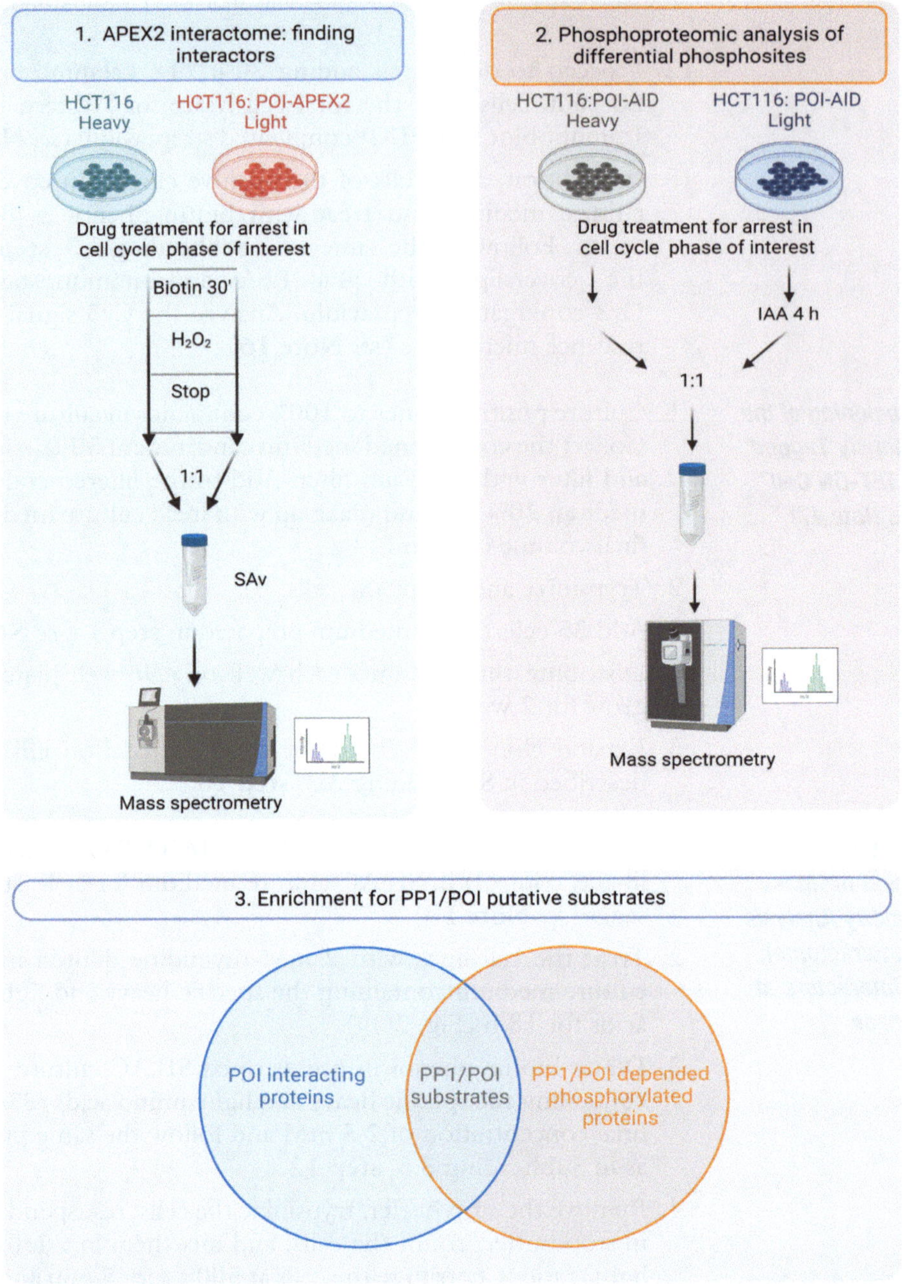

Fig. 2 Schematic workflow illustrating the procedure followed for PP1/POI substrate identification. (**1**) Schematic representation of POI interactor identification. HCT116 TET-ON and HCT116: POI-APEX2 cell lines cultured in SILAC medium. Cells were blocked at the desired cell cycle phase, treated with biotin for 30 min and with H_2O_2 for 1 min, washed 3 times with stop buffer, and mixed in 1:1 ratio. The biotinylated proteins were purified with streptavidin beads and subjected to mass spectrometry. (**2**) Schematic representation of phosphoproteomic analysis of differential phosphosites. HCT116: POI-AID cell line cultured in SILAC medium. Cells were blocked at the desired cell cycle phase, POI was degraded for 4 h after the addition of IAA, control and IAA-treated cells were mixed 1:1, and the total protein lysate was subjected to phospho–mass spectrometry. (**3**) Schematic representation of enrichment for PP1/POI putative substrates. In silico overlap of POI interacting proteins and PP1/POI depended phosphorylated proteins results in the enrichment of PP1/POI substrates. (Illustration created with BioRender.com)

9. Centrifuge at 500× *g* for 5 min at 4 °C. Wash the sample three times with IGEPAL lysis buffer for 10 min at 4 °C while rotating. Then, centrifuge at 500× *g* for 5 min at 4 °C.
10. Add 100 μL of 2× Laemmli buffer to the sample. Heat the sample to 95 °C for 20 min. The sample can be stored at −20 °C until MS analysis for validation of interactors (Fig. 2(1)).

3.6 Identification of Protein Interactors at Mitotic Exit (See Note 21)

1. For the identification of interactors at mitotic exit, treat the HCT116 TET-ON APEX2 cells with 2 mM thymidine diluted in culture medium for 18 h (Fig. 2(1)).
2. Wash three times with PBS to release the cells from thymidine. Incubate for 10 min incubation with culture medium (thymidine-free).
3. Grow cells in fresh culture medium for 8 h. Then, treat with 10 μM RO-3306 (a Cdkl inhibitor) for 18 h (Fig. 2(1)).
4. Wash the cells three times with PBS to release the cells from RO-3306. Incubate 10 min incubation with culture medium (RO-3306-free).
5. After the release, treat the cells with 0.06 μg/mL nocodazole for 2–4 h to arrest the cells in prometaphase (Fig. 2(1)).
6. Isolate mitotic cells by shaking the plate (mitotic shake off) (*see* **Note 22**).
7. Centrifuge the mitotic cells 200× *g* for 5 min and resuspend the pellet in medium containing 0.06 μg/mL nocodazole and 20 μM MG132.
8. Prepare the polylysine-coated 6-well plate by adding 2 mL of the polylysine solution in each well and incubate for 2 h at room temperature, and then remove the polylysine solution and wash three times with PBS.
9. Seed at least 1×10^6–2×10^6 cells in. Incubate the cells for 1 h.
10. Wash three times with PBS to release the cells from nocodazole and MG132 and add fresh medium only containing 20 μM MG132, to arrest the cells in metaphase. Incubate the cells for 2 h (*see* **Note 23**) (Fig. 2(1)).
11. Wash three times with PBS to release the cells from MG132 and add culture medium (drug-free). About 80 min after the release, the cells reach anaphase (*see* **Note 24**).
12. After 130 min, the cells have progressed in telophase/cytokinesis (*see* **Note 25**).
13. After 180 min time, the cells have progressed in early G1 (*see* **Note 26**).

In all the steps, the cells are mixed 1:1 (TET-ON (heavy)/APEX2 (light)) and follow the same steps as in Subheading 3.5 (*see* **Notes 27** and **28**) (Fig. 2(1)).

3.7 Phosphoproteomic Analysis of Differential Phosphosites in G1-/S-Phase

1. To perform the phosphoproteomics analyses in cells depleted of the POI in G1-/S-phase, culture the HCT116 TET-ON AID cell line in SILAC culture medium (heavy chain:control, light chain:IAA) (Fig. 2(2)).
2. Treat the cells with 2 mM thymidine diluted in culture medium for 18 h to synchronize the cells at G1-/S-phase boundary (Fig. 2(2)).
3. Treat the cells with 2 μg/mL doxycycline diluted in culture medium 24 h prior to the IAA treatment. Degrade the POI by treating with 1000 μM IAA for 4 h (Fig. 2(2)).
4. Then trypsinize the cells and mix control/IAA in a ratio 1:1 (Fig. 2(2)).
5. Centrifuge the cells and resuspend the pellet in phosphoproteomics lysis buffer (~100 μL/10^7 cells). Incubate for 30 min on ice.
6. After the incubation, sonicate the cells (*see* **Note 20**) and centrifuge the cell lysate at 20,000× *g* for 30 s at 4 °C.
7. Keep the supernatant for subsequent phosphoproteomics analysis (*see* **Note 29**) (Fig. 2(2)).

3.8 Phosphoproteomic Analysis of Differential Phosphosites on Mitotic Chromatin

1. To perform the differential phosphoproteomics analysis in mitosis, follow the protocol steps in Subheading 3.7 until **step 4**. Instead of blocking in thymidine as described in Subheading 3.7, synchronize the cells with 0.06 μg/mL nocodazole for 18 h to block them in prometaphase.
2. Isolate mitotic cells by shaking the plate (mitotic shake off).
3. Centrifuge the mitotic cells 200× *g* for 5 min
4. Supplement hypotonic buffer with 0.15 mM spermidine (*see* **Note 30**).
5. Resuspend the pellet in hypotonic buffer and incubate the cells for 30 min on ice.
6. Pre-cool the Dounce homogenizer and the pestle on ice for at least 5 min. Transfer the cell suspension to the pre-chilled Dounce homogenizer and disrupt the cell membrane using ~40 strokes of the tight-fitting pestle (*see* **Note 31**).
7. Transfer the sample to a 1.5 mL tube and centrifuge at 200× *g* for 5 min at 4 °C.
8. Carefully transfer the supernatant containing the DNA to a new tube.

9. To separate the cell cytoplasm from the DNA, centrifuge the supernatant at 1600× *g* for 30 min at 4 °C.
10. Carefully transfer the supernatant containing the cytoplasm to a new tube. Wash the pellet containing the DNA three times with hypotonic buffer and centrifuge at 1600× *g* for 30 min at 4 °C.
11. Centrifuge the fraction containing the cytoplasm fraction at 20,000× *g* for 30 min at 4 °C. Collect carefully the supernatant containing cytoplasmatic proteins.
12. Add the phosphoproteomics lysis buffer and incubate it for 30 min on ice.
13. Sonicate the fraction containing the DNA (*see* **Note 32**) and centrifuge the cell lysate at 20,000× *g* for 30 s at 4 °C.
14. Keep the supernatant for subsequent phosphoproteomic analysis.

3.9 Phosphoproteomic Analysis of Differential Phosphosites at Mitotic Exit

1. To perform the phosphoproteomic analysis during mitotic exit from cells where the POI is degraded, follow the protocol steps in Subheading 3.7 until **step 3**. For this phosphoproteomic approach, the synchronization time with 0.06 μg/mL nocodazole for 18 h to block them in prometaphase. Collect mitotic cells after this synchronization step by mitotic shake off and transfer them into another polylysine-coated cell culture dish.
2. After 4 h of IAA treatment, release the cells from nocodazole by washing three times with PBS and incubate them 10 min with culture medium (drug-free) (Fig. 2(2)).
3. Incubate the cells with culture medium containing 2 mM thymidine for 18 h.
4. Trypsinize the cells and mix them in a 1:1 ratio (control/IAA). Follow the protocol from Subheading 3.7 from **step 4** onward (Fig. 2(2)).

Phosphoproteomic analysis of differential phosphosites in different cell cycle phase (*see* **Note 33**).

3.10 Enrichment for PP1/POI Putative Substrates

1. Find the overlap in silico of the APEX2 interactor proteome and phosphoproteome (Fig. 2(3)).
2. The overlap list should contain the PP1 putative substrates (*see* **Note 34**) (Fig. 2(3)).

4 Notes

1. Make sure that all buffers during for APEX2 biotinylation and the phosphorproteomic experiments are containing fresh HALT inhibitor cocktail (sodium fluoride, sodium

orthovanadate, β-glycerophosphate, sodium pyrophosphate, aprotinin, bestatin, E64, leupeptin, and EDTA). This is to make sure that none of the proteins is degraded and the phosphorylations are maintained. The buffers that must contain the HALT inhibitor cocktail are lysis buffer for APEX2 biotinylation reaction, hypotonic buffer, and phosphoproteomics lysis buffer.

2. Store at 4 °C.
3. Intronic sequence must be included.
4. We chose three guideRNAs quite close to the STOP codon (no further than 150 bp either way from the STOP codon); however, the chosen amount of guideRNAs is user dependent. In addition, the location of the guideRNAs can vary depending on specificity, efficiency, and off-target rate calculated by the web page CRISPOR. Although some guideRNAs might have a high off-target rate, the gene editing success depends on mismatched nucleotides in the guideRNA sequence. Therefore, even if many off-targets are calculated, the guideRNA might be still good and effective if many mismatches are predicted.
5. We chose either pGEM-T easy or pBluescript SKII (+) as plasmid vectors.
6. 10 units of BsaI can be used to avoid contaminations when isolating the inserts. This increases the efficiency of subcloning the insert to the recipient plasmid. We strongly recommend this digestion.
7. We started with a 3:1 insert:vector ratio. If this wasn't successful, we increased the ratio to 5:1 or 8:1.
8. Antibiotic selection of the clones depends on the resistance cassette within the CRISPR Cas9 plasmids. We used the following antibiotics concentrations: 1 μg/mL puromycin, 700 μg/mL geneticin G-418, and 100 μg/mL hygromycin B. The concentration of the antibiotics can be increased when using only one resistance cassette plasmid. We have successfully conducted a single selection with hygromycin B and edited HCT116 TET-ON using hygromycin B concentration of 300 μg/mL. However, the efficiency of obtaining a homozygote clone is lower.
9. Single colonies will be visible by eye after 10 days of transferring to 10 cm dishes. For single clone selection, we mostly considered large colonies as we experienced that some of the smaller ones which took longer to grow didn't have a successfully edited GOI.
10. Single colonies can be trypsinized and transferred by using a metal cloning cylinder or a pipette. For the latter, we washed the cells with PBS and added a small amount of trypsin to cover

the cells. We removed the trypsin immediately and incubated the cells at 37 °C (5% CO_2 in a humidity incubator). We let the cells incubate for no longer than 5 min. Be careful not to shake or move the dish too much when moving to and from the incubator, to avoid mixing the colonies. For transferring, we used 10 μL of media and pipette the small volume carefully onto the colony. Do this step carefully to avoid clones cross-contamination. The colony is transferred in the 10 μL of media in the pipette tip. Once, the colony is transferred make sure to resuspend the colony well; otherwise, the cells are clumping together, and it will take longer for the cells to grow.

11. Our POIs were essential for the cell cycle progression, and we did not obtain any colonies with the HCT116-CMV cell line.
12. Doxycycline final concentrations might vary depending on the efficiency of protein degradation. Our suggested range is 1–2 μg/mL depending on the efficiency of POI degradation.
13. IAA final concentrations depend on how efficiently the protein target is degraded. The concentration can be increased up to 1000 μM IAA.
14. Enough to cover the cells. For a 6-well plate, we suggest at least 2 mL and for a 10 cm dish at least 5 mL.
15. We observed that the signal should increase with biotin concertation, indicating the successful incorporation of biotin in proximate proteins.
16. The streptavidin Cy3 antibody was in areas which are occupied by the POI during the cell cycle.
17. We found that for the AID-tagged cell lines, we always needed to do subsequent subcloning of the parental gene-edited cell line. This was to achieve a homogenous population of cells that responded to the IAA treatment with the same efficiency. Subcloning can be also performed when the gene-edited cell line is not homogenous.
18. Add the volume of the suspended cells which equals 35 cells.
19. HCT116 TET-ON unedited cell line as control cultured in heavy tagged amino acid SILAC medium, and HCT116 TET-ON POI-APEX2 cultured in light tagged amino acid SILAC medium.
20. We used the following sonication settings: 50% amplitude, total time of 3 min, 30 s on, and 30 s off. However, the sonication settings might need optimization depending on the sonicator machine.
21. Make sure to incubate the cells with 2.5 mM biotin–phenol 30 min before each time point.

22. At this point, you can centrifuge the mitotic cells 200× *g* for 5 min and resuspend the pellet in medium containing 0.06 μg/mL nocodazole. Seed the cells on polylysine-coated 6-well plate. After the cells attach, incubate the cells with 2.5 mM biotin–phenol for 30 min and proceed to APEX2 reaction for interactors in prometaphase.
23. You can do the H_2O_2-APEX2 reaction for interactors in metaphase.
24. You can do the H_2O_2-APEX2 reaction for interactors in anaphase.
25. You can do the H_2O_2-APEX2 reaction for interactors in cytokinesis/nuclear reformation.
26. You can do the H_2O_2-APEX2 reaction for interactors in early G1-phase.
27. The length of the cell cycle varies between cell lines. The user should be aware that the timings we used to get samples of mitotic cells is specific to the HCT116 TET-ON cell line.
28. Identification of protein interactors in different cell cycle phases: for the identification of interactors during S-phase, cells are treated with 2 mM thymidine for 18 h, and the cells are released within a 3–6 h time window; the same protocol as in Subheading 3.5 was then followed. For the identification of interactors in G2/M, cells were treated with 10 μM RO-3306 for 18 h; the same protocol as in Subheading 3.5 was then followed.
29. In our MS analysis, two lists were obtained: the "total protein" and the "phospho-enrichment." The normalized heavy/light values from the total protein list are used for the normalization of the heavy/light values from the phospho-enrichment list. The differential phosphorylated proteins for the control were obtained from the proteins with >1.2 heavy chain/light chain (normalized to the total protein values) and for IAA > 0.8 heavy chain/light chain. To identify phosphorylated residues, a comparison is performed of the identified peptide with the protein sequence from UniProt (uniport.org). Functional enrichment can be analyzed using STRING (string-db.org) and analysis of the properties of each phosphorylation through research of the literature.
30. Gently resuspend the cell pellet using a P1000 pipette with the tip cut (or wide orifice pipette tips) to avoid disrupting the cellular membranes.
31. Before continuing with the experiment, check if the membrane disruption was successful. Add 1–5 μL of the sample on a slide and check it under the microscope, if the cells are not disrupted repeat homogenizing the cells.

32. We used these settings: sonication 50%, 30 s on, 30 s on for 4 cycles. However, the sonication settings might need optimization depending on the used sonicator machine.
33. Phosphoproteomic analysis of differential phosphosites in different cell cycle phases: for S-phase progression, treat with 2 mM thymidine for 18 h and release for a time period of 3 to 6 h, then follow the same protocol as in 3.7. For G2/M, treat cells with 1 mM CDK1 inhibitor RO-3306 for 18 h and follow the same protocol as in Subheading 3.7.
34. We suggest validating the possible substrates by using FRET, proximity ligation assays (PLA), and in vitro assays.

Acknowledgments

We want to thank Dr Florentin Huguet (Brunel University London, Vagnarelli Lab) for substantial contribution toward the development and validation of the presented method. The Vagnarelli Lab is supported by the Wellcome Trust Investigator award 210742/Z/18/Z to Paola Vagnarelli.

References

1. Wang B, Wu H, Hu C et al (2021) An overview of kinase downregulators and recent advances in discovery approaches. Sig Transduct Target Ther 6:1–19. https://doi.org/10.1038/s41392-021-00826-7
2. Novak B, Kapuy O, Domingo-Sananes MR, Tyson JJ (2010) Regulated protein kinases and phosphatases in cell cycle decisions. Curr Opin Cell Biol 22:801–808. https://doi.org/10.1016/j.ceb.2010.07.001
3. Morgan DO (2007) The cell cycle: principles of control. New Science Press, London
4. Manning G, Whyte DB, Martinez R et al (2002) The protein kinase complement of the human genome. Science (1979) 298:1912–1934
5. Li X, Wilmanns M, Thornton J, Köhn M (2013) Elucidating human phosphatase-substrate networks. Sci Signal 6(275). https://doi.org/10.1126/scisignal.2003203
6. Nasa I, Kettenbach AN (2018) Coordination of protein kinase and phosphoprotein phosphatase activities in mitosis. Front Cell Dev Biol 6:30
7. Hochegger H, Takeda S, Hunt T (2008) Cyclin-dependent kinases and cell-cycle transitions: does one fit all? Nat Rev Mol Cell Biol 9:910–916. https://doi.org/10.1038/nrm2510
8. Bollen M, Gerlich DW, Lesage B (2009) Mitotic phosphatases: from entry guards to exit guides. Trends Cell Biol 19:531–541. https://doi.org/10.1016/j.tcb.2009.06.005
9. Olsen JV, Vermeulen M, Santamaria A et al (2010) Quantitative phosphoproteomics reveals widespread full phosphorylation site occupancy during mitosis. Sci Signal 3:ra3. https://doi.org/10.1126/scisignal.2000475
10. Domingo-Sananes MR, Kapuy O, Hunt T, Novak B (2011) Switches and latches: a biochemical tug-of-war between the kinases and phosphatases that control mitosis. Philos Trans R Soc Lond Ser B Biol Sci 366:3584–3594. https://doi.org/10.1098/rstb.2011.0087
11. Bouchoux C, Uhlmann F (2011) A quantitative model for ordered Cdk substrate dephosphorylation during mitotic exit. Cell 147:803–814. https://doi.org/10.1016/j.cell.2011.09.047
12. Hendrickx A, Beullens M, Ceulemans H et al (2009) Docking motif-guided mapping of the interactome of protein phosphatase-1. Chem Biol 16:365–371. https://doi.org/10.1016/J.CHEMBIOL.2009.02.012
13. Moorhead GBG, Trinkle-Mulcahy L, Ulke-Lemée A (2007) Emerging roles of nuclear protein phosphatases. Nat Rev Mol Cell Biol

8:234–244. https://doi.org/10.1038/nrm2126

14. Bollen M, Peti W, Ragusa MJ, Beullens M (2010) The extended PP1 toolkit: designed to create specificity. Trends Biochem Sci 35: 450–458. https://doi.org/10.1016/j.tibs.2010.03.002
15. Trinkle-Mulcahy L, Sleeman JE, Lamond AI (2001) Dynamic targeting of protein phosphatase 1 within the nuclei of living mammalian cells. J Cell Sci 114:4219–4228. https://doi.org/10.1242/JCS.114.23.4219
16. Egloff MP, Johnson DF, Moorhead G et al (1997) Structural basis for the recognition of regulatory subunits by the catalytic subunit of protein phosphatase 1. EMBO J 16:1876–1887. https://doi.org/10.1093/EMBOJ/16.8.1876
17. Ceulemans H, Bollen M (2004) Functional diversity of protein phosphatase-1, a cellular economizer and reset button. Physiol Rev 84: 1–39. https://doi.org/10.1152/PHYSREV.00013.2003/ASSET/IMAGES/LARGE/9J0140278009.JPEG
18. Wakula P, Beullens M, Ceulemans H et al (2003) Degeneracy and function of the ubiquitous RVXF motif that mediates binding to protein phosphatase-1. J Biol Chem 278: 18817–18823. https://doi.org/10.1074/jbc.M300175200
19. Heroes E, Lesage B, Görnemann J et al (2013) The PP1 binding code: a molecular-lego strategy that governs specificity. FEBS J 280:584–595. https://doi.org/10.1111/j.1742-4658.2012.08547.x
20. Cori GT, Green AA (1943) Crystalline muscle phosphorylase: II. Prosthetic group. J Biol Chem 151:31–38. https://doi.org/10.1016/S0021-9258(18)72111-X
21. Shimizu H, Ito M, Miyahara M et al (1994) Characterization of the myosin binding subunit of smooth muscle myosin phosphatase. J Biol Chem 269:30407–30411
22. Shirazi A, Iizuka K, Fadden P et al (1994) Purification and characterization of the mammalian myosin light chain phosphatase holoenzyme. The differential effects of the holoenzyme and its subunits on smooth muscle. J Biol Chem 269:31598–31606
23. Alessi D, MacDougall LK, Sola MM et al (1992) The control of protein phosphatase-1 by targeting subunits. The major myosin phosphatase in avian smooth muscle is a novel form of protein phosphatase-1. Eur J Biochem 210: 1023–1035. https://doi.org/10.1111/J.1432-1033.1992.TB17508.X
24. Okubo S, Ito M, Takashiba Y et al (1994) A regulatory subunit of smooth muscle myosin bound phosphatase. Biochem Biophys Res Commun 200:429–434. https://doi.org/10.1006/bbrc.1994.1467
25. Pato MD, Kerc E (1985) Purification and characterization of a smooth muscle myosin phosphatase from turkey gizzards. J Biol Chem 260: 12359–12366. https://doi.org/10.1016/S0021-9258(17)39033-6
26. Gratecos D, Fischer EH, Detwiler TC, Hurd S (1977) Rabbit muscle phosphorylase phosphatase. 1. Purification and chemical properties. Biochemistry 16:4812–4817. https://doi.org/10.1021/BI00641A009/ASSET/BI00641A009.FP.PNG_V03
27. Vagnarelli P, Ribeiro S, Sennels L et al (2011) Repo-man coordinates chromosomal reorganization with nuclear envelope reassembly during mitotic exit. Dev Cell 21:328–342. https://doi.org/10.1016/J.DEVCEL.2011.06.020
28. Wu D, De Wever V, Derua R et al (2018) A substrate-trapping strategy for protein phosphatase PP1 holoenzymes using hypoactive subunit fusions. J Biol Chem 293:15152–15162. https://doi.org/10.1074/jbc.RA118.004132
29. Huguet F, Gokhan E, Foster HA et al (2022) Repo-Man/protein phosphatase 1 SUMOylation mediates binding to lamin A and serine 22 dephosphorylation. Open Biol 12:220017. https://doi.org/10.1098/rsob.220017
30. Carrara M, Sigurdardottir A, Bertolotti A (2017) Decoding the selectivity of eIF2α holophosphatases and PPP1R15A inhibitors. Nat Struct Mol Biol 24:708. https://doi.org/10.1038/NSMB.3443
31. Fedoryshchak RO, Přechová M, Butler AM et al (2020) Molecular basis for substrate specificity of the Phactr1/PP1 phosphatase holoenzyme. elife 9:e61509. https://doi.org/10.7554/eLife.61509
32. Huet G, Rajakylä EK, Viita T et al (2013) Actin-regulated feedback loop based on Phactr4, PP1 and cofilin maintains the actin monomer pool. J Cell Sci 126:497–507. https://doi.org/10.1242/jcs.113241
33. Morris JH, Knudsen GM, Verschueren E et al (2014) Affinity purification–mass spectrometry and network analysis to understand protein-protein interactions. Nat Protoc 9:2539. https://doi.org/10.1038/NPROT.2014.164
34. Chang IF (2006) Mass spectrometry-based proteomic analysis of the epitope-tag affinity purified protein complexes in eukaryotes. Proteomics 6:6158–6166. https://doi.org/10.1002/PMIC.200600225

35. Westermarck J, Ivaska J, Corthals GL (2013) Identification of protein interactions involved in cellular signaling. Mol Cell Proteomics 12: 1752. https://doi.org/10.1074/MCP.R113.027771
36. Budayeva HG, Cristea IM (2014) A mass spectrometry view of stable and transient protein interactions. Adv Exp Med Biol 806:263. https://doi.org/10.1007/978-3-319-06068-2_11
37. Schwinn MK, Steffen LS, Zimmerman K et al (2020) A simple and scalable strategy for analysis of endogenous protein dynamics. Sci Rep 10:1–14. https://doi.org/10.1038/s41598-020-65832-1
38. Moriya H (2015) Quantitative nature of overexpression experiments. Mol Biol Cell 26: 3932–3939. https://doi.org/10.1091/mbc.E15-07-0512
39. Lackner DH, Carré A, Guzzardo PM et al (2015) A generic strategy for CRISPR-Cas9-mediated gene tagging. Nat Commun 6: 10237. https://doi.org/10.1038/ncomms10237
40. Roux KJ, Kim DI, Raida M, Burke B (2012) A promiscuous biotin ligase fusion protein identifies proximal and interacting proteins in mammalian cells. J Cell Biol 196:801–810. https://doi.org/10.1083/jcb.201112098
41. Trinkle-Mulcahy L (2019) Recent advances in proximity-based labeling methods for interactome mapping. F1000Res 8(135). https://doi.org/10.12688/F1000RESEARCH.16903.1
42. Kim DI, Jensen SC, Noble KA et al (2016) An improved smaller biotin ligase for BioID proximity labeling. Mol Biol Cell 27:1188–1196. https://doi.org/10.1091/mbc.E15-12-0844
43. Kubitz L, Bitsch S, Zhao X et al (2022) Engineering of ultraID, a compact and hyperactive enzyme for proximity-dependent biotinylation in living cells. Commun Biol 5:657. https://doi.org/10.1038/s42003-022-03604-5
44. Hall MP, Unch J, Binkowski BF et al (2012) Engineered luciferase reporter from a deep sea shrimp utilizing a novel imidazopyrazinone substrate. ACS Chem Biol 7:1848. https://doi.org/10.1021/CB3002478
45. Dixon AS, Schwinn MK, Hall MP et al (2016) NanoLuc complementation reporter optimized for accurate measurement of protein interactions in cells. ACS Chem Biol 11:400–408. https://doi.org/10.1021/ACSCHEMBIO.5B00753/ASSET/IMAGES/CB-2015-007533_M004.GIF
46. Claes Z, Jonkhout M, Crespillo-Casado A, Bollen M (2019) The antibiotic robenidine exhibits guanabenz-like cytoprotective properties by a mechanism independent of protein phosphatase PP1:PPP1R15A. J Biol Chem 294: 13478–13486. https://doi.org/10.1074/jbc.RA119.008857
47. Lalonde S, Ehrhardt DW, Loqué D et al (2008) Molecular and cellular approaches for the detection of protein–protein interactions: latest techniques and current limitations. Plant J 53: 610–635. https://doi.org/10.1111/J.1365-313X.2007.03332.X
48. Rizzo MA, Davidson MW, Piston DW (2009) Fluorescent protein tracking and detection: applications using fluorescent proteins in living cells. Cold Spring Harb Protoc 4(12). https://doi.org/10.1101/PDB.TOP64
49. Gibson TJ, Seiler M, Veitia RA (2013) The transience of transient overexpression. Nat Methods 10:715–721. https://doi.org/10.1038/nmeth.2534
50. Voeltz GK, Prinz WA, Shibata Y et al (2006) A class of membrane proteins shaping the tubular endoplasmic reticulum. Cell 124:573–586. https://doi.org/10.1016/J.CELL.2005.11.047
51. Harder Z, Zunino R, McBride H (2004) Sumo1 conjugates mitochondrial substrates and participates in mitochondrial fission. Curr Biol 14:340–345. https://doi.org/10.1016/J.CUB.2004.02.004
52. Qian J, Beullens M, Lesage B, Bollen M (2013) Aurora B defines its own chromosomal targeting by opposing the recruitment of the phosphatase scaffold repo-man. Curr Biol 23:1136–1143. https://doi.org/10.1016/j.cub.2013.05.017
53. Cheng YL, Chen RH (2010) The AAA-ATPase Cdc48 and cofactor Shp1 promote chromosome bi-orientation by balancing Aurora B activity. J Cell Sci 123:2025–2034. https://doi.org/10.1242/JCS.066043
54. Peggie MW, MacKelvie SH, Bloecher A et al (2002) Essential functions of Sds22p in chromosome stability and nuclear localization of PP1. J Cell Sci 115:195–206. https://doi.org/10.1242/JCS.115.1.195
55. Pedelini L, Marquina M, Ariño J et al (2007) YPI1 and SDS22 proteins regulate the nuclear localization and function of yeast type 1 phosphatase Glc7. J Biol Chem 282:3282–3292. https://doi.org/10.1074/jbc.M607171200
56. Eiteneuer A, Seiler J, Weith M et al (2014) Inhibitor-3 ensures bipolar mitotic spindle attachment by limiting association of SDS22 with kinetochore-bound protein phosphatase-1. EMBO J 33:2704–2720. https://doi.org/10.15252/EMBJ.201489054

57. Lakowicz JR (2006) Principles of fluorescence spectroscopy, pp 1–954. https://doi.org/10.1007/978-0-387-46312-4/COVER
58. Förster T (2012) Energy migration and fluorescence. J Biomed Opt 17:11002. https://doi.org/10.1117/1.JBO.17.1.011002
59. Algar WR, Hildebrandt N, Vogel SS, Medintz IL (2019) FRET as a biomolecular research tool – understanding its potential while avoiding pitfalls. Nat Methods 16:815–829. https://doi.org/10.1038/s41592-019-0530-8
60. Trinkle-Mulcahy L, Chusainow J, Lam YW et al (2007) Visualization of intracellular PP1 targeting through transiently and stably expressed fluorescent protein fusions. Methods Mol Biol 365:133–154. https://doi.org/10.1385/1-59745-267-X:133
61. de Los Santos C, Chang CW, Mycek MA, Cardullo RA (2015) FRAP, FLIM, and FRET: detection and analysis of cellular dynamics on a molecular scale using fluorescence microscopy. Mol Reprod Dev 82:587. https://doi.org/10.1002/MRD.22501
62. Xue Y, Ren J, Gao X et al (2008) GPS 2.0, a tool to predict kinase-specific phosphorylation sites in hierarchy. Mol Cell Proteomics 7:1598–1606. https://doi.org/10.1074/MCP.M700574-MCP200
63. Iakoucheva LM, Radivojac P, Brown CJ et al (2004) The importance of intrinsic disorder for protein phosphorylation. Nucleic Acids Res 32:1037–1049. https://doi.org/10.1093/nar/gkh253
64. Blom N, Gammeltoft S, Brunak S (1999) Sequence and structure-based prediction of eukaryotic protein phosphorylation sites1 1Edited by F. E. Cohen. J Mol Biol 294: 1351–1362. https://doi.org/10.1006/jmbi.1999.3310
65. Ingrell CR, Miller ML, Jensen ON, Blom N (2007) NetPhosYeast: prediction of protein phosphorylation sites in yeast. Bioinformatics 23:895–897. https://doi.org/10.1093/bioinformatics/btm020
66. Tang Y-R, Chen Y-Z, Canchaya CA, Zhang Z (2007) GANNPhos: a new phosphorylation site predictor based on a genetic algorithm integrated neural network. Protein Eng Des Sel 20:405–412. https://doi.org/10.1093/protein/gzm035
67. Wang Y, Liu Z, Cheng H et al (2014) EKPD: a hierarchical database of eukaryotic protein kinases and protein phosphatases. Nucleic Acids Res 42:D496–502. https://doi.org/10.1093/NAR/GKT1121
68. Liberti S, Sacco F, Calderone A et al (2013) HuPho: the human phosphatase portal. FEBS J 280:379–387. https://doi.org/10.1111/J.1742-4658.2012.08712.X
69. Duan G, Li X, Köhn M (2015) The human DEPhOsphorylation database DEPOD: a 2015 update. Nucleic Acids Res 43:D531. https://doi.org/10.1093/NAR/GKU1009
70. Mitchell CJ, Kim MS, Zhong J et al (2016) Unbiased identification of substrates of protein tyrosine phosphatase ptp-3 in C. elegans. Mol Oncol 10:910. https://doi.org/10.1016/J.MOLONC.2016.03.003
71. Feng S, Zhou H, Dong H (2019) Using deep neural network with small dataset to predict material defects. Mater Des 162:300–310. https://doi.org/10.1016/j.matdes.2018.11.060
72. Zhao W (2017) Research on the deep learning of the small sample data based on transfer learning. AIP Conf Proc 1864. https://doi.org/10.1063/1.4992835
73. Chaudhari M, Thapa N, Ismail H et al (2021) DTL-DephosSite: deep transfer learning based approach to predict dephosphorylation sites. Front Cell Dev Biol 9:662983. https://doi.org/10.3389/FCELL.2021.662983/FULL
74. Holder J, Mohammed S, Barr FA (2020) Ordered dephosphorylation initiated by the selective proteolysis of cyclin B drives mitotic exit. elife 9:1–33. https://doi.org/10.7554/ELIFE.59885
75. Touati SA, Kataria M, Jones AW et al (2018) Phosphoproteome dynamics during mitotic exit in budding yeast. EMBO J 37(10). https://doi.org/10.15252/EMBJ.201798745
76. Hoermann B, Kokot T, Helm D et al (2020) Dissecting the sequence determinants for dephosphorylation by the catalytic subunits of phosphatases PP1 and PP2A. Nat Commun 11:1–20. https://doi.org/10.1038/s41467-020-17334-x
77. Godfrey M, Touati SA, Kataria M et al (2017) PP2ACdc55 phosphatase imposes ordered cell-cycle phosphorylation by opposing threonine phosphorylation. Mol Cell 65:393–402.e3. https://doi.org/10.1016/J.MOLCEL.2016.12.018
78. Cundell MJ, Hutter LH, Bastos RN et al (2016) A PP2A-B55 recognition signal controls substrate dephosphorylation kinetics during mitotic exit. J Cell Biol 214:539. https://doi.org/10.1083/JCB.201606033
79. Hein JB, Hertz EPT, Garvanska DH et al (2017) Distinct kinetics of serine and threonine dephosphorylation are essential for mitosis.

Nat Cell Biol 19:1433–1440. https://doi.org/10.1038/ncb3634

80. McCloy RA, Parker BL, Rogers S et al (2015) Global phosphoproteomic mapping of early mitotic exit in human cells identifies novel substrate dephosphorylation motifs. Mol Cell Proteomics 14:2194–2212. https://doi.org/10.1074/MCP.M114.046938
81. Waldman T, Kinzler KW, Vogelstein B (1995) p21 is necessary for the p53-mediated G1 arrest in human cancer cells. Cancer 55:5187–5190
82. Natsume T, Kiyomitsu T, Saga Y, Kanemaki MT (2016) Rapid protein depletion in human cells by IAA-inducible degron tagging with short homology donors. Cell Rep 15:210–218. https://doi.org/10.1016/J.CELREP.2016.03.001
83. Hung V, Udeshi ND, Lam SS et al (2016) Spatially resolved proteomic mapping in living cells with the engineered peroxidase APEX2. Nat Protoc 11:456–475. https://doi.org/10.1038/nprot.2016.018
84. Nishimura K, Fukagawa T, Takisawa H et al (2009) An IAA-based degron system for the rapid depletion of proteins in nonplant cells. Nat Methods 6:917–922. https://doi.org/10.1038/nmeth.1401
85. ten Have S, Boulon S, Ahmad Y, Lamond AI (2011) Mass spectrometry-based immuno-precipitation proteomics – the user's guide. Proteomics 11:1153. https://doi.org/10.1002/PMIC.201000548

Chapter 4

Dissecting the Multiple Functions of the Polo-Like Kinase 1 in the *C. elegans* Zygote

Griselda Velez-Aguilera, Batool Ossareh-Nazari, and Lionel Pintard

Abstract

Plk1 (polo-like kinase 1) is an evolutionarily conserved serine/threonine kinase instrumental for mitotic entry and progression. Beyond these canonical functions, Plk1 also regulates cell polarization and cell fate during asymmetric cell divisions in *C. elegans* and *D. melanogaster.* Plk1 contains a specialized phosphoserine–threonine binding domain, the polo-box domain (PBD), which localizes and concentrates the kinase at its various sites of action within the cell in space and time. Here we present protocols to express and purify the *C. elegans* Plk1 kinase along with biochemical and phosphoproteomic approaches to interrogate the PBD interactome and to dissect Plk1 substrate interactions. These protocols are most suitable for the identification of Plk1 targets in *C. elegans* embryos but can be easily adapted to identify and study Plk1 substrates from any source."

Key words Polo-like kinase 1, *C. elegans* embryos, Polo-box domain, Phosphoproteomics

1 Introduction

Evolutionarily conserved from yeast to humans, the polo-like kinase 1 is pivotal for mitotic entry and progression [1–4]. *Drosophila* Polo, mammalian Plk1, *Xenopus* Plx1 and *Caenorhabditis elegans* PLK-1 are clear orthologues at the sequence and functional levels (here, we refer to them collectively as Plk1, except when we use PLK-1 in the material and methods to clearly specify the *C. elegans* protein). Besides a canonical serine/threonine kinase domain in the N-terminal part, Plk1 contains a C-terminal polo-box domain (PBD), which mediates Plk1 interaction with substrates [5, 6]. The PBD is a phosphopeptide-binding domain that recognizes specific substrates primed by phosphorylation on the so-called polo-docking site characterized by the sequence Ser-[pSer/pThr]-[Pro/X], where pSer/pThr indicates phosphorylated serine or

Griselda Velez-Aguilera and Batool Ossareh-Nazari, have equally contributed to this chapter.

Anna Castro and Benjamin Lacroix (eds.), *Cell Cycle Control: Methods and Protocols*, Methods in Molecular Biology, vol. 2740, https://doi.org/10.1007/978-1-0716-3557-5_4,

threonine, respectively, and X can be any residue [6, 7]. When X is a proline residue, the Ser or Thr are phosphorylatable by proline-directed kinases including cyclin B-Cdk1, the master mitotic kinase [8]. Thus, by phosphorylating these motifs, cyclin B-Cdk1 generates binding sites for Plk1 (non-self-priming) [9–11]. Through this mechanism, cyclin B-Cdk1 dynamically directs the localization of Plk1 to different subcellular structures and sites of action during mitosis. When X is not a proline residue and the surrounding sequence is optimal for Plk1 phosphorylation [12, 13], Plk1 itself can phosphorylate the motif (self-priming) [14, 15]. Self-priming contributes to the localization and concentration of Plk1 at its various sites of action, particularly after anaphase when cyclin B-Cdk1 activity declines [16].

Plk1 regulates an impressive array of processes, including mitotic entry, centrosome assembly and maturation, spindle assembly and function, nuclear envelope breakdown, chromosome condensation, removal of sister chromatid cohesion, kinetochore functions, and cytokinesis [1–3, 17]. Plk1 also regulates cell polarity and cell fate during asymmetric cell divisions, both in *Drosophila* neuroblasts and in *C. elegans* embryos [18, 19].

Because of its numerous substrates in various processes at different times and places, dissecting Plk1 function is challenging. For instance, the complete depletion of Plk1 activity in *C. elegans* causes a penetrant sterility phenotype. Because of this, most embryonic functions of Plk1 reported so far in the literature reflect the phenotype of embryos only partially depleted of Plk1 kinase activity. One experimental option is temporally inhibiting Plk1 activity using an analog-sensitive Plk1 allele [20]. Alternatively, instead of inactivating Plk1 itself, a viable approach is to surgically prevent Plk1 docking to a given substrate by disrupting the interaction of Plk1-PBD with specific polo-docking sites. However, this approach requires the exhaustive identification of polo-docking sites on substrates and the analysis of the consequence of their mutation in vivo.

This chapter describes methods for dissecting Plk1 function in the early *C. elegans* embryo. We describe our protocol to produce and purify the *C. elegans* PLK-1 kinase from insect Sf9 cells and the isolated human Plk1 PBD domain in *E. coli*. Then we provide a protocol to identify Plk1 PBD interacting proteins from embryonic extracts and to map the phosphorylated polo-docking sites on substrates responsible for the interaction. Finally, we describe our approach to dissect Plk1 substrate interaction using Far Western blots.

2 Materials

2.1 Purification of the C. elegans PLK-1 and the Human Plk1 PBD Domain

2.1.1 Infection of SF9 Cell with Baculovirus

1. Baculovirus expressing pFasTBAC Hta PLK-1.
2. SF9-ESF *S. frugiperda* insect cell line.
3. Insect-XPRESS™ Protein-free Insect Cell Medium with L-glutamine and antibiotics (50 U penicillin and 50 μg streptomycin per mL medium).
4. 1× PBS.
5. Sterile flasks 75 cm^2.
6. 5 and 10 mL pipettes.
7. 15 and 50 mL conical centrifuge tubes.
8. Scrapper.
9. Centrifuge.
10. Incubator.
11. Inverted microscope.

2.1.2 Purification of 6xHis-PLK-1

1. 10× phosphate buffer saline (PBS) stock solution: dissolve in 800 ml ddH_2O 80 g NaCl, 2 g KCl, 14.4 g Na_2HPO_4, and 2.4 g KH_2PO_4. Adjust the final volume to 1 L with ddH_2O and the pH to 7.4 with HCl.
2. 1× PBS: dilute 100 mL of 10× PBS in 900 mL ddH_2O. Adjust pH to 7.4 with HCl.
3. Lysis buffer: 250 mM NaCl, 30 mM imidazole in 1× PBS, pH 7.2.
4. Wash buffer: 250 mM NaCl in 1× PBS, pH 7.2.
5. Elution buffer: 250 mM NaCl, 1 M imidazole, in 1× PBS, pH 7.2.
6. 1× storage buffer: 125 mM NaCl, 25% glycerol, in 1× PBS, pH 7.2.
7. 2× storage buffer: 250 mM NaCl, 50% glycerol, in 1× PBS, pH 7.2.
8. 10× phosphatase inhibitor cocktail stock solution: in a 1.5 mL microtube, dissolve one tablet in 1 mL of ddH_20. Store at −20 °C.
9. 50× protease inhibitor cocktail: in a 1.5 mL microtube, dissolve one tablet in 1 mL of ddH_20. Store at −20 °C.
10. 100 mM $NiSO_4$.
11. 1 mL HiTrap chelating HP.
12. 0.45 μm syringe filter.
13. Fast protein LC ÄKTA System.

2.1.3 Purification of GST-Plk1 PBD Wild Type (GST-PBD^{WT}) and GST- Plk1 PBD$^{H538A/K540M}$ (GST-PBD^{mut})

1. Competent *E.coli* BL21 DE3 pLyS.
2. Plasmids: pGEX-6p1 GST-Plk1 PBD *H. s* and pGEX-6p1 GST-Plk1 PBD H538A/K540M *H. s.*
3. LB (Luria–Bertani) liquid medium: 10 g Tryptone, 10 g NaCl, 5 g yeast extract, and 950 mL of ddH$_2$O. Combine the reagents and shake until the solutes have dissolved. Adjust pH to 7.0 with 5 N NaOH. Adjust the final volume to 1 L with ddH$_2$O. Sterilize by autoclaving for 20 min at 15 psi.
4. Ampicillin: 100 mg/mL dissolved in ddH$_2$0.
5. LB ampicillin (50 μg/mL) plates.
6. 1 M IPTG (*isopropyl β-*D*-thiogalactoside*) stock solution.
7. Lysis buffer: 10 mM Tris pH 8.0, 150 mM NaCl, 1 mM EDTA, 5 mM DTT, 0.05% NP40, 1× phosphatase inhibitor cocktail, 1× protease inhibitor cocktail.
8. Binding buffer: 10 mM Tris pH 8.0, 150 mM NaCl, 1 mM EDTA, 5 mM DTT.
9. Elution buffer: 10 mM Tris pH 8.0, 150 mM NaCl, 1 mM EDTA, 5 mM DTT, 20 mM Glutathione. Adjust the pH to 8.0 (*see* **Note 1**).
10. 10× phosphatase inhibitor cocktail.
11. 50× protease inhibitor cocktail.
12. 20 mL syringe.
13. 0.45 μm syringe filter.
14. GSTrap 4B 1 mL columns.
15. Dialysis bag.
16. Centrifugal concentrator.
17. SDS-PAGE.
18. Shaking incubator.
19. Sonicator.
20. Peristaltic pump.
21. Centrifuge.

2.2 Delineating the C. elegans Plk1 PBD Interactome Using Affinity Capture and Phosphoproteomics

2.2.1 Preparation of Embryonic Cryolysate

1. N2 strain worms.
2. M9 (minimal medium): 5 g NaCl, 6 g Na2HPO$_4$, 3 g KH2PO$_4$, add ddH$_2$O to 1 L and autoclave. Add 1 mL of 1 M MgSO$_4$.
3. M9-T: M9 plus 0.01% Tween-20 or Triton X100.
4. Bleach solution: 50 mL of 30% NaOH, 50 mL bleach, 400 mL ddH$_2$O; mix and keep for up to 2 weeks.
5. S-medium (for 1 L): 100 mL S-basal (10×), 873 mL ddH$_2$0, 1 mL 5 mg/mL cholesterol, 10 mL 1 M potassium citrate

pH 6.0, 10 mL trace metal (100×), 3 mL 1 M $CaCl_2$, 3 mL 1 M $MgSO_4$.

6. 10× S-basal (for 1 L): 58.4 g NaCl, 10 g $K2HPO_4$ (or 15 g K2HPO4.3H2O), 60 g $KH2PO_4$. Dissolve in 800 mL ddH_2O. Make up to 1 L with ddH_2O. Sterilize by autoclaving.
7. Cholesterol: 5 mg/mL in 100% ethanol.
8. 1 M potassium citrate pH 6.0 (for 1 L): 20 g citric acid monohydrate, 294 g tri-potassium citrate monohydrate. Dissolve in 700 mL of ddH_2O. Make up to 1 L with ddH_2O. Sterilize by autoclaving.
9. 100× trace metals (for 500 mL): 930 mg disodium EDTA, 345 mg $FeSO_4$.7H2O, 100 mg $MnCl_2.4H_2O$, 145 mg $ZnSO_4.7H_2O$, 12.5 mg $CuSO_4$.5H2O. Dissolve in 500 mL ddH_2O. Sterilize by filtration. Protect from light in a bottle wrapped with aluminum foil.
10. 1 M $CaCl_2$.
11. 1 M $MgSO_4$.
12. 100 mM NaCl.
13. 60% (w/v) sucrose (for 500 mL): dissolve 300 g of sucrose in 450 mL ddH_2O, make up to 500 mL with ddH_2O. Sterilize by filtration on 0.22 μm. Store at 4 °C.
14. Bacteria: *E. coli* HB101.
15. TB (terrific broth).
16. LB (Luria–Bertani) agar plates with streptomycin (30 μg/mL).
17. 15 mL and 50 mL conical tubes.
18. NITEX 31 μm mesh 102.
19. Perforated funnel.
20. Wooden clamp.
21. Thick-walled polystyrene box.
22. Liquid nitrogen.
23. Stereomicroscope.
24. Mixer Mill (RETSCH MM 301).
25. Shaking incubator.

2.2.2 GST-PBD^{WT} or GST-PBD^{mut} (Negative Control) Pull-Downs

1. Binding buffer: 50 mM Tris pH 8.0, 500 mM NaCl.
2. 1× lysis buffer: 25 mM Tris pH 7.5, 100 mM NaCl, 2 mM $MgCl_2$, 1 mM DTT, 1× phosphatase inhibitor, 1× protease inhibitor cocktail.
3. 1.5× lysis buffer: 37.5 mM Tris pH 7.5, 100 mM NaCl, 3 mM $MgCl_2$, 1.5 mM DTT, 1.5× phosphatase inhibitor, 1.5× protease inhibitor cocktail.

4. 10× phosphatase inhibitor cocktail (*see* Subheading 2.1.2, **step 8**).
5. 50× protease inhibitor cocktail (*see* Subheading 2.1.2, **step 9**).
6. Benzonase nuclease.
7. Protein quantification kit.
8. Purified GST-PBD^{WT} and GST-PBD^{mut} (1.5 mg/mL).
9. Glutathione Sepharose 4B: store at 4 °C.
10. 1.5 mL Protein LoBind tubes.
11. Centrifuge.
12. Liquid nitrogen.

2.2.3 Phosphopeptides Enrichment for Mass Spectrometry Analysis

1. Ammonium bicarbonate (NH_4HCO_3).
2. Trypsin/Lys-C Mix.
3. Thermomixer.
4. Methanol (MeOH).
5. Acetonitrile (ACN).
6. Trifluoroacetic acid (TFA).
7. Formic acid (FA), LC-MS.
8. Water (H_2O), LC-MS.
9. Sep-Pak Classic C18 cartridges, waters.
10. Speed vacuum.
11. Pierce High Select Fe-NTA Phosphopeptide Enrichment Kit, including binding/wash buffer and elution buffer.
12. Centrifuge.
13. Vortex.

2.3 Probing Plk1 PBD Substrate Interaction by Far Western Blot

2.3.1 Protein Purification

The prey protein (aka Plk1 substrate) full length or fragments encompassing the polo-docking sites are produced and purified from *E. coli* tagged with MBP, GST, or STREP. GST-Plk1 PBD wild type or the negative control GST-Plk1 PBD^{mut}are used as bait proteins.

2.3.2 In Vitro Kinase Assay

1. 10× kinase buffer stock solution: 500 mM HEPES pH 7.6, 100 mM $MgCl_2$. To prepare this buffer, add 25 mL water to a 100 mL glass beaker. Weigh 5.96 g HEPES and transfer it into the glass beaker. Add 5 mL of 1 M $MgCl_2$. Add water to a volume of 45 mL. Mix and adjust pH at 7.6 with HCl. Make up to 50 mL with water. Make 50 aliquots of 1 mL, and store them at −20 °C.
2. 100 mM adenosine triphosphate (ATP) stock solution (50×).

3. 10× Phosphatase inhibitor cocktail (*see* Subheading 2.1.2, **step 8**)
4. 50× Protease inhibitor cocktail (*see* Subheading 2.1.2, **step 9**)
5. Kinase: 200 ng of the kinase 6xHis-PLK-1 per reaction is required (*see* Subheading 3.1.2).
6. Protein to be phosphorylated: 1 μg of purified protein is required.
7. 6× Laemmli buffer stock solution: 375 mM Tris-HCl (pH 6.8), 9% SDS (w/v), 50% Glycerol (v/v), 9% β-mercaptoethanol (v/v), 0.03% bromophenol blue (w/v). Make aliquots of 1 mL. Leave one aliquot at 4 °C and store the remaining aliquots at −20 °C.

2.3.3 Immunoblotting

1. SDS-PAGE.
2. Polyvinylidene fluoride (PVDF) blotting membrane 0.45 μm (*see* **Note 2**).
3. Western blot transfer buffer: 25 mM Tris base, 192 mM glycine, 20% ethanol (*see* **Note 3**).
4. 10× Tris-buffered saline (TBS): 1.5 M NaCl, 0.1 M Tris-HCl, pH 7.4.
5. 1× TBS containing 0.1% Tween-20 (TBST).
6. Blocking buffer: 4% milk in TBST. Store at 4 °C (*see* **Note 4**).
7. Transfer system (Mini Trans-Blot).
8. Whatman gel blotting sheets.
9. ECL detection kit.
10. ChemiDoc Imaging System.
11. Vertical electrophoresis apparatus.
12. Power pack.
13. Anti-human Plk1, mouse monoclonal.
14. Anti-mouse IgG-peroxidase antibody, produced in goat.

3 Methods

3.1 Purification of the C. elegans PLK-1 and the Human Plk1 PBD Domain

We engineered baculoviruses expressing 6xHis-PLK-1 to produce and purify the full-length active *C. elegans* kinase. For the PBD domain, we expressed and purified the human version from *E. coli*, fused to the glutathione-S-transferase (GST) tag. The Plk1 PBD, composed of two polo boxes (PB1 and PB2), specifically recognizes the phosphate group on substrates at least in part via the His-538 and Lys-540 residues in PB2. Accordingly, substitution of these two residues abrogates PBD binding to phosphorylated substrates

[6]. We thus employ both the wild-type PBD and the PBD mutated on these "phosphopincers," H538A/K540M as a negative control.

Insect cells are capable of carrying out many of the complex posttranslational modifications including phosphorylation, carboxymethylation, and N- and O-glycosylation, necessary for proper processing and production of active proteins. This is particularly important for protein kinases such as Plk1, which require activating phosphorylation of their T-loop [21–23].

3.1.1 Sf9 Cell Culture Conditions

The optimal temperature for cell growth and infection of Sf9 insect cells is 27 °C. These cells are typically cultivated in monolayers attached to a plastic or glass surface (T-flasks or roller flasks) or in suspension where densities can reach from 2×10^6 cells/mL to $>10 \times 10^6$ cells/mL before viability begins to decline. The cells are sensitive to density, so they must be seeded at a concentration of 10^6 cells/mL and passed when they reach 90% confluence. For bacmid preparation and virus amplification, we followed the step-by-step protocol provided by Invitrogen (Bac-to-Bac Expression System). Here, we just describe our protocol for Sf9 cell infection using baculovirus P3 stocks, to produce and purify the active 6xHis-PLK-1 kinase.

3.1.2 Infection of SF9 Cells with Baculovirus Expressing pFasTBAC Hta PLK-1

We infect 2×10^8 cells in five 75 cm^2 flasks, to have enough purified kinase for several experiments, typically ~2.5 mL of the protein at a concentration of 1 mg/mL (1.3 μM).

1. Prepare five 75 cm^2 flasks.
2. Inoculate each flask with 20 mL of Insect-XPRESS™ Medium with antibiotic containing 2×10^6 cells/mL.
3. Let the cells attach to the flask for 20 min.
4. Infect the cells with 2 mL of baculovirus.
5. Incubate at 27 °C for 3 days.
6. After 3 days, aspirate the media and add 10 mL of PBS.
7. Take off the cells using a scraper and transfer them into a 15 mL conical sterile tube.
8. Centrifuge at 200 g for 5 min at room temperature (RT).
9. Discard the supernatant.
10. Resuspend the pellet with 10 mL of PBS.
11. Centrifuge at 200 g for 5 min at RT.
12. Discard the supernatant.
13. Resuspend each pellet in 1 mL of lysis buffer.
14. Flash freeze in liquid nitrogen and keep at −80 °C.

3.1.3 Purification of 6xHis-PLK-1

Before starting with the protocol, prepare the 1 mL HiTrap chelating HP column using a peristaltic pump as follows:

1. Wash the column with 5 mL of distilled water.
2. Load 5 mL of 100 mM $NiSO_4$.
3. Wash excess with 5 mL of distilled water.
4. Eluate with 2 mL 50% of imidazole elution buffer.
5. Load 5 mL of 100 mM $NiSO_4$.
6. Wash with 5 mL of distilled water.
7. Load 5 mL of lysis buffer.

All the steps of purification should be done at 4 °C, on ice, or in the cold room; prepare all reagents and cool down the centrifuge.

1. Add 4 mL of cold lysis buffer containing phosphatase and protease inhibitors to each frozen cell suspension.
2. Once the cell suspension is partially thawed, place the tube on the rotator wheel for 15 min in the cold room.
3. Check if the cells are fully thawed and resuspended.
4. Lyse cell suspensions by passing 30 times through a 21-gauge syringe needle.
5. Check cell lysis under a microscope.
6. Centrifuge the cell lysate at 16,000 g for 10 min.
7. Collect and pool all the supernatants ~25 mL.
8. Filter the supernatant through a 0.45 μm syringe filter (*see* **Note 5**).
9. Save 5 μL of the clear supernatant for SDS-PAGE (soluble fraction).
10. Load the supernatant on the HiTrap chelating HP column using a peristaltic pump (flow rate of 0.5 mL/min (*see* **Note 6**).
11. Save 5 μL (flow-through fraction) for SDS-PAGE to check for any loss of unbound target protein.
12. Wash the column with 10 mL of wash buffer (10 column volumes) using a peristaltic pump with a flow rate of 0.5 mL/min.
13. Elute the protein with elution buffer containing imidazole using fast protein liquid chromatography (FPLC, ÄKTA System), flow rate 0.5 mL/min with a gradient from 0 to 1 M imidazole in 30 min.
14. Collect the eluted fractions, 250 μL per fraction, in a 96-well microplate.

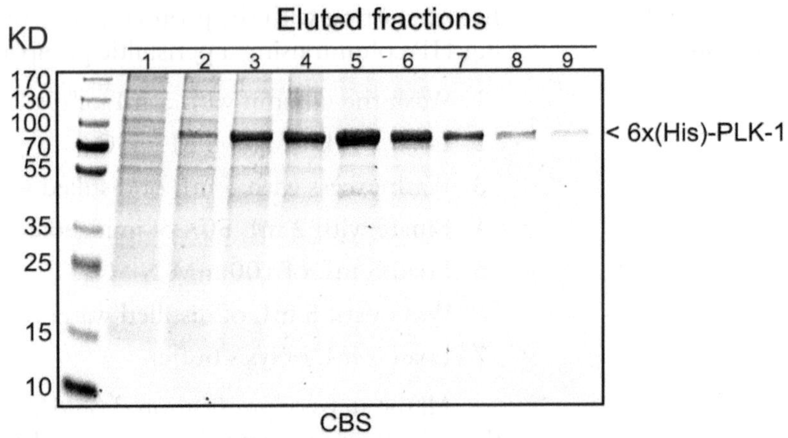

Fig. 1 Coomassie blue stained (CBS) gel after SDS-PAGE of the 6xHis-PLK-1 *C. elegans* eluted fractions from nickel column. In this example, we selected and pooled the fractions 4–9

15. Run a 10% SDS-PAGE to verify the presence and check the purity of the 6xHis-PLK-1 in the eluted fractions. Load 15 μL of each fraction with 3 μL of 6× Laemmli buffer (Fig. 1).
16. Pool the fractions containing the protein.
17. Dilute the eluted protein volume to volume in 2× storage buffer or dialyze against 1× storage buffer.
18. Concentrate the eluted protein with a centrifugal concentrator by centrifugation at 16,000 g for 15 min.
19. Quantify the protein concentration using the Pierce BCA protein assay kit, Bradford, or the method employed in your laboratory.
20. Adjust the protein concentration to 1 mg/mL.
21. Prepare single-use 6 μL 6xHis-PLK-1 aliquots to pre-chilled PCR tubes, flash freeze in liquid nitrogen, and store at −80 °C.
22. Regenerate the column by washing it with 3–5 column volumes of binding buffer (*see* **Note 7**).

3.1.4 Purification of GST-PBD^{WT} and GST-PBD^{mut}

1. Transform *E. coli* BL21 DE3 pLyS bacteria with pGEX-6p1 GST-Plk1 PBD *H. s* or pGEX-6p1 GST-Plk1 PBD H538A/K540M *H. s.* plasmid.
2. Spread the bacteria in a plate LB+ antibiotic and incubate at 37 °C overnight.
3. The day after, pick one colony, transfer to 25 mL of LB + antibiotic, and incubate overnight at 37 °C under agitation 200 rpm.
4. Transfer 5 mL of the overnight culture in 1 L of LB + ampicillin in a 2.8 L baffled flask. Do 2 L of each culture.

5. Incubate at 37 °C under agitation 200 rpm until the OD = 0.6–0.8. To verify the culture before induction, collect 1 mL of the culture, centrifuge 11,000 g for 1 min, remove the supernatant, resuspend the pellet in 100 μL of 1× Laemmli buffer, and boil the sample for 5 min.
6. Shift to 25 °C and induce protein expression by adding 1 mM IPTG for 3 h under agitation 200 rpm. To verify the expression of the protein after induction, collect 1 mL of the culture as done before induction. It is important to adjust the volume of 1× Laemmli buffer to have the same amounts of bacteria as before induction.
7. Harvest the bacteria by centrifugation at 5000 g for 10 min and discard the supernatant. The bacteria pellet can be stored at –80 °C.

   The next steps should be done at 4 °C, on ice, or cold room.
8. Resuspend the bacteria pellet in 15 mL of cold lysis buffer.
9. Transfer the cell suspension into a 50 mL conical tube.
10. Sonicate the cell suspension for 5 min at 40% amplitude with 5 s ON and 5 s OFF cycles.
11. Centrifuge the lysate at 16,000 g for 30 min.
12. Filtrate the supernatant (soluble fraction containing GST proteins) using a 0.45 μm syringe filter.
13. Save 5 μL of the clear supernatant (soluble fraction) for SDS-PAGE to check the solubility of the protein.
14. Equilibrate the 1 mL GSTrap 4B column with 5 mL of binding buffer (5 column volumes), using a peristaltic pump with a flow rate of 0.5 mL/min.
15. Load the filtered supernatant on the equilibrated column using a peristaltic pump with a flow rate of 0.5 mL/min.
16. Save 5 μL for SDS-PAGE (flow-through fraction) to check for any loss of the bound target protein.
17. Wash with 10 mL of binding buffer (10 column volumes) using a peristaltic pump with a flow rate of 0.5 mL/min. Save 5 μL for SDS-PAGE to check for any loss of bound target protein during the wash.
18. Elute with 5 mL of elution buffer (5 column volumes) using a peristaltic pump with a flow rate of 0.5 mL/min. Collect one fraction per 1 mL elution in a microtube (a total of 5 eluted fractions). Save 5 μL of each fraction for SDS-PAGE.
19. Regenerate the column by washing it with 3–5 column volumes of binding buffer.

20. Run 10% SPS-PAGE gel, of the samples collected during the purification: before induction, after induction, flow-through, after wash, and together with the elution fractions.
21. Pool the fractions containing your protein of interest.
22. Remove free glutathione by dialysis of the purified protein against 1 L of lysis buffer.
23. Concentrate the protein to 10 mg/mL with a centrifugal concentrator by centrifugation at 16,000 g for 15 min.

3.2 Delineating the C. elegans *PLK-1 PBD Interactome Using Affinity Capture and Phosphoproteomics*

To delineate the PBD interactome, we incubated *C. elegans* embryonic extracts on an affinity matrix containing either the GST-PBD^{WT}, or the negative control GST-PBD^{mut} [5–7]. A similar approach has been used previously to delineate the Plk1 interactome from human osteosarcoma U2OS cells [24]. However, while Lowery et al. eluted bound proteins using an optimal PBD-binding phosphopeptide, we digested the retained proteins on the affinity matrix with trypsin and purified the phosphopeptides on FeNTA affinity chromatography (FeNTA) before identification by liquid chromatography–tandem mass spectrometry (LC-MS/MS). We recently used this method to identify Plk1 targets at the nuclear pore complexes in *C. elegans* embryos [25]. We describe our step-by-step protocol in the following texts.

To prepare embryonic extracts, we grow worms for one generation in liquid culture. We typically inoculate 1,500,000 L1 larvae and collect adults containing up to 15 embryos. We avoid growing worms in liquid culture for more than one generation to prevent dauer formation [26]. To obtain L1, we first cultivate worms on large nematode growth medium (NGM) agar plates (150 mM), enriched with peptone, and seeded with concentrated *E. coli* HB101 as a food source (*see* **Note 8**).

3.2.1 Preparation of the Bacterial Stock for Worm Liquid Culture

All steps should be done under sterile conditions.

1. Streak *E. coli* HB101 bacteria from a frozen stock onto an LB agar plate containing streptomycin. Grow overnight at 37 °C.
2. On day 1, inoculate into 1 L of TB in a 2.8 L baffled flask and grow the bacteria overnight at 37 °C and 250 rpm.
3. On day 2, harvest bacteria by centrifugation at 5000 g for 10 min in a 1 L centrifuge bottle (determine the mass of the bottle before adding bacteria).
4. Remove the supernatant.
5. Measure the mass of the bottle with bacteria.
6. Determine bacteria mass by subtracting the mass of the bottle with and without bacteria.

7. Resuspend the bacteria in LB to obtain a suspension at 0.1 g of bacteria/mL.
8. Transfer 50 mL of the bacteria suspension into 50 mL conical tubes. Spin for 10 min at 4000 g, at RT.
9. Remove the supernatant, and store bacterial pellets (5 g) at −20 °C for several months. For 1 L bacterial culture, we obtain 20–25 g of bacteria.

3.2.2 *Preparation of Peptone-Rich NGM Agar Plates (NGM+) Seeded with* E. coli *HB101*

1. Dissolve in a 1 L bottle 3 g NaCl, 25 g agar, and 20 g peptone, and add ddH_2O to 1 L.
2. Autoclave and cool down to 50 °C.
3. Add 1 mL of 1 M $CaCl_2$, 1 mL of 1 M $MgSO_4$, 25 mL of 1 M phosphate buffer pH 6.0, and 1 mL of cholesterol 5 mg/mL and mix.
4. Pour 75 mL per 15 cm plate.
5. Let cool down and dry overnight at RT.
6. Store the plates at 4 °C in the cold room. NGM+ agar plates can be stored for several weeks in the cold room.
7. Repeat **steps 1–4** in Subheading 3.2.1. Resuspend the bacterial pellet in 40 mL LB to achieve 25× bacterial suspension (it can be stored at 4 °C for up to 1 month).
8. Add 1 mL of 25× bacteria suspension to the NGM+ plates.
9. Spread bacteria homogeneously with a bacterial spreader.
10. Let dry the plates for 48 h, at RT. During this time, bacteria will grow and form a bacterial lawn (hereafter NGM++ plates).
11. Store the plates at 4 °C. The NGM++ plates can be stored for several weeks. One NGM++ plate allows growing 150,000 worms from L1 to gravid adult.

3.2.3 *Amplifying Worms on NGM++ Plates*

1. Add 150,000 L1 to a NGM++ plate.
2. Let the worms grow at 20 °C until they become gravid.
3. Collect the worms by washing the plates with 10 mL of M9-T in a 50 mL conical tube, and fill up the tube with M9-T. Mix by inverting the tube 3–5 times (*see* **Note 9**).
4. Centrifuge for 2 min at 1800 g at RT.
5. Remove the supernatant, resuspend the worm pellet in 50 mL M9-T, and mix by inverting the tube 3–5 times.
6. Repeat **steps 4–5** twice until the supernatant becomes clear and no longer cloudy with bacteria.
7. Aliquot worms in 50 mL conical tubes to have a maximum of 2 mL settled worms in each tube.

8. Add 50 mL of the bleach solution for 7–12 min. Mix on a rotating wheel or by shaking vigorously by hand to disrupt the worms (*see* **Note 10**). After 5 min, monitor worm lysis under the stereomicroscope. Be careful about the time, since over-bleaching will kill the embryos.
9. As soon as adult bodies are broken open, centrifuge for 2 min at 1800 g.
10. Remove the supernatant and add 50 mL M9-T. Shake vigorously by hand to dissociate the egg pellet.
11. Centrifuge for 2 min at 1800 g.
12. Remove the supernatant and add 50 mL M9-T. Shake vigorously by hand to dissociate the embryo pellet.
13. Repeat **steps 11**–**12** three times.
14. By using a micropipette, resuspend the egg pellet in 1 mL M9-T and transfer into a 250 mL Erlenmeyer flask containing 49 mL M9.
15. Place the Erlenmeyer flask in a shaking incubator overnight at 20 °C and 70 rpm to let the embryos hatch.
16. Using one NGM++ plate containing 150,000 gravid adults, we obtain more than 1,500,000 L1 larvae that can be used to inoculate a liquid culture.

3.2.4 Growth of the Worms in Liquid Medium

A liquid culture is inoculated with 2500 arrested L1 larvae in 1 mL of S-medium containing 25 mg of *E. coli* HB101 bacteria. The following protocol is for 500,000 L1 larvae.

1. In a 2.8 L baffled flask, resuspend a 5 g bacterial pellet in 200 mL of S-medium.
2. Add 500,000 L1 larvae.
3. Place the flask in a shaking incubator for 3 days at 20 °C and 70 rpm to let the worms grow.
4. When worms become adults containing up to 15 embryos, harvest them by filtration through a 31 μm mesh which retains the worms, while the residual bacteria and the salt crystals that eventually precipitate in the liquid culture are eliminated. To do this, cut a piece of a 31 μm mesh, and place it over a perforated funnel installed on a bottle. Pass the culture through the mesh. Transfer the worms from the mesh using cold M9-T into a 50 mL conical tube.
5. Centrifuge the tube for 2 min at 1800 g at 4 °C.
6. Remove the supernatant and add the cold M9-T to 50 mL.
7. Centrifuge for 2 min at 1800 g at 4 °C.
8. Remove the supernatant, and resuspend pelleted worms by adding 25 mL of cold M9-T. Add 25 mL cold 60% sucrose

(if you have several tubes, pool the worms in one tube before adding sucrose). Mix and centrifuge immediately for 5 min at 1500 g at 4 °C.

9. After centrifugation, adult worms form a layer at the top of the tube, while the residual bacteria and salt crystals precipitate at the bottom. Recover the adults with a 5 mL pipette and transfer into one or two (depending on the quantity of the worms) new 50 mL conical tubes containing 30 mL of cold M9-T, and add cold M9-T to 50 mL (*see* **Note 11**).
10. Centrifuge for 2 min at 1800 g at 4 °C.
11. Remove the supernatant and add the cold M9-T to 50 mL.
12. Repeat **steps 10–11** twice.
13. Bleach the worms according to Subheading, 3.2.3, **steps 7–13** except use cold M9-T.
14. For the last wash, transfer the embryos into a 15 mL conical tube, and add cold M9-T to 15 mL.
15. Centrifuge for 2 min at 1800 g at 4 °C.
16. Remove the supernatant and resuspend the embryo pellet in an equal volume of cold 100 mM NaCl (or the desired buffer).
17. Freeze the embryos in liquid nitrogen. To do this, cut the end of a 100 μL micropipette tip with a sterile scalpel and drip 25 μL embryo suspension into a beaker containing liquid nitrogen to obtain embryo beads (*see* **Note 12**). Just before freezing, put a drop of eggs on a seeded NGM plate to control for embryo hatching and viability.
18. Collect the beads in a 15 mL or 50 mL conical tube prechilled in liquid nitrogen.
19. Let the liquid nitrogen evaporate, close the cap, and store the beads at −80 °C.

3.2.5 Cryogenic Grinding of the Embryos

1. Immerse the grinding jars containing two balls and the tube containing frozen embryo beads standing in a tube holder in a thick-walled polystyrene box filled with liquid nitrogen (*see* **Note 13**).
2. Transfer the frozen embryo beads into the jar containing one ball. For an efficient grinding, the volume of the beads should not extend beyond one-third of the jar volume.
3. Put the second ball, close it, and transfer the jars to the liquid nitrogen bath for a few minutes.
4. Pick up the jars with a wooden clamp and install them on the ball mill.
5. Tighten the jar clamps.

6. Grinding is done in four cycles of 3 min at 30 hz. Remove the jars between each grinding cycle and immerse them in the liquid nitrogen bath. Wait until the bath boiling subsides to continue the next cycle. You can take a sample of ground embryos between cycles or in the end for evaluation of the grinding process under the microscope.
7. Transfer the ground embryos (powder) to a 50 mL conical tube with a spatula. The tube and spatula should be pre-chilled in the liquid nitrogen bath.
8. Store the powder at −80 °C.
9. Clean the jars and balls with water and ethanol.

3.2.6 Preparation of the GST-PBD Affinity Matrix

Use 150 μL of packed Glutathione Sepharose beads and 165 μg of GST-PBD^{WT} or GST-PBD^{mut} (both at 1.5 mg/mL) diluted in 790 μL of binding buffer.

1. In a 1.5 mL Protein LoBind tube (*see* **Note 14**), add 188 μL of Glutathione Sepharose beads (80% slurry in ethanol) to obtain 150 μL of packed beads. Prepare two tubes, one for GST-PBD^{WT} and one for GST-PBD^{mut}.
2. Wash the beads with 500 μL of binding buffer.
3. Spin in a microcentrifuge for 1 min, at 200 g.
4. Remove the supernatant.
5. Repeat **steps 2–4** three times.
6. Dilute 165 μg (110 μL) of GST-PBD^{WT} or GST-PBD^{mut} in 790 μL of binding buffer to obtain a total volume of 900 μL. Save a 10 μL aliquot for SDS-PAGE analysis of the input.
7. Incubate the diluted proteins from **step 6** with the washed beads obtained after **step 5** for 1 h in the cold room on a rotating wheel. During this time, start to prepare embryonic extracts (Subheading 3.2.7)
8. Spin in a microcentrifuge for 1 min at 200 g.
9. Remove the supernatant (flow-through, unbound fraction). Save a 10 μL aliquot for SDS-PAGE analysis of the unbound fraction.
10. Wash the beads with 900 μL of binding buffer for 5 min in the cold room on a rotating wheel.
11. Spin in a microcentrifuge for 1 min at 200 g, and remove the supernatant.
12. Repeat **steps 10–11** twice.
13. Equilibrate the affinity matrix in lysis buffer. To do this, add 900 μL of lysis buffer and mix by inverting the tube 3–5 times.
14. Centrifuge for 1 min at 200 g and remove the supernatant.

15. Repeat **steps 13–14** twice.
16. Save a 5 μL aliquot of affinity matrix for SDS-PAGE analysis of the bound fraction.

At the end of this step, we obtain the affinity matrix for the pull-down.

3.2.7 Preparation of the Embryonic Extracts

1. Place the tube containing embryo powder (from **step 8** in Subheading 3.2.5) in liquid nitrogen.
2. Weigh 0.5 g of worm powder and transfer it to a 50 mL conical tube on ice. Use a spatula prechilled in the liquid nitrogen.
3. Add 1 mL of 1.5× cold lysis buffer.
4. Let the powder thaw in the lysis buffer.
5. Vortex the tube vigorously to resuspend the powder in the lysis buffer until it becomes homogenous.
6. Add 1600 U of benzonase nuclease for 30 min on ice. Every 5 min, shake the tube by hand.
7. Transfer the lysate into a 1.5 mL Protein LoBind tube.
8. Centrifuge at 13,000 g for 30 min at 4 °C.
9. Transfer the supernatant into a 1.5 mL Protein LoBind tube.
10. Centrifuge at 13,000 g for 15 min at 4 °C.
11. Transfer the supernatant into a 1.5 mL Protein LoBind tube. At the end of this step, you obtain the embryonic extracts.
12. Quantify the protein concentration using Pierce BCA protein assay kit or using the method employed in your laboratory. The protein concentration of the embryonic extracts is around 3 mg/mL.

3.2.8 Loading Embryonic Extracts on the GST-PBD Affinity Matrix

1. Incubate 500 μL (3 mg/mL) of the embryonic extract with the affinity matrix, prepared in **step 16** in Subheading 3.2.6, for 4 h in the cold room on a rotating wheel (*see* **Note 15**). During this time analyze the GST-PBD affinity matrix by SDS-PAGE (Subheading 3.2.9).
2. Spin in a microcentrifuge for 2 min at 200 g.
3. Remove the supernatant.
4. Wash the beads with 900 μL of binding buffer for 5 min in the cold room on a rotating wheel.
5. Spin in a microcentrifuge for 1 min at 200 g, and remove the supernatant.
6. Repeat **steps 4–5** twice.
7. Add 900 μL of ddH_20, and mix by inverting the tube 3–5 times.

8. Spin in a microcentrifuge for 1 min at 200 g, and remove the supernatant.
9. Repeat **steps 7–8** once.
10. Add 300 μL of ddH_20.
11. At the end of this step, the samples are ready to be trypsinized (*see* Subheading 3.2.10).
12. For phosphoproteomic analysis of the input, add 400 μL cold acetone to 50 μL (3 mg/mL) of embryonic extracts to precipitate proteins. These samples are ready to be trypsinized (*see* Subheading 3.2.10).

3.2.9 SDS-PAGE Analysis of the GST-PBD Affinity Matrix

To analyze the proteins that specifically interact with the Plk1 PBD, it is important to have the same amount of GST-PBD^{WT} and GST-PBD^{mut} (negative control) immobilized on the resin (Fig. 2).

Prepare the following samples:

1. Input: 10 μL (Subheading 3.2.6, **step 6**) add 10 μL 2× Laemmli.
2. Unbound fraction (FT): 10 μL (Subheading 3.2.6, **step 9**) add 10 μL 2× Laemmli.
3. Bound fraction: 5 μL beads (Subheading 3.2.6, **step 16**) add 10 μL 2× Laemmli.
4. Heat samples from **steps 1–3** for 3 min at 95 °C and spin.
5. Transfer the supernatant of the bound fraction into a new tube.
6. Load 6 μL input, 6 μL FT, and 10 μL bound fraction into a 10% polyacrylamide gel.

3.2.10 Tryptic Digestion and Desalting of the Samples

Continue from Subheading 3.2.8, **step 11**.

1. Remove water and add 100 μL NH_4HCO_3.
2. Add 4 μg (4 μL) of trypsin (1 μg/μL). Put the tube in a thermomixer for overnight at 37 °C, at 400 rpm.
3. On day 2, stop the digestion by adding trifluoroacetic acid (TFA) pH 3.0 and proceed to the desalting and lyophilization of the samples as follows.
4. Activate the C18 column (Sep-Pak Cartridge) by adding 2 mL of methanol.
5. Wash the column with 2 mL of 70% acetonitrile (ACN), 1% TFA.
6. Equilibrate the column with 2 mL volume of 1% TFA in water.
7. Add the sample from **step 3** to the column.
8. Wash the column with 1 mL of 1% formic acid (FA). Repeat this step once.

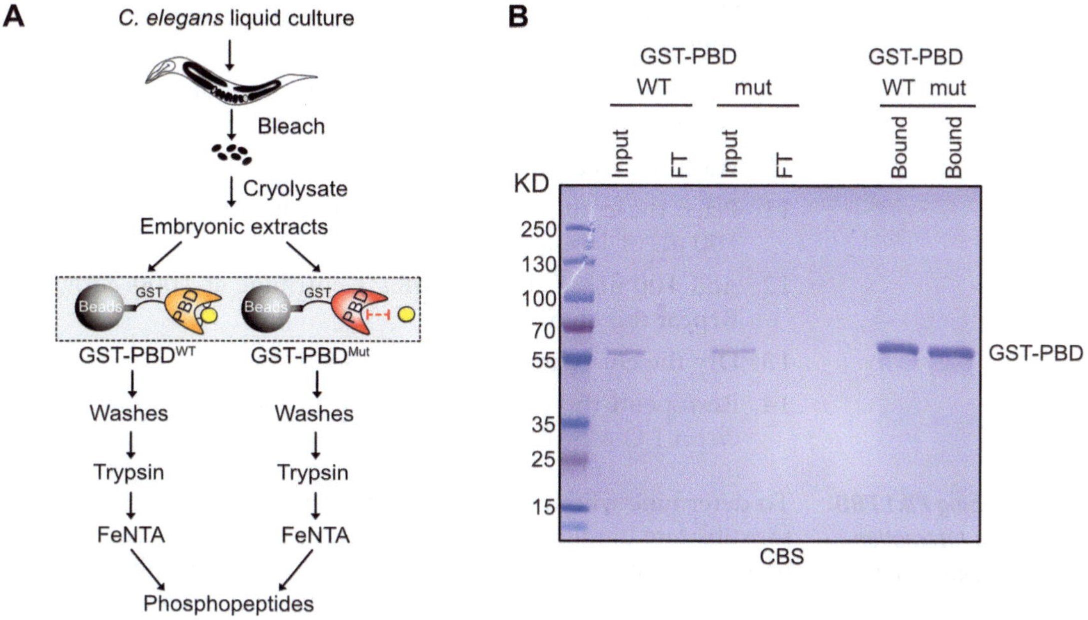

Fig. 2 Delineating the Plk1 PBD interactome. (**a**) Flowchart of the approach to identify Plk1 PBD-binding proteins from *C. elegans* embryonic extracts. After liquid cultures, worms at the adult stage are bleached to released embryos. Then a cryolysate is prepared and embryonic extracts are loaded on an affinity matrix consisting of the GST-PBD wild type or mutated on the phosphopincers (GST-PBD^{mut}). After several washing steps, retained proteins are digested with trypsin and phosphopeptides are purified (FeNTA) before identification by LC-MS/MS. (**b**) Coomassie blue-stained gel (CBS) after SDS-PAGE of GST-PBD wild type (WT) or mutant (mut) before and after preparation of the GST-PBD affinity matrix

9. Eluate the peptides with 500–750 μL of 70% ACN and 1% FA. Repeat this step once.
10. Lyophilize the sample by using speed vacuum and keep them at −80 °C.

3.2.11 Purification of the Phosphopeptides

1. Remove the bottom of the spin column and loosen the screw cap.
2. Place the column in a 2 mL microcentrifuge tube. Centrifuge at 1000 *g* for 30 s to remove the storage buffer.
3. Remove the screw cap and add 200 μL of binding/wash buffer. Centrifuge at 1000 *g* for 30 s. Repeat this step once. The flow-troughs are discarded.
4. Cap the bottom of the column with a Luer plug. Place the column with the plug in a 2 mL microcentrifuge tube.
5. Resuspend the lyophilized sample in 200 μL of binding/wash buffer and add it to the column.
6. Mix the resin with the sample softly for 30 min.
7. Remove the bottom plug and the screw cap.

8. Place the column in a 2 mL microcentrifuge tube. Centrifuge at 1000 *g* for 30 s.
9. Wash the column by adding 200 μL of binding/wash buffer. Centrifuge at 1000 *g* for 30 s. Repeat this step three times.
10. Wash the column with 200 μL of water.
11. Place the column in a new microcentrifuge tube containing 100 μL of 10% formic acid.
12. Add 100 μL of elution buffer. Centrifuge at 1000 *g* for 30 s. Repeat this step once.
13. Dry the eluate immediately in a speed vacuum.
14. Resuspend the dried eluate in 60 μL of 0.1% formic acid for direct LC-MS/MS.

3.3 Probing Plk1 PBD Substrate Interaction by Far Western Blot

To determine whether phosphorylated polo-docking sites account for substrate binding to the Plk1 PBD, different approaches can be used. For instance, purified potential substrates phosphorylated by priming kinases can be incubated with the Plk1 PBD and analyzed by analytic size exclusion chromatography (SEC) to assess complex assembly [27]. Alternatively, the interaction between phosphorylated substrates and the PBD can be probed by Far Western blotting [11, 28–30] (Fig. 3). This method is relatively simple and allows the simultaneous analysis of a combination of variants with alanine substitutions in polo-docking sites.

Far Western blot is a simple method to probe protein–protein interactions, especially when the interaction depends on a short linear motif (e.g., the phosphorylated polo-docking site). In this experiment, the purified protein of interest (POI), wild type or mutated on polo-docking sites, is phosphorylated by priming kinases and separated by SDS-PAGE. After transfer to PVDF, the membrane is incubated with the purified GST-PBD^{WT}, GST-PBD^{mut}, and the phosphorylated polo-docking site–PBD interaction is detected by immunoblot using a Plk1 antibody. The main advantage of this method is that several variants of the prey, mutated on various polo-docking sites, can be tested in parallel (Fig. 3).

3.3.1 Priming Phosphorylation of the Prey In Vitro

In the first step, the purified prey, wild type or substituted with non-phosphorylatable alanine at polo-docking sites, is primed by phosphorylation by cyclin-Cdk1 or Plk1 itself (Fig. 4). To this end, we perform in vitro kinase assays including appropriate controls.

Prepare the two reactions as follows:

Kinase assay reaction control (without kinase): In 1.5 mL microtube, add 3 μL of 10× kinase buffer, 3 μL of 10× of phosphatase inhibitor cocktail, 0.6 μL of 50× protease inhibitor cocktail, 0.6 μL of 100 mM ATP, and 0.5–1 μg of prey protein, and adjust to a total volume of 30 μL with ddH_20 (*see* **Note 16**).

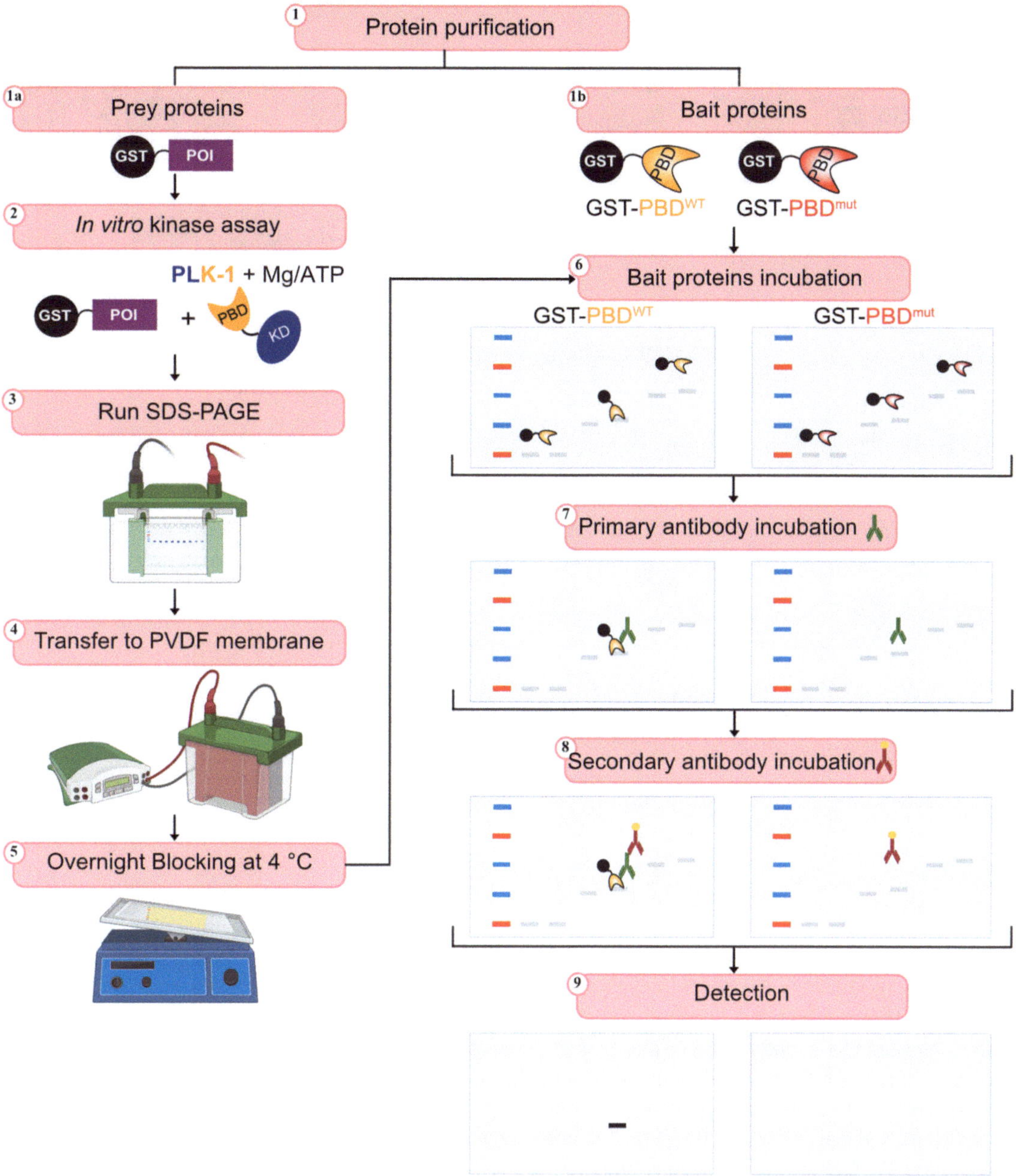

Fig. 3 Flowchart of the approach to reconstitute and detect the interaction between the Plk1 PBD and phosphorylated substrates by Far Western blot

Kinase assay reaction: In 1.5 mL microtube, add 3 μL of 10× kinase buffer, 3 μL of 10× phosphatase inhibitor cocktail, 0.6 μL of 50× protease inhibitor cocktail, 0.6 μL of 100 mM ATP, 300 ng of purified 6xHis-PLK-1, and 0.5–1 μg of prey protein, and adjust to a total volume of 30 μL with ddH_20.

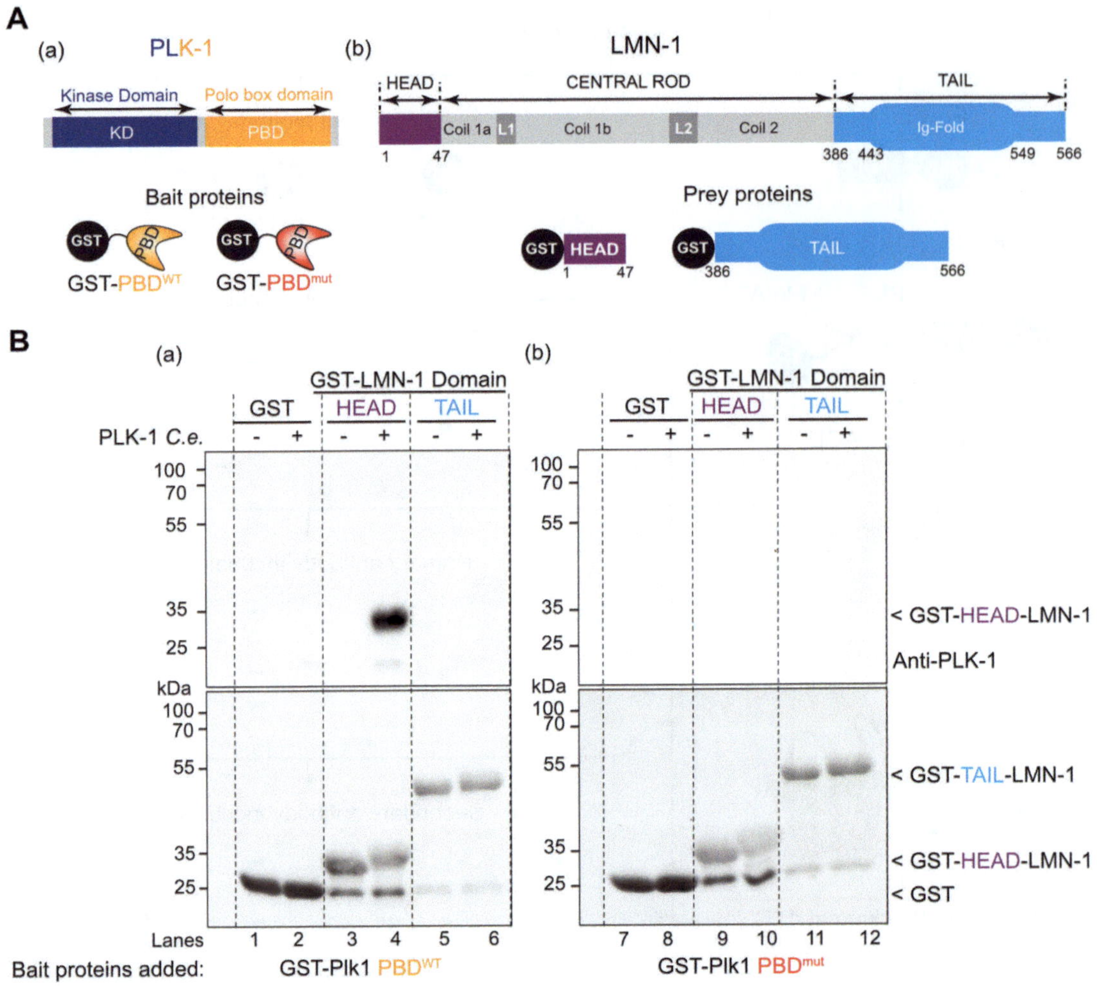

Fig. 4 (**A**) Schematic representation of *C. e* PLK-1 (*a*) and LMN-1 (lamina) with their functional domains (*b*). The bait proteins used for the Far Western blot assay were GST-PBD^{wt} and GST-PBD^{mut} (negative control), while the prey proteins were LMN-1 head and tail domains fused to GST. (**B**) In vitro kinase assay was performed with *C. elegans* PLK-1 and the GST-LMN-1 head [H] or tail [T] fragments as substrates. The samples were subjected to SDS-PAGE, followed by a Far Western ligand-binding assay using GST-PBD wild type (*a*) or the corresponding phosphate pincer mutant (GST-PBD^{mut}) (*b*). Only a direct interaction between the phosphorylated GST-HEAD domain and the GST-PBD^{wt} is observed (Lane 4). The bottom panel shows the stain-free blot (ChemiDoc, Bio-Rad) of the same membrane (*c*, *d*). GST is present as an impurity from the production of GST-LMN-1 fragments

1. Place the tubes in a shaking incubator (400 rpm) for 40 min at 30 °C.
2. Add 30 μL of 2× Laemmli buffer (final 1×).
3. Boil the samples for 5 min and set them on ice right after.

3.3.2 SDG-PAGE and Immunoblot

Prepare two gels: one for the bait protein GST-Plk1 PBD^{WT} and the second for the control GST-Plk1 PBD^{mut}.

1. Load 10 μL of the samples into the SDS-PAGE gel with the percentage according to the protein size, along with the controls and the protein ladder.
2. Run the gel at 80 V for 20 min, and then increase it to 125 V for a period depending on the size of the prey protein, which is monitored by a pre-stained protein ladder.
3. Transfer the proteins from the gel to the PVDF membrane by using liquid transfer as follows.
4. Soak the PVDF membranes in absolute methanol for 2 min; wash with distilled water, and soak with transfer buffer, before assembling the western transfer sandwich.
5. Prechill the 1 L of transfer buffer.
6. Use the traditional sandwich method to transfer the proteins from the gel to the membrane at 90 V for 90 min.
7. Block the membrane with blocking buffer overnight at 4 °C.

3.3.3 Incubation of the Membrane with GST-PBD^{WT} Wild Type or GST-PBD^{mut} Proteins

1. Remove the blocking buffer.
2. Incubate the membrane with 2 μg GST-PBD^{WT} or the control GST-PBD^{mut}, diluted in 5 mL of blocking buffer, for 5 h at 4 ° C (*see* **Note 17**).
3. Wash the membrane extensively with blocking buffer for 3 h while changing it every 15 min at 4 °C.
4. Incubate the membrane with the primary Plk1 antibody (dilution 1:1000) overnight at 4 °C.
5. Wash the membrane with the TBST buffer three times, each for 10 min.
6. Incubate the membrane with a secondary antibody for 1 h at RT. We used an HRP-conjugated anti-mouse IgG antibody at 1:10,000 dilution.
7. Wash the membrane with the TBST buffer three times, each for 10 min.
8. Perform chemiluminescent detection of the bound bait protein on the membrane using an ECL kit, according to the manufacturer's instructions.

   If there is a direct interaction between the phosphorylated protein of interest (prey protein) and the GST-PBD^{WT} (bait protein), the proteins will be located in the same spots (Fig. 4. *Lane 4*). To confirm that this interaction is phosphorylation-dependent, in the membrane that was incubated with the control GST-PBD^{mut}, there must be no detection of this bait protein (Fig. 4. *Lane 10*).

4 Notes

1. Prepare freshly and be sure to adjust the pH to 8.0 with 5 N NaOH.
2. For the Far Western blot, use PVDF and not nitrocellulose membranes.
3. The buffer should be freshly prepared and kept at 4 °C before use.
4. The blocking buffer will be used during the experiment as blocking, washing, and diluting buffer. Prepare 1 L and keep it at 4 °C.
5. The filtration is necessary to have a clear supernatant prior to loading on the column; otherwise, cell debris in the sample may clog the column.
6. Optional: collect the flow-through and reserve until the procedure has been successfully completed.
7. The reuse of the HisTrap and GSTrap 4B *columns* depends on the nature of the sample and should only be performed with identical samples to prevent cross-contamination.
8. We use the *E. coli* HB101 rather than other strains of *E. coli* since it forms a thick lawn on NGM plates and it clumps less in liquid culture [31].
9. If there are many residual bacteria in the plate that are collected with the worms, leave the tube on ice for up to 5 min to allow precipitation of the worms at the bottom, remove the supernatant, add 50 mL M9-T, mix, and proceed to **step 4**.
10. The incubation time varies depending on the freshness of the bleach solution and the concentration of the worms.
11. Do not leave the worms in sucrose for too long, and wash them rapidly after collection from the top of the tube.
12. Do not hold the tip very close to the liquid nitrogen as the worms may freeze in the tip.
13. Precautions should be taken when handling liquid nitrogen, including wearing goggles and cryo-gloves.
14. The 1.5 mL Protein LoBind tubes are designed specifically to maximize the recovery of proteins and prevent nonspecific binding to the plastic.
15. If the protein concentration is more or less than 3 mg/mL, adjust the volume to load 1.5 mg of protein.
16. This control reaction is critical for determining whether the kinase is able to phosphorylate the prey protein.
17. Due to the long incubation time with the bait protein, incubation should be at 4 °C to avoid degradation of the bait protein.

Acknowledgments

We thank P. Moussounda for media preparation and Véronique Legros and Guillaume Chevreux, from the Proteomic platform, for phosphopeptide purification and tandem mass spectrometry analysis. We thank R. Karess for critical reading of the manuscript. We thank all members of the team for stimulating discussions. Work in LP laboratory is supported by grants from Agence Nationale pour la Recherche (ANR), France, ANR-22-CE13-0022 (LP), La Ligue Nationale Contre le Cancer, Equipe Labellisée, France, and Idex Université Paris Cité, ANR-18-IDEX-0001. Griselda Velez-Aguilera is supported by a postdoctoral fellowship from "Secretaría de Educación, Ciencia, Tecnología e Innovación de la Ciudad de México," CMSECTEI/201/2022.

References

1. Archambault V, Glover DM (2009) Polo-like kinases: conservation and divergence in their functions and regulation. Nat Rev Mol Cell Biol 10:265–275
2. Zitouni S, Nabais C, Jana SC et al (2014) Polo-like kinases: structural variations lead to multiple functions. Nat Rev Mol Cell Biol 15:433–452
3. Combes G, Alharbi I, Braga LG, Elowe S (2017) Playing polo during mitosis: PLK1 takes the lead. Oncogene 36:4819–4827
4. Pintard L, Archambault V (2018) A unified view of spatio-temporal control of mitotic entry: polo kinase as the key. Open Biol 8: 180114. https://doi.org/10.1098/rsob.180114
5. Cheng KY, Lowe ED, Sinclair J et al (2003) The crystal structure of the human polo-like kinase-1 polo box domain and its phosphopeptide complex. EMBO J 22:5757–5768
6. Elia AE, Rellos P, Haire LF et al (2003) The molecular basis for phosphodependent substrate targeting and regulation of Plks by the polo-box domain. Cell 115:83–95
7. Elia AE, Cantley LC, Yaffe MB (2003) Proteomic screen finds pSer/pThr-binding domain localizing Plk1 to mitotic substrates. Science 299:1228–1231
8. Nigg EA (2001) Mitotic kinases as regulators of cell division and its checkpoints. Nat Rev Mol Cell Biol 2:21–32
9. Elowe S, Hümmer S, Uldschmid A et al (2007) Tension-sensitive Plk1 phosphorylation on BubR1 regulates the stability of kinetochore microtubule interactions. Genes Dev 21: 2205–2219
10. Qi W, Tang Z, Yu H (2006) Phosphorylation- and polo-box-dependent binding of Plk1 to Bub1 is required for the kinetochore localization of Plk1. Mol Biol Cell 17:3705–3716
11. Chan EH, Santamaria A, Sillje HH, Nigg EA (2008) Plk1 regulates mitotic Aurora A function through betaTrCP-dependent degradation of hBora. Chromosoma 117:457–469
12. Nakajima H, Toyoshima-Morimoto F, Taniguchi E, Nishida E (2003) Identification of a consensus motif for Plk (Polo-like kinase) phosphorylation reveals Myt1 as a Plk1 substrate. J Biol Chem 278:25277–25280
13. Kettenbach AN, Schweppe DK, Faherty BK et al (2011) Quantitative phosphoproteomics identifies substrates and functional modules of Aurora and Polo-like kinase activities in mitotic cells. Sci Signal 4:rs5
14. Neef R, Preisinger C, Sutcliffe J et al (2003) Phosphorylation of mitotic kinesin-like protein 2 by polo-like kinase 1 is required for cytokinesis. J Cell Biol 162:863–875
15. Kang YH, Park JE, Yu LR et al (2006) Self-regulated Plk1 recruitment to kinetochores by the Plk1-PBIP1 interaction is critical for proper chromosome segregation. Mol Cell 24:409–422
16. Park JE, Soung NK, Johmura Y et al (2010) Polo-box domain: a versatile mediator of polo-like kinase function. Cell Mol Life Sci 67: 1957–1970
17. Saurin AT (2018) Kinase and phosphatase cross-talk at the kinetochore. Front Cell Dev Biol 6:62
18. Noatynska A, Tavernier N, Gotta M, Pintard L (2013) Coordinating cell polarity and cell cycle

progression: what can we learn from flies and worms? Open Biol 3:130083

19. Kim AJ, Griffin EE (2020) PLK-1 regulation of asymmetric cell division in the Early C. elegans embryo. Front cell. Dev Biol 8:632253
20. Taylor SJP, Bel Borja L, Soubigou F et al (2023) BUB-1 and CENP-C recruit PLK-1 to control chromosome alignment and segregation during meiosis I in C. elegans oocytes. Elife 12:e84057
21. Endicott JA, Noble ME, Johnson LN (2012) The structural basis for control of eukaryotic protein kinases. Annu Rev Biochem 81:587–613
22. Macurek L, Lindqvist A, Lim D et al (2008) Polo-like kinase-1 is activated by aurora A to promote checkpoint recovery. Nature 455: 119–123
23. Tavernier N, Noatynska A, Panbianco C et al (2015) Cdk1 phosphorylates SPAT-1/Bora to trigger PLK-1 activation and drive mitotic entry in C. elegans embryos. J Cell Biol 208: 661–669
24. Lowery DM, Clauser KR, Hjerrild M et al (2007) Proteomic screen defines the polo-box domain interactome and identifies Rock2 as a Plk1 substrate. EMBO J 26:2262–2273
25. Nkombo Nkoula S, Velez-Aguilera G, Ossareh-Nazari B, Van Hove L, Ayuso C, Legros V, Chevreux G, Thomas L, Seydoux G, Askjaer P, Pintard L (2023) Mechanisms of nuclear pore complex disassembly by the mitotic Polo-like kinase 1 (PLK-1) in C. elegans embryos. Sci Adv 9:eadf7826
26. Stiernagle T (2006) Maintenance of C. elegans. WormBook 1–11
27. Singh P, Pesenti ME, Maffini SC, et al (2021) BUB1 and CENP-U, primed by CDK1, are the main PLK1 kinetochore receptors in mitosis. Mol Cell 81:67–87.e9
28. Wu Y, Li Q, Chen XZ (2007) Detecting protein-protein interactions by far western blotting. Nat Protoc 2:3278–3284
29. Martino L, Morchoisne-Bolhy S, Cheerambathur DK et al (2017) Channel nucleoporins recruit PLK-1 to nuclear pore complexes to direct nuclear envelope breakdown in C. elegans. Dev Cell 43:157–171.e7
30. Velez-Aguilera G, Nkombo Nkoula S, Ossareh-Nazari B et al (2020) PLK-1 promotes the merger of the parental genome into a single nucleus by triggering lamina disassembly. elife 9:e59510
31. Brooks KK, Liang B, Watts JL (2009) The influence of bacterial diet on fat storage in C. elegans. PLoS One 4:e7545

Chapter 5

Artificial Modulation and Rewiring of Cell Cycle Progression Using Synthetic Circuits in Fission Yeast

Akanksha Jain, Pei-Yun Jenny Wu, and Damien Coudreuse

Abstract

Cell cycle control is a central aspect of the biology of proliferating eukaryotic cells. However, progression through the cell cycle relies on a highly complex network, making it difficult to unravel the core design principles underlying the mechanisms that sustain cell proliferation and the ways in which they interact with other cellular pathways. In this context, the use of a synthetic approach to simplify the cell cycle network in unicellular genetic models such as fission yeast has opened the door to studying the biology of proliferating cells from unique perspectives. Here, we provide a series of methods based on a minimal cell cycle module in the fission yeast *Schizosaccharomyces pombe* that allows for an unprecedented artificial control of cell cycle events, enabling the rewiring and remodeling of cell cycle progression.

Key words Fission yeast, Synthetic cell cycle, Minimal cell cycle network, Cell cycle rewiring, Artificial cell cycle, Cyclin-dependent kinase

1 Introduction

Cell cycle progression is controlled by an intricate regulatory network. At the heart of this system are cyclin-dependent kinases (CDKs), which associate with various cyclin subunits to sequentially drive the different phases of the cell cycle [1, 2]. While the core mechanisms underlying cell proliferation are conserved among eukaryotic species, unicellular genetic models such as fission yeast remain instrumental for deciphering the processes that sustain cell cycle progression. Indeed, the simplicity of their cell cycle control system and their amenability to state-of-the-art genetic manipulation allow for unraveling the basic architecture of the cell proliferation regulatory network [1, 2]. Remarkably, using a synthetic biology approach, the fission yeast cell cycle network was artificially simplified, with the endogenous system being replaced by a single module consisting of a fusion between CDK1/Cdc2 and cyclin B/Cdc13 (Fig. 1a, b). Surprisingly, cells operating with this

Anna Castro and Benjamin Lacroix (eds.), *Cell Cycle Control: Methods and Protocols*, Methods in Molecular Biology, vol. 2740,
https://doi.org/10.1007/978-1-0716-3557-5_5,

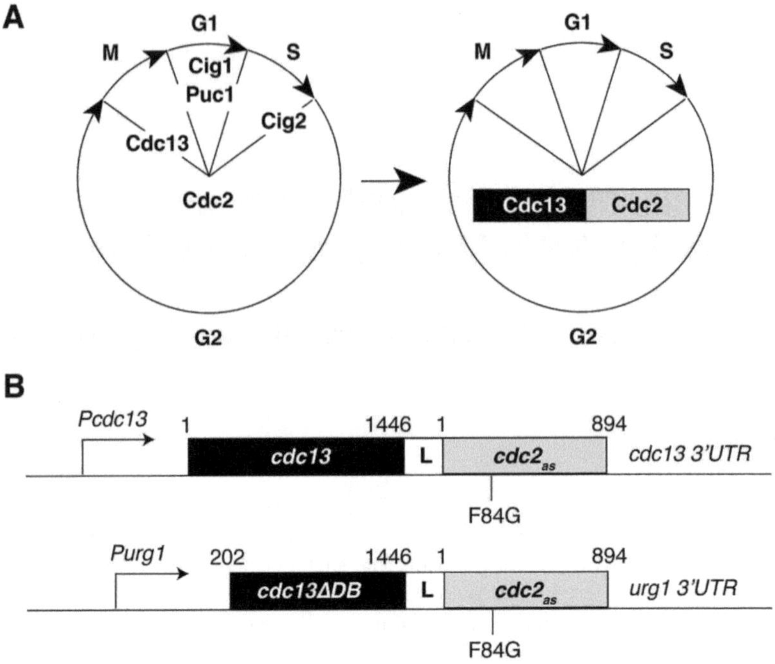

Fig. 1 (**a**) Left: the wild-type fission yeast cell cycle relies on the activity of Cdc2/CDK, which associates with various cyclins at different phases of the cell cycle (G1: Cig1 and Puc1; S: Cig2; G2 and M: Cdc13). Right: the complex endogenous cell cycle network can be replaced by a simple fusion of Cdc13 and Cdc2 in the absence of the other cell cycle cyclins and of the endogenous copy of *cdc2*. Strains operating with this minimal cell cycle network (MCN) have a normal cell cycle [3]. (**b**) Top: schematic representation of the Cdc13-Cdc2 fusion cassette in MCN cells. *Pcdc13*: *cdc13* promoter. The F84G alteration ($cdc2_{as}$) makes the synthetic module sensitive to dose-dependent inhibition by non-hydrolysable ATP analogs [3, 14]. L: linker (GGGSGGGSGGGS). Bottom panel: the MCN cassette was modified for artificially driving the entire cell cycle [3]. *Purg1*: uracil-inducible *urg1* promoter [16]. The cyclin destruction box was deleted to make the protein insensitive to APC-dependent degradation at mitotic exit (*cdc13ΔDB*) [17]. Strains carrying this cassette are referred to as MCN*ΔDB*. Numbers are open reading frame coordinates. Both cassettes are integrated at the *leu1* locus

minimal cell cycle network (MCN) were shown to have a normal cell cycle [3]. This demonstrated that cell cycle progression in fission yeast solely relies on the oscillation of a single qualitative CDK activity between two major thresholds at the G1/S and G2/M transitions. Importantly, this approach paved the way for using the MCN system to rewire and modulate cell cycle progression in fission yeast. Indeed, this system made it possible not only to manipulate cell cycle events and reveal the logic of the cell cycle network, but also to investigate the interplay between cell cycle progression, cell cycle phases, CDK function, and other cell biological processes [3–7]. The capacity to externally impose a complete remodeling of the temporal sequence of the cell cycle, to alter the duration of any cell cycle phase, or even to drive the entire fission yeast cell cycle through simply modulating the activity of the MCN CDK module represents a unique approach for a broad range of studies beyond cell cycle research. Here we describe a set of key protocols, from cell preparation to cell cycle rewiring to fully artificial cell cycle control. This provides a basis for the external modulation of cell cycle progression in MCN cells and their derivatives. These methods can easily be combined or adjusted, and the strains described subsequently can be further modified to address other specific questions.

2 Materials

1. Fission yeast cells operating with the minimal cell cycle network (MCN) [3]. MCN: *leu1Δ::Pcdc13::cdc13Lcdc2as::cdc13UTR::ura4+ cdc2Δ::KanMX6 cdc13Δ::NatMX6 cig1Δ::HygMX6 cig2Δ::KanMX6 puc1Δ::leu2+ ura4-D18 h+*(DC450) [4, 8].
2. MCN *nda3-km311* cells: *leu1Δ::Pcdc13::cdc13Lcdc2as::cdc13UTR::ura4+ cdc2Δ::KanMX6 cdc13Δ::NatMX6 cig1Δ::HygMX6 cig2Δ::KanMX6 puc1Δ::leu2+ nda3-km311 ura4-D18* (DC510) [7]. The mutation *nda3-km311* in the tubulin-encoding gene *nda3* allows for blocking cells in mitosis with high CDK activity upon shift to 18 °C [9].
3. MCN*ΔDB* cells: *leu1Δ::Purg1::cdc13ΔDB-L-cdc2as::urg13'UTR::ura4+ cig2Δ::natMX6 cdc2-33 ura4*-D18 *h-* (DC348) [3].
4. 50X salt stock: 53.5 g/L $MgCl_2*6H_2O$, 0.74 g/L $CaCl_2*2H_2O$, 50 g/L KCl, 2 g/L Na_2SO_4.
5. 1000X Vitamin stock: 1 g/L calcium pantothenate, 10 g/L nicotinic acid, 10 g/L myo-inositol, 10 mg/L biotin. Do not autoclave - filter sterilize.
6. 10,000X Mineral stock: 5 g/L boric acid, 4 g/L $MnSO_4$, 4 g/L $ZnSO_4*7H_2O$, 2 g/L $FeCl_2*6H_2O$, 0.4 g/L molybdic acid, 1 g/L KI, 0.4 g/L $CuSO_4*5H_2O$, 10 g/L citric acid. Do not autoclave - filter sterilize.
7. Minimal supplemented medium EMM6S [10, 11]: 3 g/L potassium hydrogen phthalate, 4.2 g/L Na_2 HPO_4*7H_2O, 20 g/L dextrose, 20 mL/L salt stock, 0.1 mL/L mineral stock, 1 mL/L vitamin stock, 5 g/L NH_4Cl, 0.225 g/L adenine, 0.225 g/L histidine, 0.225 g/L leucine, 0.225 g/L uridine, 0.225 g/L lysine, and 0.225 g/L arginine. EMM5S: EMM6S without uridine. Sterilize by filtration and pre-warm or precool to the appropriate temperature prior to usage.

8. YE4S [10] agar plates containing antibiotics: 5 g/L Bacto yeast extract, 30 g/L dextrose, 20 g/L agar, 0.225 g/L adenine, 0.225 g/L histidine, 0.225 g/L leucine, 0.225 g/L uridine, 200 μg/mL hygromycin B, and/or 100 μg/mL G418 sulfate. Autoclave and add antibiotics after the medium has cooled to ~55 °C prior to pouring the plates.
9. 50 and 125 mL glass Erlenmeyer flasks, sterilized.
10. Static incubators at 25 °C and 32 °C, shaking incubators at 18 °C, 25 °C, and 32 °C.
11. Non-hydrolyzable ATP analog 3-MBPP1 (Toronto Research Chemicals, A602960; *see* **Note 1**) stock solution: 10 mM in DMSO. Alternative ATP analogs may be used, although their efficiencies and the concentrations that are required for the provided protocols would require calibration.
12. Transmitted light benchtop microscope with a 40X objective, standard microscopy-grade glass slides and coverslips.
13. Filtering system and 1.2 μm filters. The system must be efficient to allow for fast filtration of the cultures, a key point for most of the provided protocols.
14. 1000X DAPI (2-(4-Amidinophenyl)-1*H*-indole-6-carboxamidine) solution: 50% glycerol, 0.1 M Tris pH 8, 1 mg/mL DAPI.
15. 1 mg/mL Blankophor (in water).
16. Cold 70% ethanol (4 °C).
17. 50 mM sodium citrate.
18. 10 mg/mL RNase A (100X).
19. 1 mg/mL propidium iodide.
20. Benchtop sonicator.
21. Flow cytometer.
22. Benchtop hot plate that can heat to ~70 °C.
23. Fluorescence microscope equipped for transmitted and reflected (for DAPI and Blankophor; 405 nm wavelength) light imaging equipped with a 60X objective.

3 Methods

Schematics for all of the methods and protocols presented in the following sections are provided in Figs. 2, 3, 4, and 5. Except otherwise indicated, all assays should be performed at 32 °C in a shaking incubator.

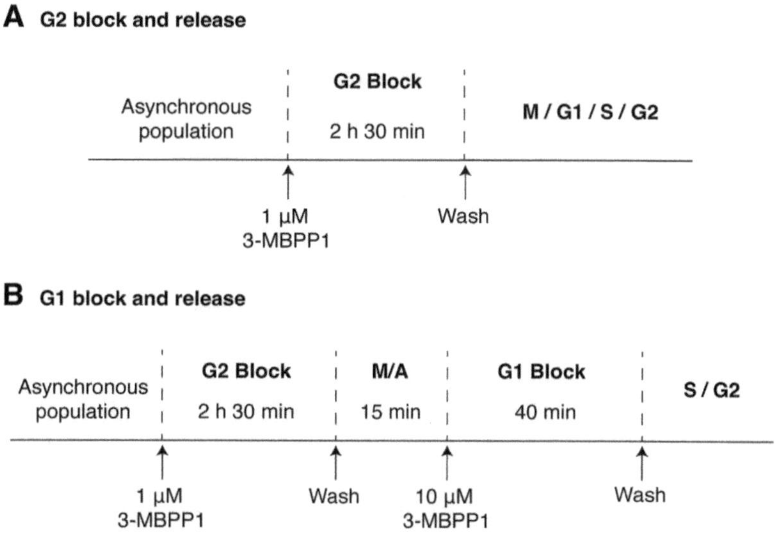

Fig. 2 (**a**) Experimental design for the G2 block and release experiment. (**b**) Experimental design for the G1 block and release experiment. M/A: metaphase to anaphase transition

3.1 Monitoring Cell Cycle Progression

For all of the specific protocols that are presented here to rewire the cell cycle using the synthetic minimal cell cycle network, the quality of each step and the behavior of the cells should be evaluated by monitoring essential cycle markers, such as changes in DNA content (S-phase progression) and percentage of bi-nucleated cells (completion of mitosis). This will ensure that cells have responded efficiently to the artificial modulation of their cell cycle regulatory system.

3.1.1 DNA Content Analysis of Fission Yeast Cell Populations

1. Sample 1 mL of culture (*see* Subheading 3.2) in a 1.5 mL microtube.
2. Centrifuge at 8000 rpm for 30 s, discard the supernatant, and resuspend the cells in 1 mL cold 70% ethanol (vortex for 5 s).
3. Keep samples at 4 °C for a minimum of 1 h (we recommend incubating overnight to ensure full cell fixation). *See* **Note 2**.
4. Wash each sample with 1 mL of 50 mM sodium citrate and resuspend in 500 μL sodium citrate +0.1 mg/mL RNase A (vortex for 5 s).
5. Incubate for a minimum of 3 h at 37 °C (we recommend incubating overnight).
6. Add 500 μL sodium citrate +4 μg/mL propidium iodide (2 μg/mL final concentration) and vortex for 5 s.
7. Sonicate the samples using a standard benchtop sonicator. *See* **Note 3**.
8. Process the samples using a standard flow cytometer. *See* **Note 4**.

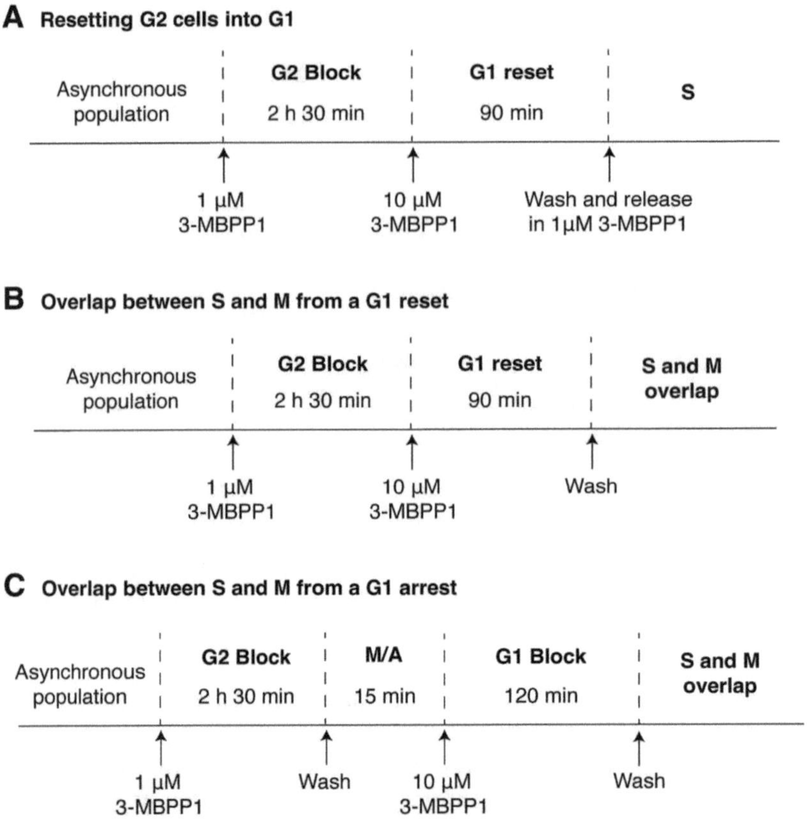

Fig. 3 (**a**) Experimental design for resetting G2 cells into G1 without an intervening mitosis. (**b**) Experimental design for inducing an overlap between S and M from G2 cells that have been reset in G1. (**c**) Experimental design for inducing an overlap between S and M from cells blocked in G1. M/A: metaphase to anaphase transition

3.1.2 Monitoring Nuclear and Cell Division

1. Prepare a hot plate at 70 °C.
2. Sample 500 μL of culture (*see* Subheading 3.2) in a 1.5 mL tube.
3. Centrifuge at 8000 rpm for 30 s, discard the supernatant, and resuspend the cells in 5–10 μL of medium.
4. Deposit 2 well-separated drops of 3–5 μL of cells on a microscope slide. *See* **Note 5**.
5. Put the slide on the 70 °C hot plate for 2 min. *See* **Note 6**.
6. Put a 2–3 μL drop of DAPI solution on the heat-fixed cells and cover them with a microscopy-grade coverslip. *See* **Note 7**.
7. Determine the percentage of binucleated cells in the population using a microscope equipped for DAPI imaging. *See* **Note 8**.

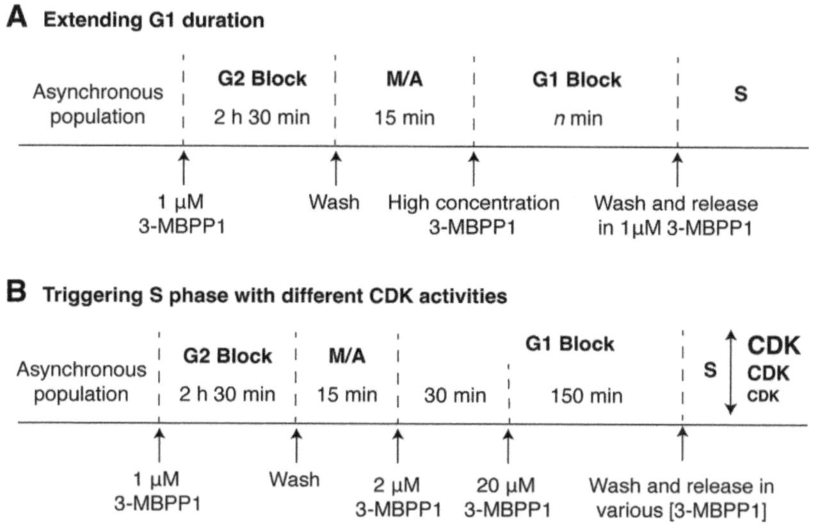

Fig. 4 (**a**) Experimental design for extending G1 duration. After the metaphase to anaphase transition, cultures are treated with a high concentration of the 3-MBPP1 inhibitor to arrest cells in G1 for various periods of time (*n* min). Different concentrations should be used depending on the target G1 extension (*see* main text). M/A: metaphase to anaphase transition. (**b**) Experimental design for triggering S phase with different CDK activities. At the end of the extended G1 block, cells have accumulated a high level of the MCN module. Release into various concentrations of 3-MBPP1 allows for cells to enter S phase with a range of CDK activities. M/A: metaphase to anaphase transition

3.2 Preparation of MCN or MCN nda3-km311 *Cultures for Cell Cycle Manipulation*

The initial preparation of the cultures is a key step that should be strictly followed, as this is essential for the quality of the subsequent cell cycle manipulations. The following protocol is the starting point for all of the methods presented in this article.

1. Plate cells from a frozen stock on YE4S plates and grow at 32 °C for 2–3 days until colonies are visible.
2. Day 0: inoculate a single colony in 10 mL EMM6S in a 50 mL flask and incubate this starter culture overnight at 32 °C without shaking.
3. Day 1: in the morning, dilute the starter culture in 40 mL EMM6S in a 125 mL flask and grow cells at 32 °C in a shaking incubator. The target optical density (OD_{595}) after dilution should be calculated to reach an OD_{595} of ~0.5–0.6 at the end of the day (*see* **Note 9**).
4. Dilute cells to the appropriate OD_{595} in 40 mL EMM6S and grow overnight. The target OD_{595} after dilution should be calculated so that the culture is at an OD_{595} of ~0.5–0.6 the next morning.

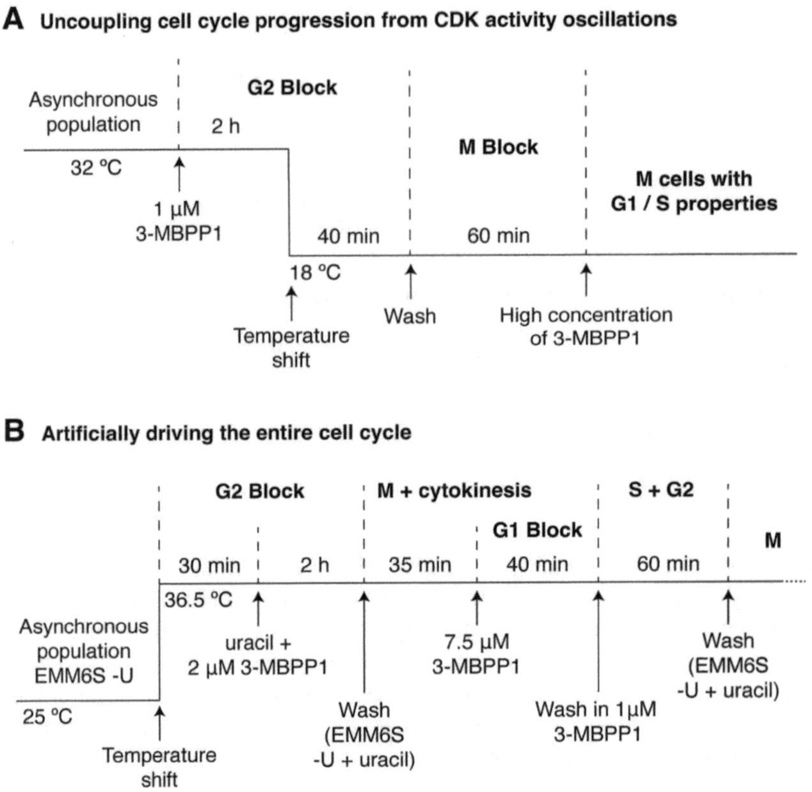

Fig. 5 (**a**) Experimental design for uncoupling cell cycle progression from CDK activity oscillations. Representative assay using the *nda3-km311* cold-sensitive mutation, which blocks cell cycle progression independently of CDK activity. Modulating CDK activity using high concentrations of 3-MBPP1 in M-arrested MCN *nda3-km311* cells allows for triggering G1 and S events in cells that remain in a mitotic state. (**b**) Experimental design for artificially driving an entire cell cycle using the MCN*ΔDB* background. Constitutive expression of the nondegradable cassette prevents cell growth without external control of the activity. The strain is therefore designed to rely on endogenous Cdc2/CDK for growth prior to the experiment. The use of the *cdc2-33* temperature-sensitive allele to inactivate Cdc2, together with the *Purg1* promoter to induce the expression of the MCN*ΔDB* cassette, makes it possible to switch from the endogenous to the synthetic system by shifting to the restrictive temperature of 36.5 °C while adding uracil to the medium. Once cells are dependent on the externally controlled MCN*ΔDB* module, the experiment must be maintained at 36.5 °C to prevent interference from endogenous Cdc2

5. Day 2: in the morning, check cells under a standard benchtop microscope with transmitted light to ensure that the culture is not contaminated and that cells have a normal morphology. Dilute cells to an OD_{595} of ~0.1 in 40 mL EMM6S (this volume should be adjusted depending on the subsequent assays to be performed). Grow cells for one generation until they reach an OD_{595} of ~0.2.
6. All experiments using MCN and MCN *nda3-km311* presented below start from this point.

3.3 G2 Block and Release of MCN Cells (Fig. 2a)

Mitotic entry requires high CDK activity. A modest overall reduction of this activity allows cells to progress through G1, S, and G2 but prevents mitotic entry, thus arresting cells at G2/M.

1. Add 3-MBPP1 to the culture of MCN cells to a final concentration of 1 μM. This reduces CDK activity to a level that prevents mitotic entry, arresting cells in G2. Collect a control sample from the exponentially growing culture prior to treatment with the inhibitor as a reference for monitoring DNA content and assessing binucleated cells. *See* **Note 10**.
2. After 2.5 h, check the cells under a standard benchtop microscope. All cells should be significantly elongated and there should be no visible division septa in the population (*see* **Note 11**). The quality of the block is critical for the quality of the subsequent steps.
3. Collect a sample for DNA content analysis and binucleated count prior to the release. The G2 block can also be used to study the physiological status of the cell at the G2/M.
4. Release the cells from the 3-MBPP1 block: filter cultures on a 1.2 μm pore membrane and wash three times with 100 mL of pre-warmed EMM6S (*see* **Note 12**).
5. Inoculate the released population in 40 mL pre-warmed EMM6S. Cells will synchronously enter mitosis and proceed through the cell cycle.
6. The next steps depend on the goal of the specific experiment (*see* **Note 13**). Culture volumes depend on the specific application (*see* **Note 14**).

3.4 G1 Block and Release of MCN Cells (Fig. 2b)

Blocking fission yeast cells in G1 requires a strong inhibition of CDK activity below the S-phase threshold [3, 11–13]. Using the MCN system, this is achieved using high concentrations of 3-MBPP1 [3]. However, in asynchronous populations of *S. pombe* cells, the large majority of cells are in G2. Treatment of such cultures with 3-MBPP1 thus leads to most cells blocking in G2. Arresting cells in G1 therefore requires a pre-synchronization of the culture in G2.

1. Synchronize the population in G2 using the protocol described in **steps 1–5** of Subheading 3.3.
2. 15 min after the release from the G2 block, add 3-MBPP1 to the culture at a final concentration of 10 μM (*see* **Note 15**). After ~40 min, most cells undergo cytokinesis without genome duplication, resulting in accumulation of mononucleated 1C cells.
3. Collect samples for DNA content analysis and binucleated count. The G1 block can also be used to study the physiology of cells in G1.

4. Release cells from the G1 block using the same filtration method as for the G2 block and release protocol. Inoculate the released population in 40 mL pre-warmed EMM6S. Cells will synchronously enter S phase and proceed through the cell cycle. Depending on the length of the G1 block, S-phase entry can be rapid, occurring within 5 min of the release.
5. Cells can also be allowed to eventually "leak through" into S phase (modulation of the leak through timing can be achieved by blocking cells in G1 with different 3-MBPP1 concentrations in **Step 2**; *see* **Note 16**).
6. The following steps depend on the goal of the specific experiment. Nevertheless, it is essential to evaluate the quality of the procedure as for the G2 block and release (*see* **Note 13**). The same restrictions as for the G2 block and release apply in terms of scaling up the assay (*see* **Note 14**).

3.5 Rewiring the Cell Cycle

In fission yeast, cell cycle progression solely relies on the oscillation of single qualitative CDK activity. Using the synthetic MCN background, artificial modulation of the activity of the minimal CDK module therefore allows for extensive rewiring of the cell cycle.

3.5.1 Bypassing Mitosis: Resetting G2 Cells into G1 (Fig. 3a)

This protocol allows for resetting G2 cells into a G1-like state without an intervening mitosis.

1. Block cells in G2 as in **Steps 1** and **2** of Subheading 3.3.
2. Add 3-MBPP1 to the culture at a final concentration of 10 μM. This lowers CDK activity to a G1-level, resetting G2 cells into G1 (*see* **Note 17**).
3. After 90 min, release cells in EMM6S + 1 μM 3-MBPP1. Cells enter S phase ~20 min after the release, without an intervening mitosis. This can be observed by a shift from a 2C (G2) to 4C DNA content. The release in 3-MBPP1 prevents an overlap between S- and M-phases (*see* Subheading 3.5.2).

3.5.2 Inducing an Overlap Between S and M Phases from a G1 Reset (Fig. 3b)

The use of the synthetic MCN network allowed us to demonstrate that mitosis and DNA replication are not fundamentally exclusive and can overlap if both processes are triggered simultaneously by the appropriate levels of CDK activity [3]. This makes it possible to investigate how these two processes, which are normally separated by checkpoint functions, interact in the context of deregulated cell cycles. Inducing an overlap between S and M phases can be achieved by two different protocols, which allows for studying this deleterious event in cells with distinct physiological statuses and DNA contents (also *see* Subheading 3.5.3).

1. Perform a G1 reset from G2 through **Step 2** (*see* Subheading 3.5.1).

2. After 90 min in EMM6S + 10 μM 3-MBPP1, release cells in EMM6S without 3-MBPP1. Cells enter M and S (DNA content shifts from 2C to 4C) simultaneously within 10–20 min.
3. The overlap between S and M can be followed by combining DNA content analysis with DAPI staining (*see* **Note 18**).

3.5.3 Inducing an Overlap Between S and M Phases from a G1 Arrest (Fig. 3c)

1. Perform a G1 block using 10 μM 3-MBPP1 (*see* Subheading 3.4, **steps 1–3**).
2. Maintain the G1 block for 120 min.
3. Release cells in EMM6S without 3-MBPP1. A similar nuclear phenotype as in Subheading 3.5.2 is then observed but with a ~10 min delay. DNA content shifts from 1C to 2C.

3.5.4 Modulating the Duration of G1 (Fig. 4a)

CDK activity is required during the early phases of the cell cycle, playing a key role in core functions such as DNA replication. MCN cells can be used to investigate the importance of G1 timing as well as the role of CDK activity levels (*see* Subheading 3.5.5) for the onset and organization of DNA replication [4].

1. Follow the protocol from Subheading 3.4 with the following modifications. For short G1 extensions (<20 min), 2 μM 3-MBPP1 is sufficient to maintain cells in G1. For extensions up to 90 min, use 10 μM 3-MBPP1; for longer extensions, 20 μM 3-MBPP1 must be used to prevent leak through (*see* **Note 19**).
2. During the G1 block, proteins continue to accumulate [3, 4]. This includes the MCN module, which can be monitored by Western blot analyses (e.g., anti-Cdc13 or anti-Cdc2 antibody, both allow detection of the Cdc13-Cdc2as fusion protein).
3. Release cells by filtration in 1 μM 3-MBPP1 (allows S-phase entry but prevents mitosis) after the target extension times.

3.5.5 Triggering S Phase with Different CDK Activities (Fig. 4b)

1. Follow a modified version of the protocol from Subheading 3.5.4, with an initial G1 block of 30 min using 2 μM 3-MBPP1 and a subsequent 150 min extension of the arrest by adding 3-MBPP1 to a final concentration of 20 μM.
2. Split the culture according to the number of distinct S-phase CDK activities required for the experiment.
3. Wash and release each culture in a distinct concentration of 3-MBPP1. Concentrations from 1 to 6 μM 3-MBPP1 allow S-phase onset but prevent mitotic entry [3, 4] (*see* **Note 20**).

While cell cycle progression relies on oscillations in CDK activity, events that depend on CDK function can, in certain cases, be studied while isolated from cell cycle progression. For instance, a number of G1 and S-phase related processes can be uncoupled from cell cycle events and monitored even in mitotic cells [7]. Here, we provide an example based on MCN cells harboring *nda3-km311*, a cold-sensitive mutation of tubulin in fission yeast, which blocks cells in mitosis with high CDK activity upon switching to 18 °C [9]. External modulation of CDK activity in these cells triggers G1 and S-phase processes while cells remain in a mitotic state. This protocol can easily be modified for use with other conditional mutants.

3.5.6 Uncoupling Cell Cycle Progression from CDK Activity Oscillations (Fig. 5a)

1. Synchronize MCN *nda3-km311* cells using a modified version of the protocol described in Subheading 3.3. Specifically, to inactivate Nda3, cells are shifted from 32 °C to 18 °C 2 h after addition of 1 μM 3-MBPP1 for the G2 arrest.
2. After 40 min at 18 °C, release cells from the 3-MBPP1 inhibition in precooled (18 °C) EMM6S. This allows cells to resume their cell cycle but rapidly arrest in mitosis with high CDK activity.
3. After 1 h, cells can be treated with various concentrations of 3-MBPP1, allowing for the reset of cells in different pseudophases (e.g., G1) while they remain in mitosis. Treatment with 10 μM 3-MBPP1 at this stage triggers a switch in periodic transcription, activating post-mitotic gene clusters [7]. Interestingly, release from the temperature shift allows for evaluating mitotic events in cells that were rewired toward G1 and S.

3.6 Artificially Driving the Entire Cell Cycle (Fig. 5b)

A derivative of the MCN system (*leu1Δ::Purg1::cdc13ΔDB-L-cdc2as::urg1 3′UTR::ura4+ cig2Δ::natMX6 cdc2-33 ura4-D18* [3]; *see* Fig. 1b and **Note 21**) can be used to artificially drive the entire cell cycle, modulating the duration of each phase and the level of CDK activity at each transition. The following protocol provides a basis for such experiments and can be modified in order to alter cell cycle timing and CDK activity levels (*see* **Note 22**).

1. Cultures should be grown as described for MCN cells but at 25 °C in EMM6S -U (*see* **Note 23**). The dilution steps should take into account the slower generation time at this temperature.
2. Shift cells to 36.5 °C for the remainder of the experiment to inactivate endogenous Cdc2 and render cells solely dependent on the MCN*ΔDB* module.
3. After 30 min at 36.5 °C, add uracil (250 mg/L final concentration – this induces expression of the MCN*ΔDB* cassette) + 2 μM 3-MBPP1 (inhibits the activity of the induced MCN*ΔDB* cassette) and incubate for 2 h. Cells are blocked in late G2.

4. Release cells in EMM6S – U + uracil (maintains the *Purg1* promoter in an induced state). Cells will undergo synchronous mitotic entry.
5. After 35 min, add 3-MBPP1 to a final concentration of 7.5 μM. This corresponds to a time at which cells have passed the metaphase to anaphase transition but have not committed to S phase. The high concentration of 3-MBPP1 arrests cells in G1. Cytokinesis occurs and cells accumulate with a 1C DNA content.
6. After 40 min, release cells in EMM6S – U + uracil +1 μM 3-MBPP1. In these conditions, cells have sufficient CDK activity to proceed through S phase and G2 but not mitosis. This maintains the temporal sequence of cell cycle events.
7. After 60 min, release cells in EMM6S – U + uracil to trigger entry into the next mitosis.
8. Repeat the sequence from **step 5** to drive another cycle.

This protocol is the most advanced method for artificial control of the entire cell cycle in a eukaryotic cell, making it possible to investigate the interplay between cell cycle dynamics and a variety of cell cycle processes.

4 Notes

1. We have tested a range of suppliers and found variable efficiencies for the ATP analog between manufacturers. The concentrations described here have been optimized using this specific reference. Changing provider may require optimization of the concentrations used.
2. Ethanol-fixed samples can be kept at 4 °C for several months.
3. The parameters must be adjusted to the specific sonicator that will be used. As a reference, we are using a Branson 450 Digital Sonifier and a 10 s cycle at 30% power. Sonicators compatible with the simultaneous treatment of several samples can be used but would require optimization.
4. To evaluate cell cycle progression and S-phase duration, DNA content is generally plotted as a histogram showing the intensity of propidium iodide staining. In fission yeast, DNA replication occurs prior to cytokinesis. As a result, cells spend most of their cell cycle as haploids with 2 genome complements (2C). In synchronized populations, a transient switch from 2C to 4C can be observed as cells duplicate their genome, followed by a return to 2C upon completion of cytokinesis. Appearance of a 1C peak generally reflects an extension of G1/delay in S phase onset; in this case, cells proceed through

cytokinesis prior to DNA replication (such as in the Subheading 3.4).

5. Only one of the two drops of cells is processed at a time. This allows the user to have a backup sample if any of the subsequent steps fails. When both DNA content and nuclear division are being monitored, a small aliquot of the cells from the DNA content sample (prior to ethanol fixation) can be used for heat fixation and assessment of nuclear division.
6. After heat fixation, slides can be stored in a dry, dark environment for months.
7. Septum formation/cytokinesis can also be evaluated by adding Blankophor to the DAPI solution (11:1 DAPI/Blankophor). We have however found the quality of the double-staining to be variable.
8. We recommend using an objective with at least 60X magnification. Cells are considered as binucleated when the two DNA masses are completely separated and cells have not started to physically divide (septated cells that have not engaged in cytokinesis are considered as binucleated). A cell that shows any sign of cytokinesis (even a small medial invagination) is considered as two mononucleated cells. We often use simultaneous exposure to reflective and low levels of transmitted light to visualize both cell contours and DAPI staining.
9. MCN cells and their derivatives should not be diluted to an OD_{595} lower than 0.008. Over-dilution results in a growth delay or no growth. If the culture reaches an $OD_{595} > 0.6$, cells have been subjected to nutritional stress, which may affect the outcome of the experiment.
10. The MCN module harbors a mutation in the Cdc2/CDK1 moiety of the fusion protein (Fig. 1b), making it sensitive to reversible and dose-dependent inhibition by non-hydrolysable ATP analogs [3, 14].
11. 2.5 h corresponds to the doubling time of MCN cells at 32 °C, allowing cells that just exited mitosis when the inhibitor was added to reach late G2 at the end of the block. At any phase of the cell cycle (except mitosis), growth continues even if the cell cycle is arrested. This leads to cell elongation.
12. All releases and filtrations of the cultures described in the protocols that are provided are based on three successive washes with 100 mL of pre-warmed or pre-cooled medium depending on the assay. The quality of the assays is significantly improved by limiting the time it takes to filter the culture to a minimum (<1 min per washing step).
13. It is essential to evaluate the quality of the synchronization using DNA content analysis and binucleate count. At 32 °C,

we recommend taking samples every 10 min for 2.5 h. As a reference, mitosis and S-phase onset take place ~15 min and ~ 30 min following release from G2 arrest, respectively [3, 4].

14. While this method can be applied to larger cultures, volumes over 150 mL will affect the rapidity of the filtration and subsequent synchronization.
15. To block cells in G1, CDK activity must be strongly inhibited after the metaphase to anaphase transition but prior to S-phase onset. In fission yeast, this represents a very short time window. At 32 °C in EMM6S, we add a high concentration of 3-MBPP1 15 min after inoculating the G2-released cells in pre-warmed medium. This precise timing can be adjusted for different media, genetic backgrounds, and growth conditions by verifying DAPI staining of heat-fixed samples and determining the time at which the majority of cells are in anaphase but have not yet completed nuclear division.
16. Cells can be arrested in G1 using a minimum of 2 μM 3-MBPP1. However, the lower the concentration, the faster they eventually leak through into S phase. Concentrations below 2 μM failed at providing a robust G1 arrest at the population level. We have routinely used 10 μM as a standard concentration.
17. Cell status prior to the release is complex to evaluate, as cells remain 2C during the "reset." However, this approach triggers a switch in the cell cycle transcription program, and the reset can be characterized by assessing the expression of genes that belong to the S-phase gene cluster, which are activated prior to the release from 3-MBPP1 [7].
18. DAPI staining allows for monitoring abnormal segregation of the genetic material while it is being duplicated. Coupled with Blankophor staining, it makes it possible to observe cells that undergo septation without normal nuclear segregation, referred to as a "cut" phenotype [15].
19. Certain ATP analogs such as 1-NmPP1 induce nonspecific effects (e.g., septation defect) at concentrations as low as 10 μM [3, 5]. 3-MBPP1 is less prone to such off-target effects and can be reliably used above 20 μM.
20. Releasing these cells in EMM6S without 3-MBPP1 will trigger an overlap between S and M. When using concentrations above 1 μM at the release, cells enter S phase at the same time for all concentrations, but S-phase duration is increased at higher inhibitor concentrations [3, 4].
21. The *Purg1* promoter is induced in the presence of uracil [16]. *ΔDB* is a deletion of the cyclin destruction box [17];

this modification prevents degradation of the MCN module at mitotic exit, thereby making CDK activity independent of oscillations in protein level. *cdc2-33* is a temperature-sensitive allele that inactivates Cdc2 at 36.5 °C [18]. We refer to this strain as MCN*ΔDB*.

22. It is particularly critical to monitor mitosis, cytokinesis, and DNA content to ensure the success of this complex artificial control of the entire cell cycle. Note that while these experiments are performed using batch cultures, microfluidic technologies would allow for automated switching of the medium and single-cell analysis of cell cycle progression.
23. The permissive temperature for the *cdc2-33* allele is 25 °C. Higher temperatures impair cell cycle progression. EMM6S cannot be used for the initial growth of these MCN*ΔDB* cells, as the presence of uridine induces the expression of the modified MCN module. In the absence of the cyclin destruction box, CDK activity cannot be controlled by the anaphase promoting complex (APC), resulting in mitotic exit failure. Therefore, the MCN*ΔDB* module must not be constitutively present, and its expression must only be induced at a later step, as mitotic exit can then be triggered by the timely addition of 3-MBPP1.

Acknowledgments

This work was supported by grants to DC from the Agence Nationale de la Recherche (PRC eVOLve, ANR-18-CD13-0009), the Région Nouvelle Aquitaine (program CHESS, grant agreements 15963520 and 15964420), and the Ligue Contre le Cancer (Gironde, Dordogne). AJ was funded by the ANR eVOLve. PYW was supported by the Région Nouvelle Aquitaine (program CHESS, grant agreements 15963520 and 15964420) and the Ligue Contre le Cancer (Pyrénées-Atlantiques).

References

1. Morgan DO (1997) Cyclin-dependent kinases: engines, clocks, and microprocessors. Annu Rev Cell Dev Biol 13:261–291
2. Morgan DO (2007) The cell cycle: principles of control. New Science Press, London
3. Coudreuse D, Nurse P (2010) Driving the cell cycle with a minimal CDK control network. Nature 468:1074–1079
4. Perrot A, Millington CL, Gómez-Escoda B et al (2018) CDK activity provides temporal and quantitative cues for organizing genome duplication. PLoS Genet 14:e1007214
5. Swaffer MP, Jones AW, Flynn HR et al (2016) CDK substrate phosphorylation and ordering the cell cycle. Cell 167:1750–1761.e16
6. Gutiérrez-Escribano P, Nurse P (2015) A single cyclin–CDK complex is sufficient for both mitotic and meiotic progression in fission yeast. Nat Commun 6:6871
7. Banyai G, Baïdi F, Coudreuse D et al (2016) Cdk1 activity acts as a quantitative platform for coordinating cell cycle progression with periodic transcription. Nat Commun 7:11161

8. Chen T, Gomez-Escoda B, Munoz-Garcia J et al (2016) A drug-compatible and temperature-controlled microfluidic device for live-cell imaging. Open Biol 6:160156
9. Moreno S, Hayles J, Nurse P (1989) Regulation of p34cdc2 protein kinase during mitosis. Cell 58:361–372
10. Moreno S, Klar A, Nurse P (1991) [56] Molecular genetic analysis of fission yeast Schizosaccharomyces pombe. Methods Enzymol 194: 795–823
11. Gérard C, Tyson JJ, Coudreuse D et al (2015) Cell cycle control by a minimal Cdk network. PLoS Comput Biol 11:e1004056
12. Stern B, Nurse P (1996) A quantitative model for the cdc2 control of S phase and mitosis in fission yeast. Trends Genet 12:345–350
13. Fisher DL, Nurse P (1996) A single fission yeast mitotic cyclin B p34cdc2 kinase promotes both S-phase and mitosis in the absence of G1 cyclins. EMBO J 15:850–860
14. Bishop AC, Ubersax JA, Petsch DT et al (2000) A chemical switch for inhibitor-sensitive alleles of any protein kinase. Nature 407:395–401
15. Uemura T, Yanagida M (1984) Isolation of type I and II DNA topoisomerase mutants from fission yeast: single and double mutants show different phenotypes in cell growth and chromatin organization. EMBO J 3:1737–1744
16. Watt S, Mata J, López-Maury L et al (2008) urg1: a uracil-regulatable promoter system for fission yeast with short induction and repression times. PLoS One 3:e1428
17. Yamano H, Gannon J, Hunt T (1996) The role of proteolysis in cell cycle progression in Schizosaccharomyces pombe. EMBO J 15:5268–5279
18. Nurse P, Thuriaux P, Nasmyth K (1976) Genetic control of the cell division cycle in the fission yeast Schizosaccharomyces pombe. Mol Gen Genet 146:167–178

Chapter 6

Measuring Molecular Diffusion in Self-Organizing *Xenopus* Extracts by Fluorescence Correlation Spectroscopy

William Y. C. Huang, James E. Ferrell Jr., and Xianrui Cheng

Abstract

The cytoplasm is densely packed with macromolecules and organelles, displaying viscoelastic properties at various scales. How biochemical reactions function efficiently enough in a seemingly jammed environment remains elusive. Cell-free *Xenopus laevis* extracts represent a powerful system for investigating the biochemistry and biophysics of living systems. Here we present a protocol for characterizing macromolecular diffusion in self-organizing cytoplasmic extracts using fluorescence correlation spectroscopy (FCS), which measures the motions on a distance scale of ~200 nm. The method can also be used to characterize diffusion in the cytoplasm as it progresses through different phases of the cell cycle.

Key words Self-organization, Cytoplasm, Cell extracts, Diffusion, Fluorescence correlation spectroscopy

1 Introduction

Diffusion directly influences diffusion-limited reactions and diffusion-coupled processes in cells. Diffusion is affected by multiple cellular mechanisms. For example, yeast cells adjust cytoplasmic viscosity through metabolite production to maintain invariant diffusion rates at vastly different temperatures [1–4]. Although most enzymes are reaction-limited rather than diffusion-limited in vitro [5], it is possible that the crowded environment of the cell slows enzyme–substrate dissociation enough relative to catalysis to make diffusion control more common.

Xenopus egg extracts are an attractive system for probing the biochemistry and biophysics of complex living systems. *Xenopus* egg extracts are undiluted, cell-free cytoplasm isolated from frog eggs, and they can carry out important physiological processes including the cell cycle [6, 7]. In addition, they are highly manipulable and can be kept alive for many hours [8]. Cytoplasmic extracts are initially spatially disorganized, but over the course of

Anna Castro and Benjamin Lacroix (eds.), *Cell Cycle Control: Methods and Protocols*, Methods in Molecular Biology, vol. 2740, https://doi.org/10.1007/978-1-0716-3557-5_6,

~30–60 min they self-organize into macroscopic structures that resemble cells (Fig. 1) [9–12]. In this organized state, organelles are concentrated near the nuclei, and organelle-depleted borders separate neighboring cell-like compartments. Overall, the pattern consists of similarly sized polygons and resembles a sheet of epithelial cells [9, 13].

Here we present a method for carrying out fluorescence correlation spectroscopy (FCS) on self-organizing, interphase-arrested *Xenopus* egg extracts. FCS provides information about protein motions over a length scale of ~200 nm. This length scale is relevant to biochemical reactions in the cytoplasm, since an average protein with a concentration of 200 nM uniformly distributed in the cytoplasm would need to travel about 100 nm to find its nearest reaction partner [10, 14]. FCS detects single molecules diffusing through a diffraction-limited confocal spot of about 200 nm and reports a fluorescence intensity trace that contains the diffusion information [15, 16]. Diffusion characteristics are then retrieved from the fluctuations in the fluorescence intensity trace using autocorrelation analysis. This procedure yields the characteristic diffusion time and the type of diffusion motion of fluorescent molecules passing through the confocal spot [15–17]. In the example here, we monitor protein diffusion at the hundred-nanometer scale as a function of cytoplasmic self-organization.

The method readily applies to all other types of *Xenopus* egg extracts. Extracts can be manipulated to arrest at specific cell cycle phases and transition between phases [8]. Extracts that carry out as many as 20 cell cycles over the course of 10–15 h can also be produced [18, 19]. The method could therefore be used to characterize diffusion kinetics of cell cycle regulators in the different spatial architectures that emerge throughout the cell cycle.

2 Materials

2.1 *Xenopus laevis* Interphase-Arrested Egg Extracts

1. Mature female *Xenopus laevis* frogs.
2. Pregnant mare serum gonadotropin (PMSG): reconstitute in sterile water to 500 IU/mL and store at 4 °C.
3. Human chorionic gonadotropin (hCG): reconstitute in sterile water to 2000 IU/mL and store at 4 °C.
4. Egg laying buffer: 100 mM NaCl, 2 mM KCl, 1 mM $MgSO_4{\cdot}7H_2O$, 2.5 mM $CaCl_2{\cdot}2H_2O$, 0.5 mM HEPES, 0.1 mM EDTA. Prepare as a 20× stock solution, adjust to pH 7.4 with NaOH, and dilute to 1× with clean water drawn from the water system in which the frog population is housed before use.

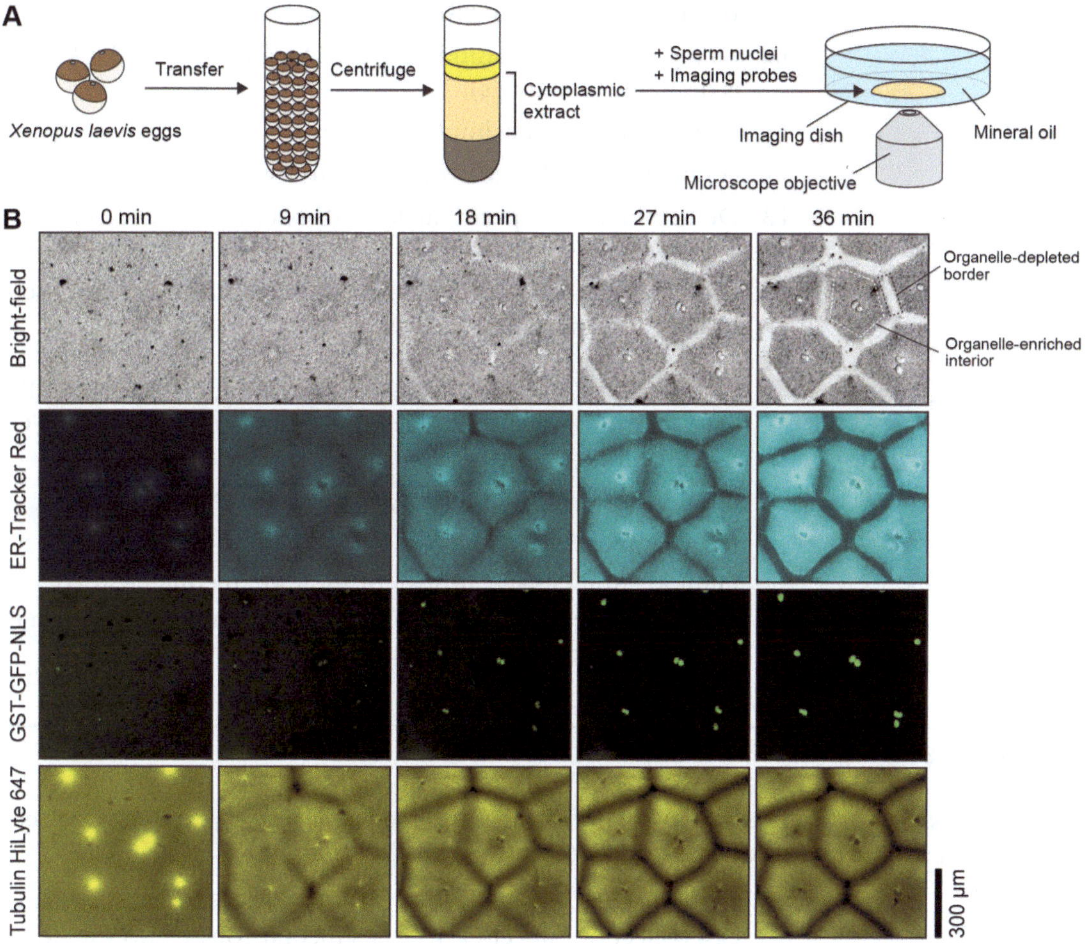

Fig. 1 Self-organizing cytoplasmic extracts. (**a**) Schematics of *X. laevis* egg extract preparation and imaging setup. The cytoplasmic extract was arrested in the interphase stage and further supplemented with demembranated sperm nuclei at approximately 27 per μL. (**b**) Visualization of the self-organization process. 1 μM ER-Tracker Red dye (cyan), 0.38 μM GST-GFP-NLS (green), and 1 μM HiLyte Fluor 647-labeled tubulin (yellow) were added to the extract to visualize the endoplasmic reticulum (ER), interphase nuclei, and microtubules, respectively. At 0 min, the homogenized extract was released from 0 °C to room temperature and allowed to self-organize

5. Egg lysis buffer (ELB): 250 mM sucrose, 10 mM HEPES, 50 mM KCl, 2.5 mM $MgCl_2$, 1 mM DTT (dithiothreitol). Prepare fresh and adjust to pH 7.7 with KOH.
6. Marc's modified Ringer's (MMR) solution: 100 mM NaCl, 2 mM KCl, 1 mM $MgCl_2$, 2 mM $CaCl_2$, 0.1 mM EDTA, 5 mM HEPES. Adjust to pH 7.8 with NaOH.
7. Dejellying solution: 2% w/v L-cysteine in water. Adjust to pH 8.0 with NaOH and prepare fresh.
8. 5 mg/mL aprotinin (Roche) in water.

9. 10 mg/mL leupeptin (Roche) in water.
10. 10 mg/mL cytochalasin B in DMSO (can be omitted if F-actin is desirable in extracts).
11. 10 mg/mL cycloheximide in water.
12. cOmplete EDTA-free protease inhibitor (Roche).
13. Demembranated sperm nuclei [20] (can be omitted, *see* **Note 1**).
14. ER-Tracker Red (or other organelle labels for monitoring self-organization such as MitoTracker dyes [9]).

2.2 Fluorescence Correlation Spectroscopy (FCS)

1. ZEISS LSM 780 laser scanning confocal microscope and ZEN data-acquisition software (or equivalent setup capable of FCS and imaging).
2. Bioinert μ-Dish 35 mm imaging chamber (No. 1.5) (ibidi).
3. Heavy mineral oil.
4. Fluorescently labeled probes of interest. In this protocol, we use bovine serum albumin (67 kDa) or dextran-70 kDa conjugated with Alexa Fluor 488 as examples [10].

3 Methods

3.1 X. laevis Interphase-Arrested Egg Extract Preparation

Procedures for preparing interphase *X. laevis* egg extracts have been described in previous work [20, 21].

1. Prime female *X. laevis* frogs. At least 3 days before the day of egg collection, inject each frog subcutaneously into the dorsal lymph sac with 100 IU of PMSG.
2. Induce egg laying 16–18 h before egg collection. Inject each primed frog with 500 IU of hCG and isolate the frog in egg laying buffer in a separate container.
3. Collect eggs that meet quality criteria (*see* **Note 2**).
4. Remove the egg jelly coat by incubating the eggs in dejellying solution with gentle swirling at intervals. Proceed to the next step within no more than 5 min (*see* **Note 3**).
5. Pour off the dejellying solution and wash the dejellied eggs a few times with 1 L of 0.25× MMR.
6. Wash the eggs a few times with 400 mL of ELB.
7. Transfer the washed eggs into a centrifuge tube and centrifuge for 15 s at 400× g in a swinging bucket rotor.
8. Remove as much buffer from the top of the packed eggs as possible (sacrifice some eggs if necessary).

9. Determine the volume of packed eggs, and add 5 μg aprotinin, 5 μg leupeptin, 5 μg cytochalasin B, and 50 μg cycloheximide per mL of eggs to the top of the packed eggs.
10. Centrifuge the eggs at 12,000× g for 15 min at 4 °C in a swinging bucket rotor.
11. Isolate the cytoplasmic extract with an 18-gauge needle attached to a syringe. Puncture the side wall of the centrifuge tube with the needle and draw slowly. Cytoplasmic extract is the translucent fraction between the yellow lipid fraction at the top and the dark yolk fraction at the bottom (Fig. 1a).
12. Supplement the extract with all the inhibitors used in **step 9** except cycloheximide. Use the same doses as in **step 9**.
13. Add the protease inhibitor cocktail, cOmplete EDTA-free (follow the reagent's instruction to attain 1× final concentration), to the extract. Incubate for 30 min on ice to maximize the inhibition (*see* **Note 4**).

3.2 Measuring Diffusion with FCS

1. Turn on the FCS microscope and warm up the instrument for at least 10 min.
2. Select an objective that is certified for FCS measurements (such as the C-APO 40× water-based objective from ZEISS). Adjust the objective collar to match the imaging chamber substrate and temperature (i.e., room temperature).
3. Select the optical path for 488 nm measurements (e.g., lasers, dichroic mirrors, filters, and detectors).
4. Mount a standard on the microscope for alignment and calibration; commonly used standards include 100 nM fluorescein (diffusion coefficient $D = 425\ \mu m^2/s$), Alexa Fluor 488 (diffusion coefficient $D = 435\ \mu m^2/s$), or Atto 488-carboxylic acid (diffusion coefficient $D = 400\ \mu m^2/s$) in water [22] (*see* **Note 5**).
5. Align the pinhole using the standard (for the ZEISS LSM 780, this can be achieved using the "adjust pinhole" function in the ZEN program); during this step, make sure the z position is 100 μm above the substrate surface such that the focus is in the solution.
6. Calibrate the spot size by taking an autocorrelation function of the standard for 5 min (or sufficient time to obtain a smooth autocorrection function; Fig. 2a, b).
7. Prior to the primary FCS measurements, add 160 per μL of demembranated *X. laevis* sperm nuclei, 25 nM fluorescently labeled molecules, and 1 μM ER-Tracker™ Red to the extract. Mix thoroughly (*see* **Note 6**).

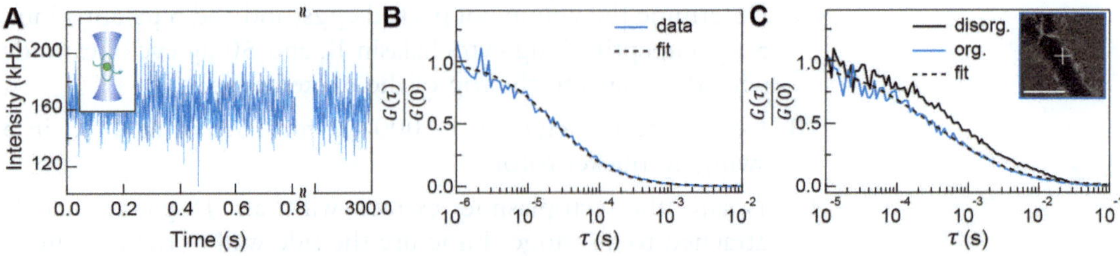

Fig. 2 FCS measurements in buffers and self-organizing extracts. (**a**) The intensity trace and (**b**) the autocorrelation function of Alexa Fluor 488 in water at room temperature. The inset in (**a**) shows a labeled molecule diffusing through a confocal spot, which gives the fluctuations observed in the intensity trace. The correlation function was calculated from 5 min of data acquisition. The dashed line shows the fit of a 3D Brownian diffusion model (Eq. 2 with $\alpha = 1$) to the data. The fitted τ_D is 24.7 μs, yielding a confocal radius of $w = 207$ nm. (**c**) Normalized autocorrelation functions of BSA-Alexa Fluor 488 in the organized (blue) and disorganized (black) cytoplasmic extracts. The inset shows the organized ER at about 50 min after the initial deposit of the extract and right before this FCS measurement; the cross shows the focus spot at the organelle-depleted border region. Scale bar, 100 μm. The dashed line shows the fit of an anomalous diffusion model (Eq. 2) to the data. The fitted values are $\alpha = 0.71$ and $\tau_D = 310$ μs in the organized cytoplasm; in contrast, the fitted values for the disorganized cytoplasm are $\alpha = 0.76$ and $\tau_D = 700$ μs (fit not shown). Note the decrease in the diffusion time in the organized cytoplasm. (See data in [10] for comparison)

8. Add a droplet (~2 μL) of the extract on an imaging dish, immediately cover the extract with mineral oil, and mount it on an inverted FCS confocal microscope (Fig. 1a). This step marks the time zero of the self-organization process.
9. Regularly take images of ER (e.g., 568 nM for ER-Tracker Red) to monitor the state of self-organization.
10. Select an *x-y* position of interest to measure the autocorrelation function (Fig. 2c). We typically fix the *z*-position such that nuclei are near the focal plane (about 30–40 μm above the substrate) (*see* **Note 7**).
11. Repeat 9 and 10 until the end of the experiment.

3.3 FCS Data Analysis

1. Retrieve the autocorrelation function from the experiment. If only raw intensity traces are available, the autocorrelation function $G(\tau)$ is calculated by [15]

$$G(\tau) = \frac{\langle \delta I(t)\delta I(t+\tau)\rangle}{\langle I(t)\rangle^2}, \tag{1}$$

where $I(t)$ is the fluorescence intensity, $\delta I(t) = I(t) - \langle I(t)\rangle$, $\langle\cdot\rangle$ denotes the average, and τ is the delay time.

2. Fit the autocorrelation function with a 3D anomalous diffusion model [15, 17]

$$G(\tau) = \frac{1}{N\left(1+\left(\frac{\tau}{\tau_D}\right)^\alpha\right)\sqrt{\left(1+\frac{1}{s^2}\left(\frac{\tau}{\tau_D}\right)^\alpha\right)}}, \tag{2}$$

where α is the diffusion-mode parameter, τ_D is the characteristic diffusion time, N is the particle number, and s is the structural parameter of the optics. α describes the dependence of mean squared displacement (MSD) on time, $MSD(t) \propto t^\alpha$. Small dye molecules diffusing in buffer solution may appear Brownian ($\alpha = 1$); in contrast, diffusion of protein molecules in the cytoplasm is often subdiffusive ($\alpha < 1$). Strictly speaking, anomalous diffusion does not have a defined diffusion coefficient. Nevertheless, it is possible to estimate an effective diffusion coefficient using the relation $D = w^2/4\tau_D$, where w is the confocal spot radius (calibrated with the standard assuming a known diffusion coefficient and using the same relation) (*see* **Note 8**).

4 Notes

1. Cell-like spatial organization forms in *X. laevis* egg extracts regardless of whether demembranated sperm nuclei are added or not [9]. When sperm nuclei are added, the spatial patterns emerge earlier, presumably due to the presence of centrosomes associated with the sperm nuclei.
2. Egg quality affects the properties of egg extracts. The quality criteria we apply for extracts used in FCS experiments are as follows: no fecal matter or other debris in the egg laying buffer; no variation in egg appearance across the population; the animal hemisphere of each egg has uniform pigment color and a single, distinct white dot near the pole region.
3. Prolonged incubation in the dejellying solution may cause the eggs to lyse. The degree of jelly coat removal can be assessed by visual inspection. Before dejellying, opaque eggs are separated by some "clear" space due to the presence of the transparent jelly coat. As the incubation goes on, the jelly coat is gradually dissolved and this "clear" space shrinks. When the jelly coat is sufficiently dissolved, this space vanishes, and eggs directly touch each other. At this point, remove the dejellying solution and start washing the eggs with 0.25× MMR.
4. For protein measurements, protease inhibitors are crucial for preventing protein degradation over the course of an experiment (~1 h). Alternatively, one may use protein-sized polysaccharides, such as dextran-70 kDa, which are more resistant to degradation.
5. For each instrument, it is important to characterize the point spread function (PSF) to ensure that the illumination is approximately Gaussian in the *x*-*y* plane and not tilted along the *z*-axis [23]. Typical calibration can be performed by taking a scanning

image of a bright 100-nm immobilized microsphere. For a well-shaped PSF, the autocorrelation function of the standard sample should give α that approaches 1 when fitted with an anomalous diffusion model, which would suggest that any anomalous diffusion measured in extracts is not due to imperfection in the optics.

6. Well-mixed cytoplasmic extracts are in a disorganized state (Fig. 1b, $t = 0$ min), which is difficult to achieve in live cell experiments. This convenience demonstrates a unique strength of extract systems.
7. Self-organizing extracts are highly dynamic; macroscopic features change significantly on minute timescales. Thus, data acquisition in self-organizing extracts should minimize the acquisition time while obtaining sufficient signals. We have chosen 1 min as the measurement time though a shorter acquisition time may be possible. For each region of interest, it is also advisable to take multiple data points from different positions due to the spatial heterogeneity of cytoplasmic extracts.
8. Model selection depends on the probe of interest. In general, the simplest model should be used to fit the data. In some cases, a Brownian diffusion model ($\alpha = 1$) may be sufficient. Importantly, it is preferrable that the measurement is taken without the substantial presence of complex photophysics; for example, triplet-state interconversion may attribute to additional correlation decay in the timescale of μs [15]. -Photobleaching should always be avoided in taking measurements.

Acknowledgments

This work was supported by the National Institutes of Health (NIH) (K99 GM143481 to W.Y.C.H. and R35 GM131792 to J. E.F.). We thank Stanford University's Cell Sciences Imaging Facility (CSIF) for microscopy support.

References

1. Persson LB, Ambati VS, Brandman O (2020) Cellular control of viscosity counters changes in temperature and energy availability. Cell 183(6):1572–85 e16. https://doi.org/10.1016/j.cell.2020.10.017
2. Parry BR, Surovtsev IV, Cabeen MT et al (2014) The bacterial cytoplasm has glass-like properties and is fluidized by metabolic activity. Cell 156(1–2):183–194. https://doi.org/10.1016/j.cell.2013.11.028
3. Molines AT, Lemiere J, Gazzola M et al (2022) Physical properties of the cytoplasm modulate the rates of microtubule polymerization and depolymerization. Dev Cell 57(4):466–79 e6. https://doi.org/10.1016/j.devcel.2022.02.001
4. Chen Y, Huang JH, Phong C et al (2023) Protein homeostasis from diffusion-dependent control of protein synthesis and degradation. bioRxiv. https://doi.org/10.1101/2023.04.24.538146

5. Bar-Even A, Noor E, Savir Y et al (2011) The moderately efficient enzyme: evolutionary and physicochemical trends shaping enzyme parameters. Biochemistry 50(21):4402–4410. https://doi.org/10.1021/bi2002289
6. Murray AW, Kirschner MW (1989) Cyclin synthesis drives the early embryonic cell cycle. Nature 339(6222):275–280. https://doi.org/10.1038/339275a0
7. Tunquist BJ, Maller JL (2003) Under arrest: cytostatic factor (CSF)-mediated metaphase arrest in vertebrate eggs. Genes Dev 17(6): 683–710. https://doi.org/10.1101/gad.1071303
8. Murray AW (1991) Cell cycle extracts. Methods Cell Biol 36:581–605
9. Cheng X, Ferrell JE Jr (2019) Spontaneous emergence of cell-like organization in Xenopus egg extracts. Science 366(6465):631–637. https://doi.org/10.1126/science.aav7793
10. Huang WYC, Cheng X, Ferrell JE Jr (2022) Cytoplasmic organization promotes protein diffusion in Xenopus extracts. Nat Commun 13(1):5599. https://doi.org/10.1038/s41467-022-33339-0
11. Mitchison TJ, Field CM (2021) Self-Organization of Cellular Units. Annu Rev Cell Dev Biol 37:23–41. https://doi.org/10.1146/annurev-cellbio-120319-025356
12. Nguyen T, Mitchison TJ, Wühr M. Immunofluorescence of microtubule assemblies in amphibian oocytes and early embryos (1940–6029 (Electronic))
13. Gires PY, Thampi M, Krauss SW et al (2023) Exploring generic principles of compartmentalization in a developmental in vitro model. Development 150(3). https://doi.org/10.1242/dev.200851
14. Huang WYC, Boxer SG, Ferrell JE (2022) Membrane localization accelerates association under conditions relevant to cellular signaling. bioRxiv. https://doi.org/10.1101/2022.10.22.513358
15. Krichevsky O, Bonnet G (2002) Fluorescence correlation spectroscopy: the technique and its applications. Rep Prog Phys 65(2):251–297. https://doi.org/10.1088/0034-4885/65/2/203
16. Lakowicz JR (2006) Principles of fluorescence spectroscopy. Springer, New York
17. Hofling F, Franosch T (2013) Anomalous transport in the crowded world of biological cells. Rep Prog Phys Phys Soc 76(4):046602. https://doi.org/10.1088/0034-4885/76/4/046602
18. Guan Y, Li Z, Wang S et al (2018) A robust and tunable mitotic oscillator in artificial cells. elife:7. https://doi.org/10.7554/eLife.33549
19. Afanzar O, Buss GK, Stearns T et al (2020) The nucleus serves as the pacemaker for the cell cycle. elife:9. https://doi.org/10.7554/eLife.59989
20. Deming P, Kornbluth S (2006) Study of apoptosis in vitro using the Xenopus egg extract reconstitution system. Methods Mol Biol 322: 379–393
21. Cheng X, Ferrell JE Jr (2021) Xenopus laevis egg extract preparation and live imaging methods for visualizing dynamic cytoplasmic organization. J Vis Exp 172. https://doi.org/10.3791/61923
22. Kapusta P (2010) Absolute diffusion coefficients: compilation of reference data for FCS calibration. Access
23. Hess ST, Webb WW (2002) Focal volume optics and experimental artifacts in confocal fluorescence correlation spectroscopy. Biophys J 83(4):2300–2317. https://doi.org/10.1016/S0006-3495(02)73990-8

Chapter 7

Mechanical Characterization of Murine Oocytes by Atomic Force Microscopy

Rose Bulteau, Lucie Barbier, Guillaume Lamour, Tristan Piolot, Elsa Labrune, Clément Campillo, and Marie-Emilie Terret

Abstract

The quality of murine and human oocytes correlates to their mechanical properties, which are tightly regulated to reach the blastocyst stage after fertilization. Oocytes are nonadherent spherical cells with a diameter over 80 μm. Their mechanical properties have been studied in our lab and others using the micropipette aspiration technique, particularly to obtain the oocyte cortical tension. Micropipette aspiration is affordable but has a low throughput and induces cell-scale deformation. Here we present a step-by-step protocol to characterize the mechanical properties of oocytes using atomic force microscopy (AFM), which is minimally invasive and has a much higher throughput. We used electron microscopy grids to immobilize oocytes. This allowed us to obtain local and reproducible measurements of the cortical tension of murine oocytes during their meiotic divisions. Cortical tension values obtained by AFM are in agreement with the ones previously obtained by micropipette aspiration. Our protocol could help characterize the biophysical properties of oocytes or other types of large nonadherent samples in fundamental and medical research.

Key words Oocyte, Biomechanics, Large nonadherent cells, Cortical tension, AFM

1 Introduction

Female gametes, called oocytes, are produced for sexual reproduction during meiosis. These cells are large, 80 μm diameter for mouse oocytes, and nonadherent, surrounded by an outer glycoprotein matrix called the zona pellucida (ZP). They are composed of several layers: a plasma membrane associated with a contractile actomyosin cortex and a cytoplasm. Their meiotic divisions, which terminate oocyte morphogenesis, last more than 10 h in the mouse model and are characterized by equal chromosome segregation but asymmetric cytoplasm partitioning in the absence of canonical centrosomes. This tour de force is ensured in part by the tight temporal regulation of cortical tension during this process. As in

Anna Castro and Benjamin Lacroix (eds.), *Cell Cycle Control: Methods and Protocols*, Methods in Molecular Biology, vol. 2740, https://doi.org/10.1007/978-1-0716-3557-5_7,

the case of a liquid droplet, increasing the area of the cell costs energy, and the cortical tension is the ratio of this energy over the area increase. This cortical tension reflects the dynamics of the cellular actomyosin cortex [1]. We and others have shown that cortical tension regulation was a key parameter to produce oocytes with a good developmental potential [2, 3]. A widely used technique to study the mechanical properties of large cells like oocytes is the micropipette aspiration technique [2–5]. In this technique, the cell is aspirated into a glass pipette with controlled pressure steps until reaching a critical pressure at which the length of the cell part aspirated into the pipette is equal to its radius. For larger aspiration pressures, the cell flows into the pipette. From the critical pressure, the cell radius, and the pipette radius (typically one quarter of the cell diameter), the cortical tension is calculated using the Laplace law [6]. This method allows to obtain the mechanical properties of the cell at a global scale with an affordable setup. However, it has a low throughput, the measurement lasting 5 min per cell, and is invasive as it implies a large cell deformation during measurements [7].

AFM is based on the physical interaction between the sample and a local probe, inducing only nanometer-scale deformation and thus representing a good alternative to micropipette aspiration [7]. This technique was developed in 1986 to study the surfaces of insulators at the atomic scale [8]. The probe is composed of a cantilever with a sharp tip that can be moved precisely in all directions by piezoelectric actuators. When the tip is close to the surface, the interaction forces between the tip and the sample cause the cantilever deflection, which is measured by a laser reflection system. AFM can be used in a large variety of modes, many of which are used for imaging. Here, we used force spectroscopy, where the tip is brought in contact to indent the sample and then retracted while the cantilever deflection, and thus the force on the tip, is measured. Such experiments yield force vs. distance curves, and the mechanical parameters of the sample are calculated from the adjustment of these curves using appropriate physical models.

AFM is compatible with physiological conditions, the preparation of the samples is nondestructive and requires no labelling, and the experimental parameters can be adjusted to induce no mechanical damage to the samples. As a result, it is used in biology to quantitatively map the mechanics of biomimetic systems [9] and living material, yielding measurements of cell elasticity [10], cortical tension [11], and viscosity [12]. Recent works have used AFM to study the mechanical properties of oocytes [13–16], focusing on the ZP. Complete mechanical characterization of the oocyte, including the measurement of its cortical tension during meiotic divisions, is hampered by technical limitations to maintain stable physiological conditions and sample immobilization simultaneously. Here, we propose an AFM protocol to perform these measurements without affecting meiotic divisions even after removal of the ZP of the oocyte. We then present an analysis of AFM force curves to extract cortical tension based on the linear model developed by [11], and we obtain values in the same range as those obtained with micropipette aspiration.

2 Materials

2.1 AFM

1. Atomic force microscope (NanoWizard IV, JPK) coupled to a widefield microscope (Axio Observer ZEISS with a Hamamatsu SCMOS Flash 4.0 camera). We perform force spectroscopy experiments in contact mode.
2. Pyramidal MLCT-C tips ($k = 0.01$ N/m), silicon nitride cantilevers (Bruker).
3. AFM experimental parameters: maximal force applied ("set point") = 0.5 nN, approach velocity = 1 μm/s (*see* **Note 1**).

2.2 Force Curve Analysis

We used the JPK Data Processing software tool developed by JPK/Bruker (version 6.3.36).

2.3 Experimental Chamber

1. TPP (Techno Plastic Products) tissue culture petri dish, Ø 40 mm, 9.2 cm^2 (lab and industry equipment, Buch & Holm).
2. Electronic microscopy thin bar grids, nickel (Ni), hexagonal mesh, gilder (diameter = 3.05 mm, thickness = 0.8 μm, hole = 73 μm) (Delta Microscopies).
3. Alcohol lamp.
4. Forceps.

2.4 Medium Circulation

1. Mounting flange to connect the dish to the syringe pump (JPK, NanoWizard II).
2. AFM head evaporation protection (JPK, NanoWizard II).
3. PVC tubing, ACF00003, inside diameter = 1.6 mm.
4. A 5 mL PP/PE syringe without needle.
5. Two programmable syringe pumps set to a flow rate of 1 mL/h (respectively for pushing and aspirating).

2.5 Oocyte Handling

1. Glass pipette attached to mouth pipette.
2. Homemade M2-BSA medium supplemented or not with 1 μM Milrinone (Sigma Aldrich – M4659-10MG) and 0.4% Pronase (P5147, Sigma).

We do not describe in this protocol oocyte preparation and manipulation (*see* **Note 2**).

3 Methods

This section describes the setup for AFM experiments (Subheadings 3.1 and 3.2) and AFM measurement (Subheadings 3.3, 3.4, and 3.5).

3.1 Holder Preparation

1. Sonicate the electronic microscopy grids in a 50% ethanol/50% acetone solution for 10 min and dry them for 1 h.
2. Place one grid (shiny side up) in the middle of a TPP tissue culture petri dish (Fig. 1a, b).
3. Heat the tip of a forceps over a flam (in our case, with an alcohol lamp). Under a binocular loupe, melt the petri dish in one point to stick the grid to the dish (*see* **Note 3**).

3.2 Medium

1. Set the two syringe pumps to a flow rate of 1 mL/h; one pushes the medium into the chamber and the other aspirates to ensure sufficient medium circulation while limiting flows in the chamber (total volume = 3 mL).
2. Connect the holder (placed under the AFM head) through a plate mounting flange, tubing, and 5 mL syringes to syringe pumps (*see* **Note 4**) (Fig. 1a).

3.3 Laser Alignment and Calibration

1. Mount the MLCT-C tip on the AFM head.
2. Focus the laser on the extremity of the tip and align it to the center of the photodiode.
3. Calibrate the cantilever (*see* **Note 5**) to determine its deflection sensitivity and spring constant to convert the electric signal of the photodiode in volts (V) into a force in newtons (N). The sensitivity of the photodiode in m/V is calculated by performing a contact between the tip and a hard surface to obtain a voltage–distance curve that reflects the pure bending of the cantilever. The cantilever spring constant in N/m is determined by the thermal tune method [17]. The force on the cantilever is calculated by Hooke's law ($F = -k*d$).

3.4 Force Curves Acquisition

1. Transfer an oocyte to the hexagonal grid with a mouth pipette under the microscope.
2. Wait for the oocyte to fall at the bottom of a hole in the grid (Fig. 2a).
3. Move the cantilever over the oocyte and lower the tip until it touches the top of the oocyte.
4. Acquire ten force curves per oocyte with the following parameters: set point = 0.5nN, approach speed = 1 μm/s (*see* **Note 6**) (Fig. 2b).

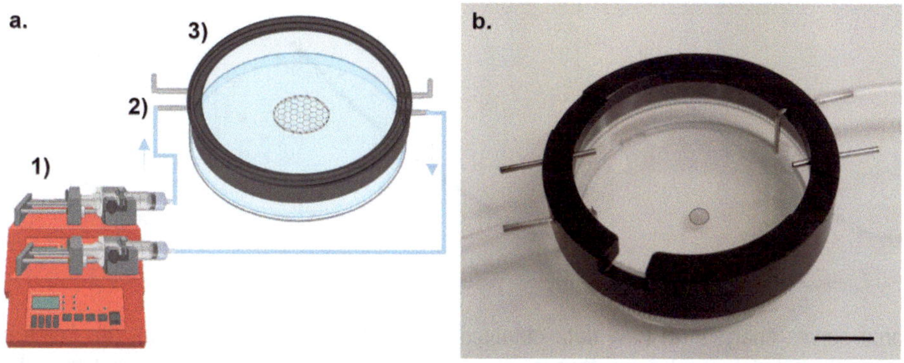

Fig. 1 Chamber for AFM experiment. (**a**). Scheme of the setup: the hexagonal electronic microscopy grid is fixed on a TPP tissue culture petri dish. Medium circulation is ensured by (1) two syringe pumps, connected to (2) the chamber with PVC tubing and (3) a mounting flange. (**b**). Image of the chamber. Scale bar = 500 μm

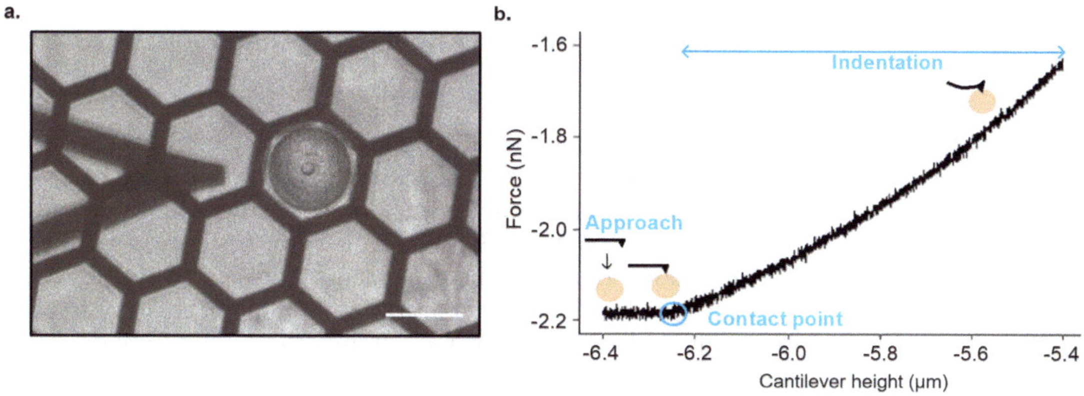

Fig. 2 Acquisition of force curves. (**a**) Top view of an oocyte in a hole of the electronic microscopy grid and MLCT C cantilever. Scale bar = 70 μm. (**b**). Force (in nN) applied by the cantilever as a function of its height (in μm). We highlighted the three regions of the curves: the probe approaching the sample (*approach*), the probe touching the sample (*contact point*), and the probe indenting the sample (*indentation*)

3.5 Force Curves Analysis

For the analysis, we use the JPK Data Processing software tool to adjust the force curves and fit them with a linear model to extract the oocyte cortical tension (Fig. 3).

1. Force curve adjustment (Fig. 3b).
   - Apply the "baseline subtraction" function of the software to obtain a perfectly straight baseline.
   - Subtract the cantilever's deflection to the piezo vertical position to get the real vertical tip position.
   - Define the contact point determined by the software as the zero height.
2. Model fitting: as a first approximation, we fit the large deformation part of the force curves (indentation >0.4 μm) with a

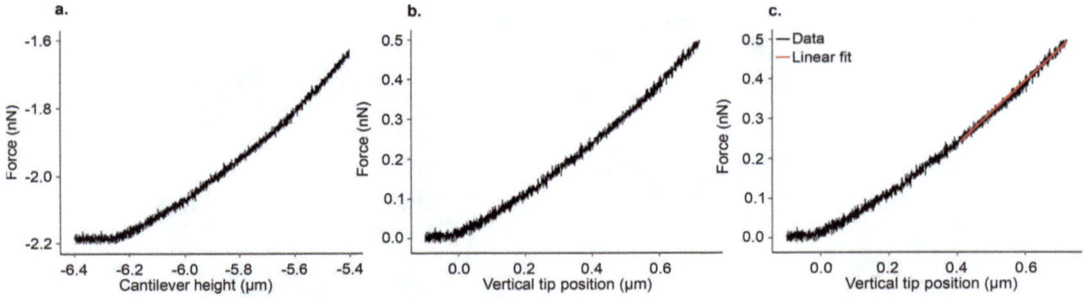

Fig. 3 Force curves analysis (**a**) Raw data obtained after AFM acquisition (Fig. 2b). (**b**). Force curve after adjustment with JPK data analysis software. The height of the cantilever is corrected by subtracting its deflection to obtain the real tip vertical position, the baseline is corrected to obtain a perfectly straight baseline, and the height is set to zero at the contact point as determined by the software. (**c**). Linear regression (red) applied to large deformations (indentation >0.4 µm) of the adjusted force curve (black) to extract cortical tension using a linear model

linear model to extract the cortical tension of the cell [11] (Fig. 3c).

3.6 Results Interpretation

From the force curve analysis presented in Fig. 3, we extract a cortical tension value of 1.2 ± 0.5 nN/µm ($n = 15$ oocytes), which is of the same order of magnitude as the value obtained by micropipette aspiration (3.6 ± 1.1 nN/µm, $n = 14$ oocytes) [18] for prophase I oocytes. The discrepancy between the values could be explained by the different spatial and temporal scales of the two methods. Indeed, the AFM measurement induces a local deformation compared to the large deformation created by the micropipette aspiration. In addition, the AFM measurement time is shorter (less than a minute per cell) than the micropipette measurement time (about 5 min per cell). For the AFM force curves analysis, we used a linear model to extract the cortical tension. However, it only applies to large deformations. An improved physical model, adding mechanical components such as elasticity and viscosity, would be needed to fully exploit the AFM data and reflect the mechanical complexity of the oocyte.

4 Notes

1. We chose the AFM experimental parameters to avoid hydrodynamic flows in the chamber during measurements.
2. For details on oocyte preparation and manipulation, see previously published protocols from our lab [19, 20]. Briefly, we extract oocytes in prophase I before meiotic divisions from mouse ovaries in homemade M2-BSA medium supplemented with 1 µM Milrinone. For outer glycoprotein matrix removal, oocytes are incubated for 5 min in M2-BSA medium

supplemented with 1 μM Milrinone and 0.4% Pronase and then washed six times with fresh M2-BSA medium supplemented with 1 μM Milrinone [21].

3. One point is sufficient; do not apply too much pressure on the grid to avoid bending it. Choose the grid according to the sample size: we chose grids with holes of 72 μm inner diameter for murine oocytes whose diameter is around 75 μm without their outer glycoprotein matrix.
4. Check the medium level in the holder and avoid any overflow that could damage the electronic head of the AFM. Plastic protection is placed on the AFM head to minimize evaporation and protect the AFM head.
5. The calibration process is necessary for the accuracy of AFM measurements, as the manufacturers do not always give the nominal value of the spring constant.
6. If the oocyte moves during the measurement, do not hesitate to make successive approaches to wedge the oocyte on the edge of the grid hole.

Acknowledgments

The AFM microscopy was performed at the Orion Platform (member of France-Bioimaging ANR-10-INBS-XX) of the Center for Interdisciplinary Research in Biology (UMR7241/U1050) of Collège de France. This work was supported by DIM ELICIT du Conseil regional d'Ile de France (DIM ELICIT-AAP-2020 to Marie-Emilie Terret), Biomedical Engineering seed grant program (BME to Clément Campillo), PSL-QLife interdisciplinary program (QLife-CdF-06-2022 to Marie-Emilie Terret), and FRM ECO contract doctoral program (ECO202206015524 to Clément Campillo, Marie-Emilie Terret, and Rose Bulteau). This work has received support from the Fondation Bettencourt Schueller, support under the program Investissements d'Avenir launched by the French Government and implemented by the Agence Nationale de la Recherche, with the following references: ANR-10-LABX-54 MEMO LIFE and ANR-11-IDEX-0001-02 PSL* Research University.

References

1. Tinevez J-Y, Schulze U, Salbreux G et al (2009) Role of cortical tension in bleb growth. Proc Natl Acad Sci 106:18581–18586. https://doi.org/10.1073/pnas.0903353106
2. Larson SM, Lee HJ, Hung P et al (2010) Cortical mechanics and meiosis II completion in mammalian oocytes are mediated by myosin-II and Ezrin-radixin-Moesin (ERM) proteins. MBoC 21:3182–3192. https://doi.org/10.1091/mbc.e10-01-0066
3. Chaigne A, Campillo C, Gov NS et al (2013) A soft cortex is essential for asymmetric spindle

positioning in mouse oocytes. Nat Cell Biol 15: 958–966. https://doi.org/10.1038/ncb2799

4. Chaigne A, Campillo C, Gov NS et al (2015) A narrow window of cortical tension guides asymmetric spindle positioning in the mouse oocyte. Nat Commun 6:6027. https://doi.org/10.1038/ncomms7027
5. Evans JP, Robinson DN (2018) Micropipette aspiration of oocytes to assess cortical tension. In: Verlhac M-H, Terret M-E (eds) Mouse oocyte development. Springer, New York, pp 163–171
6. Evans E, Yeung A (1989) Apparent viscosity and cortical tension of blood granulocytes determined by micropipet aspiration. Biophys J 56:151–160. https://doi.org/10.1016/S0006-3495(89)82660-8
7. Daza R, González-Bermúdez B, Cruces J et al (2019) Comparison of cell mechanical measurements provided by atomic force microscopy (AFM) and micropipette aspiration (MPA). J Mech Behav Biomed Mater 95:103–115. https://doi.org/10.1016/j.jmbbm.2019.03.031
8. Binnig G, Quate CF, Gerber C (1986) Atomic force microscope. Phys Rev Lett 56:930–933. https://doi.org/10.1103/PhysRevLett.56.930
9. Lamour G, Allard A, Pelta J et al (2020) Mapping and modeling the nanomechanics of bare and protein-coated lipid nanotubes. Phys Rev X 10:011031. https://doi.org/10.1103/PhysRevX.10.011031
10. Cross SE, Jin Y-S, Rao J, Gimzewski JK (2007) Nanomechanical analysis of cells from cancer patients. Nat Nanotech 2:780–783. https://doi.org/10.1038/nnano.2007.388
11. Mandriota N, Friedsam C, Jones-Molina JA et al (2019) Cellular nanoscale stiffness patterns governed by intracellular forces. Nat Mater 18:1071–1077. https://doi.org/10.1038/s41563-019-0391-7
12. Abuhattum S, Mokbel D, Müller P et al (2022) An explicit model to extract viscoelastic properties of cells from AFM force-indentation curves. iScience 25:104016. https://doi.org/10.1016/j.isci.2022.104016
13. Andolfi L, Masiero E, Giolo E et al (2016) Investigating the mechanical properties of zona pellucida of whole human oocytes by atomic force spectroscopy. Integr Biol 8:886–893. https://doi.org/10.1039/c6ib00044d
14. Choi JK, Yue T, Huang H et al (2015) The crucial role of zona pellucida in cryopreservation of oocytes by vitrification. Cryobiology 71:350–355. https://doi.org/10.1016/j.cryobiol.2015.08.012
15. Giolo E, Martinelli M, Luppi S et al (2019) Study of the mechanical properties of fresh and cryopreserved individual human oocytes. Eur Biophys J 48:585–592. https://doi.org/10.1007/s00249-019-01379-y
16. Battistella A, Andolfi L, Zanetti M et al (2022) Atomic force spectroscopy-based essay to evaluate oocyte postovulatory aging. Bioeng Transla Med 7:e10294. https://doi.org/10.1002/btm2.10294
17. Sader JE, Chon JWM, Mulvaney P (1999) Calibration of rectangular atomic force microscope cantilevers. Rev Sci Instrum 70:3967–3969. https://doi.org/10.1063/1.1150021
18. Bennabi I, Crozet F, Nikalayevich E et al (2020) Artificially decreasing cortical tension generates aneuploidy in mouse oocytes. Nat Commun 11:1649. https://doi.org/10.1038/s41467-020-15470-y
19. Almonacid M (2018) Photoactivation of actin in mouse oocyte. In: Verlhac M-H, Terret M-E (eds) Mouse oocyte development. Springer, New York, pp 145–151
20. Bennabi I, Manil-Ségalen M (2018) Laser ablation of microtubule–chromosome attachment in mouse oocytes. In: Verlhac M-H, Terret M-E (eds) Mouse oocyte development. Springer, New York, pp 153–161
21. Al Jord A, Letort G, Chanet S et al (2022) Cytoplasmic forces functionally reorganize nuclear condensates in oocytes. Nat Commun 13:5070. https://doi.org/10.1038/s41467-022-32675-5

Chapter 8

Manipulation of Embryonic Cleavage Geometry Using Magnetic Tweezers

Jing Xie, Daniel L. Levy, Nicolas Minc, and Jérémy Sallé

Abstract

The geometry of reductive divisions that mark the development of early embryos instructs cell fates, sizes, and positions, by mechanisms that remain unclear. In that context, new methods to mechanically manipulate these divisions are starting to emerge in different model systems. These are key to develop future innovative approaches and understand developmental mechanisms controlled by cleavage geometry. In particular, how cell cycle pace is regulated in rapidly reducing blastomeres and how fate diversity can arise from blastomere size and position within embryos are fundamental questions that remain at the heart of ongoing research. In this chapter, we provide a detailed protocol to assemble and use magnetic tweezers in the sea urchin model and generate spatially controlled asymmetric and oriented divisions during early embryonic development.

Key words Embryogenesis, Cleavage pattern, Magnetic tweezers, Asymmetric division, Sea urchin

1 Introduction

The choreography of cell divisions during the very first hours of embryonic development has been a source of fascination for generations of biologists. During this critical period, the one cell zygote divides rapidly in a stereotyped manner while cellular fates emerge and diversify to set the stage for more complex morphogenetic processes [1–3]. Concomitantly, maternal RNAs and proteins are progressively replaced by the products of zygotic transcription, which is almost systematically correlated with a drastic reduction in cell volume and cell cycle speed [4–6]. However, despite several major studies, the causal relationship between these different events, as well as their underlying mechanisms, are still debated. One of the main hypotheses involves the ability of blastomere cells to "sense" the ratio between their nuclear and cytoplasmic volume (N/C ratio). Changes in the N/C ratio may be in part attributed to the limited ability of the nucleus to scale with cytoplasmic volume

Anna Castro and Benjamin Lacroix (eds.), *Cell Cycle Control: Methods and Protocols*, Methods in Molecular Biology, vol. 2740, https://doi.org/10.1007/978-1-0716-3557-5_8,

based on nuclear import dynamics as well as nuclear components availability [7, 8]. As a result, the most drastic variations of the N/C ratio is a consequence of the rapid reduction of cytoplasmic volume during cleavages which are taking place in a constant volume embryo. Thus, embryonic N/C patterning can be viewed as a direct output of early cleavage geometry that acts as a developmental cue for key cellular transitions in the early embryo [5, 6, 9].

The direct impact of division geometries and consequent N/C ratio patterning, on cell cycle progression or fate specifications, quickly emerged as a fundamental question for early development that requires novel methods to manipulate cleavage. Recently, several approaches have been used successfully in different species. These methods provide the key advantage of relying on the physical manipulation of intracellular structures rather than on genetic losses of function or chemical compounds, both having high risks of causing pleiotropic effects [10–13]. Here, we describe a method to reliably alter the positioning of the division plane in living sea urchins embryos (*Paracentrotus lividus*) in a minimally invasive manner using magnetic tweezers [11, 12, 14]. This approach relies on magnetic particles which possess the unique property of being transported, in a dynein-dependent manner, toward the minus end of microtubules, and aggregate at the centrosome. Thus, through the application of an external magnetic force, microtubule structures such as interphase asters and mitotic spindles can be displaced or rotated in a controlled fashion to produce asymmetric or oriented divisions. This chapter will cover all experimental steps including the gametes collection, the magnetic tweezer fabrication, the injection needle and magnetic particles preparation, the magnetic particles injection, and finally the induction of an asymmetric division. Using this method, it has been possible, for instance, to reveal an uncoupling between nuclear volume and cellular volume during the early divisions of the sea urchin embryo [12]. We anticipate that this type of approach would benefit a large variety of studies and could be expanded to other model systems in which mechanical manipulation of cell division is currently challenging.

2 Materials

2.1 Sea Urchins

1. Adult mature purple sea urchins (*Paracentrotus lividus*).
2. Sea water salts.
3. >200 L fully equipped sea water tank.
4. 0.5 M KCl solution.
5. 55 mM plastic petri dishes.
6. 80 μm Nitex nylon mesh.

2.2 Magnetic Particles

1. 1 μm streptavidin magnetic particles (Nanolink®).
2. 1 M NaCl, 1% Tween-20 solution.
3. Biotin–Atto 488/565. 2 mg/mL stock diluted in DMSO.

2.3 Magnetic Particles and Magnetic Probe

1. 100 mM borosilicate glass capillaries (OD: 1 mM, ID: 0.75 mM).
2. Micropipette puller.
3. Microforge and microgrinder.
4. Siliconizing agent (Sigmacote®).
5. Cylindrical permanent neodymium magnets.

2.4 Particles Injection

1. MatTek glass bottom dishes (P50G-0-14-F).
2. Protamine sulfate salt.
3. Inverted fluorescence microscope equipped with a microinjection system.

3 Methods

3.1 Gametes Collection

1. Adult purple sea urchins (*Paracentrotus lividus*) are shipped from the Roscoff marine station (France) and can be stored for several weeks in artificial sea water (ASW) in 250–350 L tanks equipped with a 600–800 L/h mechanical filtration (Eheim®), a protein skimmer (Tunze®), an oxygenation air pump, and a water cooler (TECO®) to maintain the temperature at 16 °C.
2. ASW is prepared from commercial salts (Reef Crystals, Aquarium systems®) and deionized water at least 24 h before use to obtain a density of 1023–1025 kg/m^3. Every week, 10% of tank water is replaced with fresh ASW and several water parameters are controlled and maintained in the following ranges: pH: 7.8–8.4; KH: 10–12; NH3 < 0.1 ppm; NO2 < 0.2 ppm; NO3 < 0.2 ppm.
3. Spawning is induced by intracoelomic injection of 0.5–1 mL of 0.5 M KCl. After injection, female individuals are placed with the aboral side facing down on a glass beaker filled with ASW for 15 min maximum. Spawned eggs will sediment and accumulate at the bottom of the beaker and can then be transferred into a glass crystallizer with fresh ASW. Egg density is adjusted in order to avoid egg stacking by using larger and/or multiple containers and are stored at 16 °C for up to 12 h. Sperm is collected directly from the aboral side of male individuals using a micropipette and stored undiluted at 4 °C in a microtube for up to 1 week.

4. Gamete quality is assessed with a quick fertilization test. Sperm is first activated by dilution in ASW (1:1000) followed by several up and down aspirations with a plastic pipette for 30 s. Ten drops of activated sperm are immediately mixed with 10 mL of egg suspension and observed under a microscope in a 50 mM petri dish (*see* **Note 1**). The general criteria for good gamete quality are as follows: more than 90% of sperm cells are motile, nearly 100% of eggs are fertilized (elevation of the fertilization envelope is visible a couple of minutes after sperm addition), zygote nucleus is properly centered 15 min after fertilization, and first cleavages are symmetric and do not show any significant delays (60 min for the first division and ~30 min for the second and the third division). Once quality is validated, a fraction of the eggs is passed three times through an 80 μm Nitex mesh (for 90–100 μm eggs) in order to mechanically remove the jelly coat which otherwise prevents egg adhesion on the injection chamber. "Dejellied" eggs are then stored at 16 °C in a separate container and are ready for injections. More details and alternative methods for sea urchin handling are available in several technical resources [15, 16].

3.2 Magnetic Particle Preparation

1. 10 μL of magnetic particles functionalized with streptavidin (diameter 1 μm, Nanolink®) are first washed for 15–30 min in 1 M NaCl, 1% Tween-20 under gentle agitation (*see* **Note 2**). A magnetic rack is then used to exchange the washing solution with 100 μL of PBS and particles are resuspended by pipetting up and down. This step is repeated at least once to eliminate any trace of detergent and particles are finally resuspended in 20 μL of PBS (final particle concentration = 5 mg/mL). Final concentration may be adjusted from 1 to 10 mg/mL depending on the amount of particles one wants to inject and/or the aperture size of the injection needle. It is important to keep in mind that while the injection of a large amount of particles in a small volume is generally favored, higher concentrations tend to be particularly prone to needle clogging.
2. If the experiment requires fluorescently labelled particles, an extra 15 min incubation in 100 μL of 2 μg/mL Biotin–Atto 488/560/650 (Sigma) diluted in PBS can be introduced before final PBS rinses. After particle concentration, the injection solution is ready to use and can be stored on ice protected from light for 6 h. 20 μL of injection solution is sufficient to load approximatively 10–15 injection needles.

3.3 Injection Needle Preparation

1. Injection needles are pulled from 100 mM borosilicate glass capillaries (TW100-4, OD: 1 mM, ID: 0.75 mM, *Kwik-Tip*™, World Precision Instruments). To minimize glass–particle interaction and reduce the risk of needle clogging, capillaries

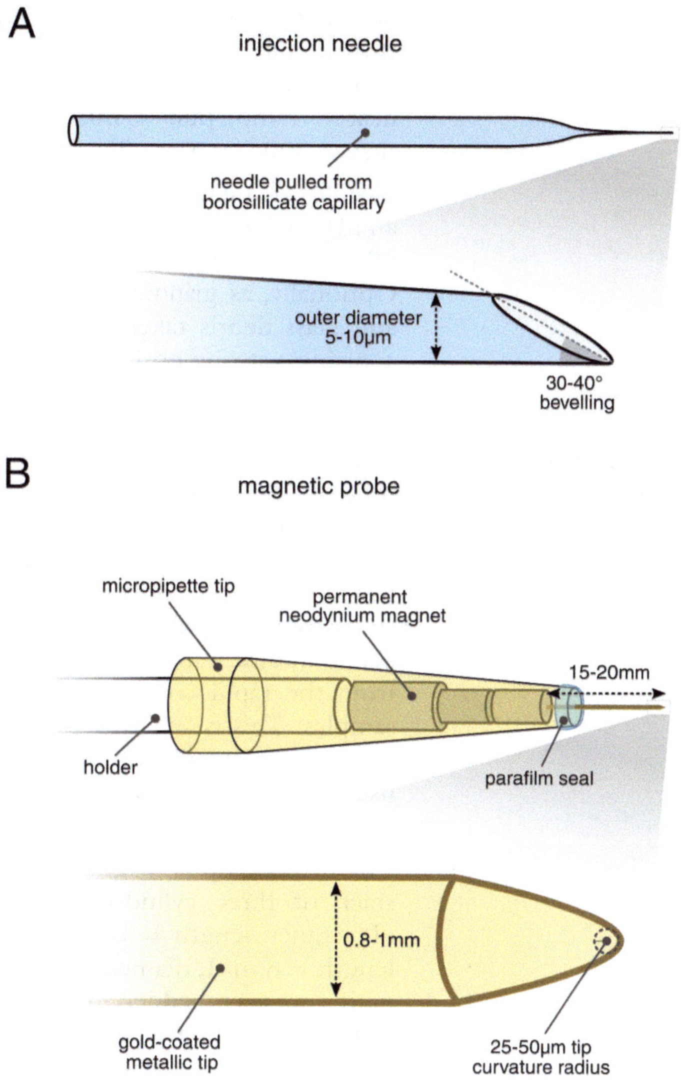

Fig. 1 (**a**) Schematic view of the needle used for the microinjection of magnetic particles in sea urchin eggs. (**b**) Schematic view of the magnetic probe used to apply forces on injected magnetic particles and generate an asymmetric division

are treated with a siliconizing reagent (Sigmacote®). Typically, 10–15 capillaries are incubated with 2 mL of reagent for 12 h at room temperature on a rotating wheel. Under the hood, the siliconizing reagent is then discarded and excess reagent is removed from the inner part of capillaries by applying an absorptive paper on both extremities. Capillaries are then left to dry under the hood for at least 6 h protected from dust, well separated to prevent them from sticking to each other. Capillaries can then be stored for months in a clean container before further use.

2. Siliconized capillaries are pulled using a P-1000 Flaming/Brown micropipette puller (Sutter Instruments®) equipped with a 2.5 × 2.5 mM box filament with the following settings: heat = ramp; pull = 80; velocity = 60; delay = 70; pressure = 200 (*see* **Note 3**).
3. Needle tips are then cut to obtain an aperture of 5–10 μm using an MF2 microforge (Narishige©) and beveled with an angle of 40° using an EG-402 microgrinder (Narishige©) (Fig. 1a). Optionally, as grinding occurs on a wet surface, trace liquid and glass debris taken up by the needle can be expelled by applying a strong positive pressure using a pneumatic injection system (IM-11-2, Narishige©).

3.4 Magnetic Tip Fabrication

1. The terminal part of the magnetic probe is made of a sharpened rod of ferromagnetic material of 15–20 mM in length, 0.8–1 mM in diameter, and with a curvature radius at the tip of 25–50 μm (Fig. 1b). Interestingly, the tip of a sewing needle has proven to be a cheap and relatively robust option compared to custom-made alternatives (*see* **Note 4**). However, since this part is in direct contact with ASW, it is necessary to preserve it from the rapid corrosion that can harm surrounding living samples. This is achieved by the deposition of a thin layer of gold on the metallic tip by electro-coating. A single tip can be used for years if carefully cleaned between experiments and protected from shocks.
2. The metallic tip is magnetized by placing it in contact with a stack of three cylindrical permanent neodymium magnets (1 magnet: length = 10 mM, diameter = 4 mM; 2 magnets: length = 8 mM, diameter = 3 mM, SuperMagnet). This magnetic core is embedded in a shortened micropipette tip so that 5–10 mM of the tip protrudes from the distal extremity. ASW infiltration in the magnetic core is prevented by sealing the micropipette tip with a piece of Parafilm® or nail polish (Fig. 1b). Finally, the "core + tip" assembly is mounted on a minutien pin holder (Fine Science Tools©) to easily replace the injection capillary holder during the experiment.

3.5 Injection and Imaging Setup

1. Injection and imaging are carried out in glass bottom dishes (MatTek, P50G-0-14-F) coated with protamine to ensure egg adhesion. The glass surface of the dish is exposed to 1% protamine sulfate (Sigma) dissolved in water for 1–2 min and thoroughly rinsed with deionized water. Coated dishes can be used immediately or dried and stored for months protected from dust.
2. The whole experiment is carried out on a standard inverted fluorescence microscope equipped with a micromanipulation (InjectMan 4, Eppendorf®) and a microinjection system

(FemtoJet, Eppendorf®). The injection axis forms a 40° angle with the plane of the microscope stage. If the experiments require a specific type of imaging (e.g., spinning disc, confocal), the dish can be carefully moved to a different system after injection.

3.6 Before Injection

1. Using a micropipette, a band of dejellied eggs is carefully deposited on the protamine-coated glass-bottom dish filled with ASW (Fig. 2a). A couple of minutes are necessary for the eggs to completely sediment and adhere to the glass surface before the dish can be moved safely to the microscope stage. The band of egg is oriented perpendicular to the injection axis and it is in general more convenient to only inject eggs located at the edge of the band from one end to the other in order to have a large empty area to move the magnet afterward.
2. The injection needle is filled with 2 μL of the magnetic particle suspension using a microloader tip (Eppendorf®). As particles tend to sediment quickly, it is essential to resuspend them by gently flicking the tube and/or pipetting up and down before each needle loading. Once the needle is loaded and connected to the pressure controller, make sure that the particle suspension fills the tip of the needle before lowering it into the ASW. If not, use the "clean" function to briefly apply a high-pressure pulse. Due to the large needle aperture, one may use the lowest compensation pressure available for the device (5 hPa). Using a zone in the dish with no eggs, one should visually check the constant flow of particles coming out from the needle before starting injections.

3.7 Particles Injection

1. The injection system is used in "step injection" mode which produces a quick axial back and forth motion of the needle (speed = 500–2000 μm/s; distance = 20–35 μm) when pressing the "inject" button. An injection pressure of 15–30 hPa is applied for 0.1–0.4 s when the needle is at its maximum distance (injection synchronization set to "limit"). To facilitate piercing of the egg membrane, it is often necessary to deform it by gently pushing the tip of needle to make a 10–20 μm inward indentation in the egg before injection (*see* **Note 5**). A brief particle flow inside the egg cytoplasm should be clearly visible during injection and a fast manual retraction, along the needle axis, might be required if the needle tip remains inside the egg at the end of the automated needle motion (*see* **Note 6**).
2. After a successful injection, the egg cytoplasm should quickly return to a homogenous state. The "coarsening" of cytoplasmic material is a sign of an irreparable egg damage. Similarly, cytoplasm leakage at the injection site tends to drastically reduce embryo survival. Elevation of the fertilization

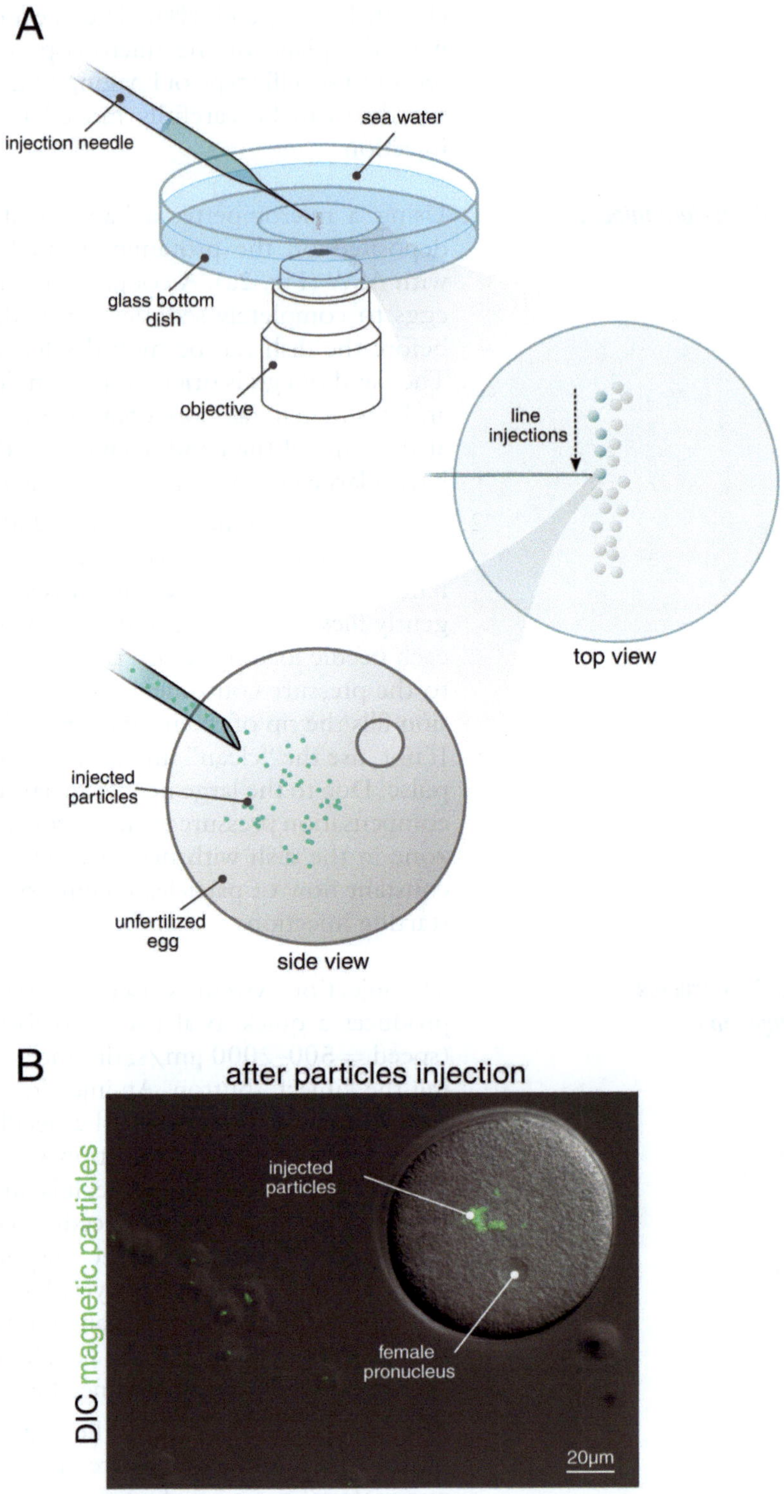

Fig. 2 (**a**) Schematized view of the microinjection setup including close-up views of the injection chamber and injected unfertilized egg. (**b**) Example of injected egg. DIC is in gray and injected magnetic particles are in green. (Adapted from Sallé et al. [11])

membrane can sometimes be triggered mechanically by the injection. A complete elevation of the fertilization membrane will prevent fertilization and should also be avoided. However, if the elevation is restricted to the injection site or ideally completely absent, the egg remains competent for fertilization (*see* **Note** 7). The amount of injected particle can also be roughly estimated at this step (Fig. 2b). Modifying the injection duration (0.1 s steps) and/or the injection pressure (5 hPa steps, up to 30 hPa) is a good option to slightly modulate the number of injected particles within the same injection series. Otherwise, adjusting the particle concentration in the injection solution remains the only way to obtain a wider range of injected particle densities.

3. Successive injections must take place in time windows of 15–20 min maximum before fertilization. As mentioned before, it is generally better to inject eggs on a 2D vertical line relative to the microscopy field (Fig. 2a). This allows one to inject 20 to 50 eggs per dish. After removing the injection needle, fertilization is triggered by adding 1–4 droplets of diluted sperm (1:1000) that have been vigorously agitated for 30 s. Successful fertilization and pronuclear migration are monitored directly under the microscope.

3.8 Artificial Asymmetric Division

1. After swapping the injection needle holder with the magnet holder on the micromanipulation arm, one must quickly position the magnetic tip at a distance of 50–100 μm from a fertilized injected zygote. In a couple of minutes, the particles will move toward the magnet and form a flattened subcortical aggregate that will be referred to as the "pulling cap" (*see* **Note 8**) (Fig. 3a, b). This step has to be carried out quickly to avoid particle capture by the growing sperm aster. Captured particles are however still able to aggregate to form a pulling cap although it might require more time (1–2 extra minutes).
2. Alternatively, the pulling cap can be assembled before fertilization (Fig. 4a). This offers the advantage of a slightly faster and "cleaner" cap formation. However, the cap formation step must now be included in the 15–20 min time window mentioned earlier originally devoted exclusively to the injection (*See* **Note 9**). In addition, because the egg membrane is significantly softer before fertilization, it is important to temporarily reduce the magnet pulling force by increasing the distance between the pulling cap and the magnet tip, to limit membrane deformation. Also, due to potential increased fragility, it is generally advised to not assemble the pulling cap too close to the injection site when possible.
3. From sperm aster centration to cell division, the pulling cap should be maintained continuously at the cortex. There is no

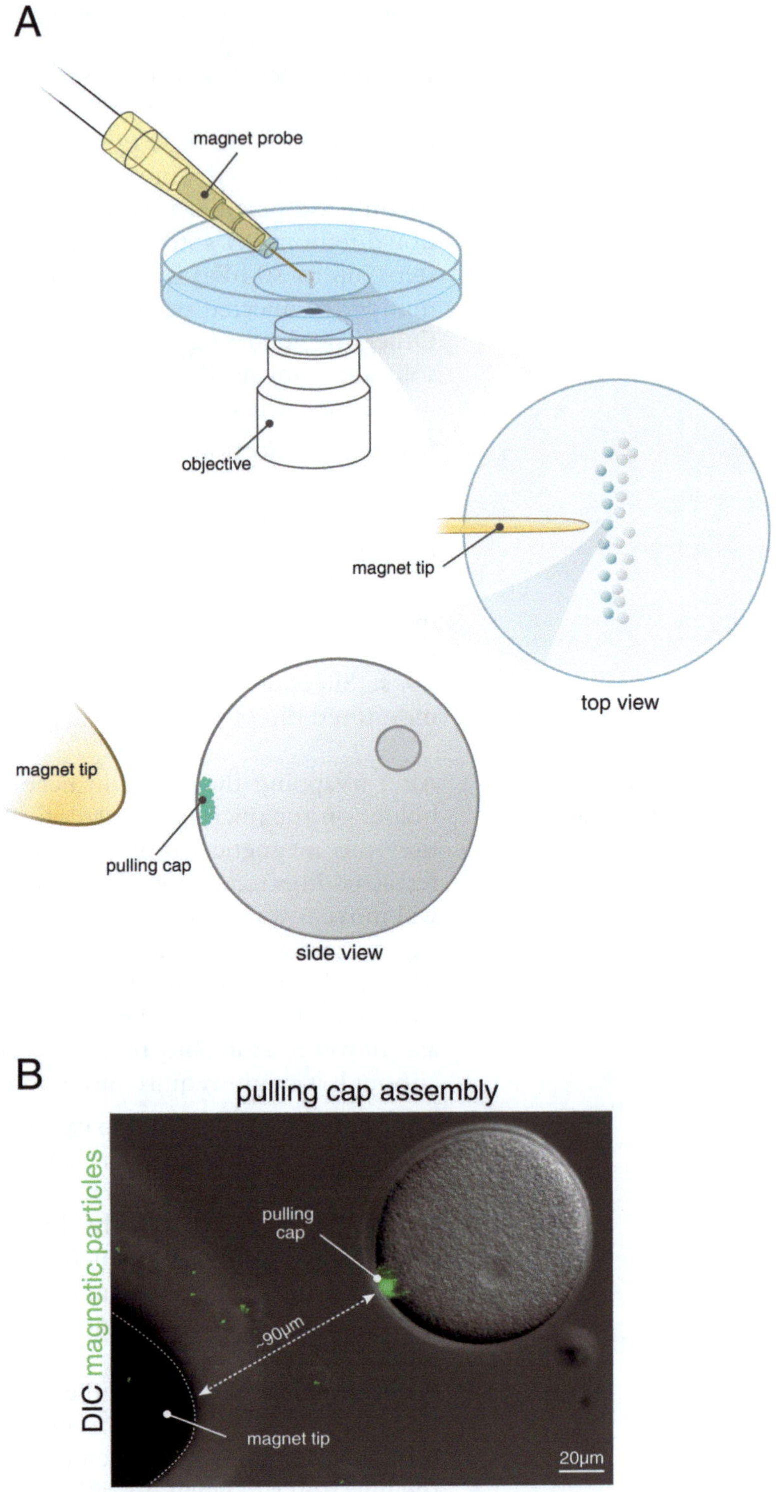

Fig. 3 (**a**) Schematized view of the magnetic tweezer setup including close-up views of the injection chamber and assembled pulling cap. (**b**) Example of pulling cap assembly. DIC is in gray and injected magnetic particles are in green. (Adapted from Sallé et al. [11])

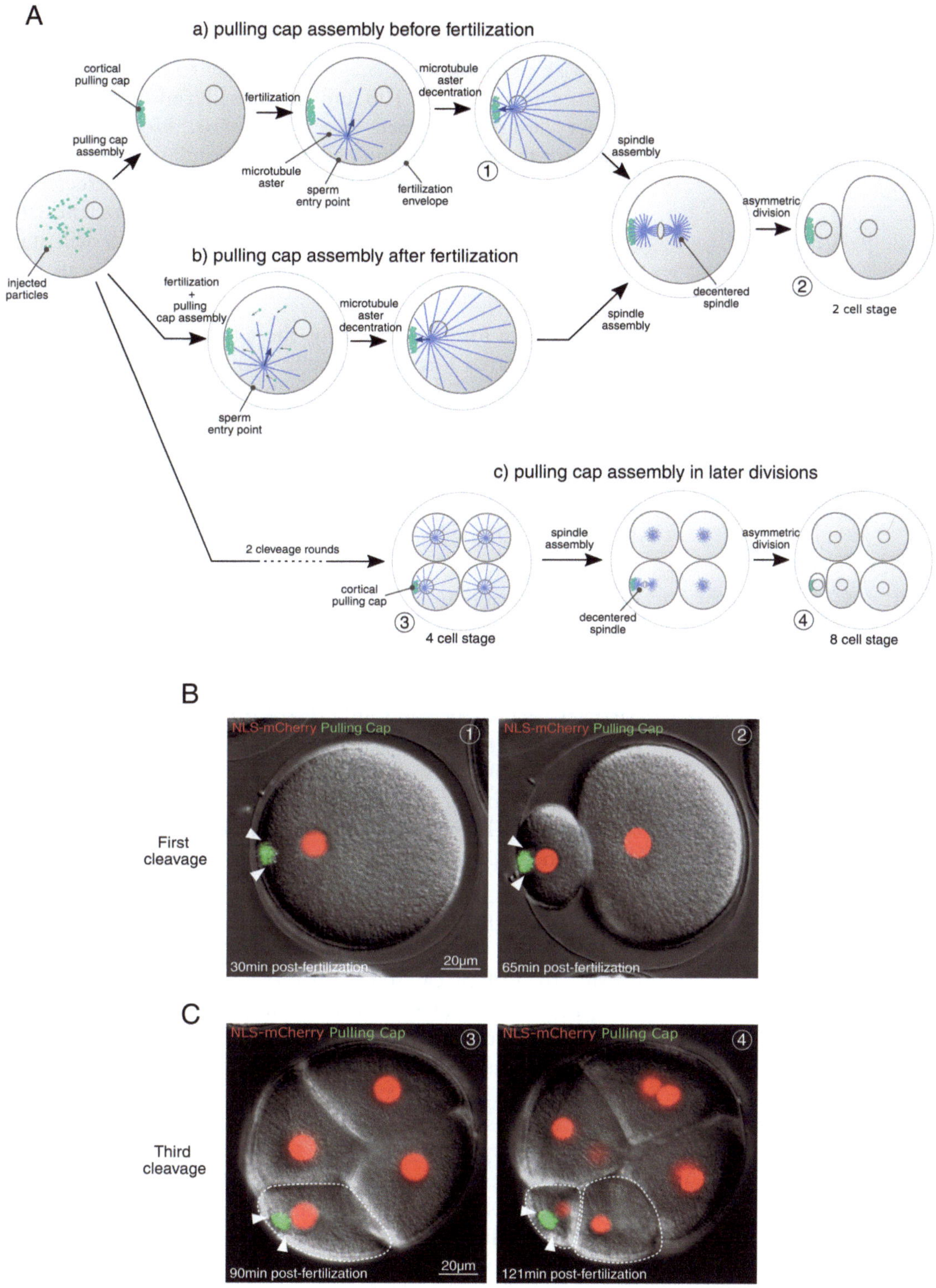

Fig. 4 (**A**) Overview of the experimental steps required to assemble the magnetic pulling cap before (*a*) or after fertilization (*b*) and generate an asymmetric division during the first embryonic cleavage (*a, b*) as well as subsequent cleavages (*c*). (**B, C**) Example of asymmetric divisions during the first (**B**) and the third (**C**) cleavage. Arrowheads indicate the position of the pulling cap. DIC is in grey, NLS-mCherry is in red, and the pulling cap is in green. (Adapted from Mukherjee et al. [12])

major risk of keeping the magnet too close to the embryo, but distances from 50 to 100 μm are usually enough with the magnet design described here (Fig. 3b). However, beads detaching from the pulling cap and accumulating at the aster center are a sign of low magnetic force that can be corrected by moving the magnet closer. As soon as the aster reaches the center of the cell, the interphasic microtubule aster will start to be susceptible to repositioning by the pulling cap (*See* **Note 10**).

4. The magnetic force maintaining the aggregate is opposed by dynein-generated forces that try to move the particles toward the aster center. Therefore, in response, the aster center is displaced away from the cell center toward the pulling cap, generally stabilizing its position 15–25 μm away from the cortex (*See* **Note 11**). After 45–50 min post-fertilization, the mitotic spindle will assemble next to the cortex forming a division axis aligned with the pulling axis.
5. Asymmetric divisions can also be generated during subsequent cleavages (up to the eighth cleavage so far in our hands) with some adjustments and precautions. First, the size of the pulling cap should be reduced accordingly by lowering the particle density of the injection solution. Second, without magnet, particles tend to be randomly distributed among blastomeres resulting in various aggregate sizes and pulling efficiencies. To circumvent this, it is possible to use several brief magnetic force applications during each round of divisions to restrict the particles to fewer blastomeres, or even a single blastomere, before assembling a proper cortical pulling cap at the desired stage (Fig. 3a, b).
6. After a successful asymmetric division, the magnet can be carefully removed from the holder once cytokinesis is complete, causing the pulling cap to rapidly move toward the aster center. A slight delay in division timing can be observed during the geometrically modified mitosis without obvious consequences up to the blastula stage. However, a significantly extended prophase (e.g., NEBD after 70 min post-fertilization for the first cleavage) should be considered as a failed attempt.

3.9 Imaging

1. Manipulated embryos can be observed for several hours after the induced asymmetric division. The amount of fluorescent illumination that embryos can tolerate is often batch dependent and the best compromise between frame intervals, number of channels, and number of z stacks should be determined beforehand. As an example, we successfully imaged a manipulated embryo for 6 h from the four-cell stage to the blastula stage on a spinning disc microscope (40× objective) with 2 fluorescent channels, 2 min frame interval, and 10 stacks per frame.

2. Another important parameter influencing the long-term viability of manipulated embryos is the number of particles remaining in the small cell generated by the asymmetric division and its progeny. Even though an oversized pulling cap will systematically ensure a strong asymmetry, an excess of particles trapped in the small blastomere is likely to interfere with mitotic spindle assembly and/or cytokinesis during subsequent cleavages. Thus, it is best to aim for the minimal pulling cap size required to achieve the desired asymmetry.
3. As a nongenetic model, live imaging in sea urchin embryos relies almost exclusively on fluorescent protein and mRNA injection as well as incubation with fluorescent probes [12, 16, 17]. While combining magnetic tweezers with chemical incubations is fairly straightforward, the combined injection of magnetic particles with other biomolecules often requires further troubleshooting. The easiest option is to co-inject proteins or mRNA with the magnetic particles. In the case of co-injection with proteins, one frequent caveat is the progressive nonspecific binding of the protein to the particles within the injection solution. This can be limited by mixing particles and proteins just before the injection. Co-injection with mRNAs can increase the viscosity of the injection solution and make the injection of high-density particle solutions almost impossible. For this type of situation, a secondary injection using a standard liquid injection needle can be performed after the particle injection either before or after fertilization.

4 Notes

1. Fertilization rate can vary slightly with this standard quantity of activated sperm. One can double the sperm amount to try to reach 100% fertilization efficiency. If it is still not sufficient, it is usually better to collect a new gamete batch. On the other hand, very active sperm is likely to cause polyspermy on dejellied eggs and its amount should therefore be reduced accordingly.
2. While several types of commercially available particles exhibit partial spontaneous minus end-directed movements [18], only Nanolink® magnetic particles have shown a systematic and reproducible transport toward the microtubule aster center in sea urchin embryos. Even though the exact nature of the coupling between the particles and the molecular motors is unclear, it has been established that these unique properties depend on dynein motors and microtubules, and these particles have been used reliably to collect both qualitative and quantitative data over the last few years [11, 14, 19].

3. Pulling parameters are only suggested and should be optimized for a given capillary type and puller/filament model. The goal remains to get a relatively long and shallow taper to easily control the opening diameter.
4. It might be tempting to use a very sharp metallic tip to gain better control over the injected particles. Unfortunately, a tip curvature radius below 10 μm will reduce the effective range of the magnet below the value of the egg radius making the manipulation of particles throughout the whole cytoplasmic volume impossible.
5. Classical microinjections in sea urchins are normally performed after fertilization (up to 15 min post-fertilization). The stiffer cortex allows for easier needle penetration and is less harmful for the cell. However, the large needle opening required for the injection of large amounts of particles is too damaging for a stiffened cortex. Here again, injection parameters are only suggested and should be modulated between and during experiments as needle size, quantity of cell debris clogging the needle, or egg quality might vary.
6. The combined use of an automated pressure controller and micromanipulator has greatly increased the throughput and the consistency of particle injections. However, it is important to note that similar injections have been successfully carried out using manual hydraulic micromanipulators and microinjectors.
7. The sensitivity at which fertilization envelope elevation can be mechanically triggered and propagated is often batch dependent. If, in spite of efforts to inject gently, contact with the needle still causes envelope elevation, it is usually better to collect new eggs.
8. Just after fertilization, the viscosity of the cytoplasm increases for a couple of minutes [20]. If the fertilization envelope is already partially elevated before sperm addition, making it difficult to assess the fertilization status of the egg, fast or slow particle motion toward the magnet is a good early indicator of unfertilized vs. fertilized eggs, respectively.
9. Practically, assembling the pulling cap after fertilization offers generally more flexibility as one can select a specific fertilization geometry among the many injected eggs or quickly switch to another egg if the one with the pre-assembled cap does not fertilize.
10. For experiments performed during the first cleavage, it is important to keep in mind that aster centration is not altered by the pulling cap as centration forces are stronger than decentering forces imparted by the cap. An efficient aster displacement is only possible once centration has been completed and centration forces reduced [11].

11. There is a threshold size for the pulling cap to be effective at generating an asymmetric division. Below this threshold (i.e., if not enough particles have been injected, or if particles have been distributed into too many blastomeres), the division will remain mostly symmetric in spite of properly executed cap formation.

Acknowledgments

This work was supported by the Centre National de la Recherche Scientifique (CNRS), the Appel Emergence en recherche ("MAG-NUC") from Université de Paris to JS, and grants from La Ligue Contre le Cancer (EL2021. LNCC/ NiM), the Fondation Bettencourt Schueller ("Coup d'Elan"), and the European Research Council (ERC CoG "Forcaster" no. 647073) to NM. JX. acknowledges the "Ecole Doctorale FIRE (Frontières de l'Innovation en Recherche et Education) – Program Bettencourt," a fellowship from the Chinese Scholarship Council (201708070046), and from the LabEx "Who am I?" (ANR-11-LABX-0071). DLL is supported by the National Institutes of Health/National Institute of General Medical Sciences (R35GM134885 and P20GM103432).

References

1. Sallé J, Minc N (2022) Cell division geometries as central organizers of early embryo development. Semin Cell Dev Biol 130:3–11. https://doi.org/10.1016/j.semcdb.2021.08.004
2. Pierre A, Sallé J, Wühr M, Minc N (2016) Generic theoretical models to predict division patterns of cleaving embryos. Dev Cell 39:667–682. https://doi.org/10.1016/j.devcel.2016.11.018
3. Wolpert L, Tickle C, Arias AM (2015) Principles of development. Oxford University Press, Oxford
4. Chen H, Einstein LC, Little SC, Good MC (2019) Spatiotemporal patterning of zygotic genome activation in a model vertebrate embryo. Dev Cell 49:852–866.e7. https://doi.org/10.1016/j.devcel.2019.05.036
5. Jukam D, Kapoor RR, Straight AF, Skotheim JM (2021) The DNA-to-cytoplasm ratio broadly activates zygotic gene expression in Xenopus. Curr Biol 31:4269–4281.e8. https://doi.org/10.1016/j.cub.2021.07.035
6. Syed S, Wilky H, Raimundo J et al (2021) The nuclear to cytoplasmic ratio directly regulates zygotic transcription in drosophila through multiple modalities. Proc Natl Acad Sci USA 118:e2010210118. https://doi.org/10.1073/pnas.2010210118
7. Chen P, Tomschik M, Nelson KM et al (2019) Nucleoplasmin is a limiting component in the scaling of nuclear size with cytoplasmic volume. J Cell Biol 218:4063–4078. https://doi.org/10.1083/jcb.201902124
8. Jevtić P, Schibler AC, Wesley CC et al (2019) The nucleoporin ELYS regulates nuclear size by controlling NPC number and nuclear import capacity. EMBO Rep 20:e47283. https://doi.org/10.15252/embr.201847283
9. Jevtić P, Levy DL (2015) Nuclear size scaling during Xenopus early development contributes to midblastula transition timing. Curr Biol 25:45–52. https://doi.org/10.1016/j.cub.2014.10.051
10. Garzon-Coral C, Fantana HA, Howard J (2016) A force-generating machinery maintains the spindle at the cell center during mitosis. Science 352:1124–1127. https://doi.org/10.1126/science.aad9745
11. Sallé J, Xie J, Ershov D et al (2019) Asymmetric division through a reduction of microtubule centering forces. J Cell Biol 218:771–782. https://doi.org/10.1083/jcb.201807102

12. Mukherjee RN, Sallé J, Dmitrieff S et al (2020) The perinuclear ER scales nuclear size independently of cell size in early embryos. Dev Cell 54:395–409.e7. https://doi.org/10.1016/j.devcel.2020.05.003
13. Jankele R, Jelier R, Gönczy P (2021) Physically asymmetric division of the C. elegans zygote ensures invariably successful embryogenesis. elife 10:10.7554/eLife.61714
14. Tanimoto H, Sallé J, Dodin L, Minc N (2018) Physical forces determining the persistency and centering precision of microtubule asters. Nat Phys 14:848–854. https://doi.org/10.1038/s41567-018-0154-4
15. Ettensohn CA, Wray G, Wessel GM (2004) Development of sea urchins, ascidians, and other invertebrate deuterostomes: experimental approaches. Gulf Professional Publishing
16. Barone V, Lyons DC (2022) Live imaging of echinoderm embryos to illuminate evo-devo 2022.08.05.503002
17. von Dassow G, Verbrugghe KJC, Miller AL et al (2009) Action at a distance during cytokinesis. J Cell Biol 187:831–845. https://doi.org/10.1083/jcb.200907090
18. Hamaguchi MS, Hamaguchi Y, Hiramoto Y (1986) Microinjected polystyrene beads move along astral rays in sand dollar eggs. Develop Growth Differ 28:461–470. https://doi.org/10.1111/j.1440-169X.1986.00461.x
19. Xie J, Najafi J, Le Borgne R et al (2022) Contribution of cytoplasm viscoelastic properties to mitotic spindle positioning. Proc Natl Acad Sci 119:e2115593119. https://doi.org/10.1073/pnas.2115593119
20. Hiramoto Y (1969) Mechanical properties of the protoplasm of the sea urchin egg. II Fertilized egg. Exp Cell Res 56:209–218. https://doi.org/10.1016/0014-4827(69)90004-4

Chapter 9

Cross Talk Between Metabolism and the Cell Division Cycle

Diana Vara-Ciruelos and Marcos Malumbres

Abstract

Cell division requires a massive rewiring of cellular pathways, including molecular routes involved in providing energy for cell survival and functionality. The energetic requirements and the metabolic opportunities for generating energy change during the different phases of the cell cycle and how these processes are connected is still poorly understood. This chapter discusses basic concepts for a coordinated analysis of cell cycle progression and metabolism and provides specific protocols for studying these two connected processes in mammalian cells.

Key words Cell cycle, Cell synchronization, Cellular energy, Metabolism

1 Introduction

Cell division is an orchestrated process by which one cell will split in two daughter cells, ensuring that the genomic information is properly transmitted from one generation to the next. During the cell division cycle, the progenitor cell needs to go through different phases to safely duplicate (S phase) and segregate (mitosis or M phase) the genetic material into the daughter cells (Fig. 1a). The first preparatory phase (Gap-1 or G1, which take places before S phase) requires a major effort of the cell to increase the amount of several organelles (i.e., ribosomes and mitochondrias) and to synthesize the enzymes that are in charge of DNA replication. Once the DNA is duplicated, the cell will go through another preparatory phase (G2) in which many organelles are also duplicated and move to future division poles in preparation for their segregation, along chromosomes, into the two daughter cells before the final cytokinesis occur. Cell cycle progression is monitored by specific sensoring molecular pathways, known as checkpoints, that control that the previous phase ended without specific aberrations. For instance, possible errors in DNA replication trigger intra-S-phase and G2 checkpoints that prevent entry into mitosis until these errors have

Anna Castro and Benjamin Lacroix (eds.), *Cell Cycle Control: Methods and Protocols*, Methods in Molecular Biology, vol. 2740,
https://doi.org/10.1007/978-1-0716-3557-5_9,

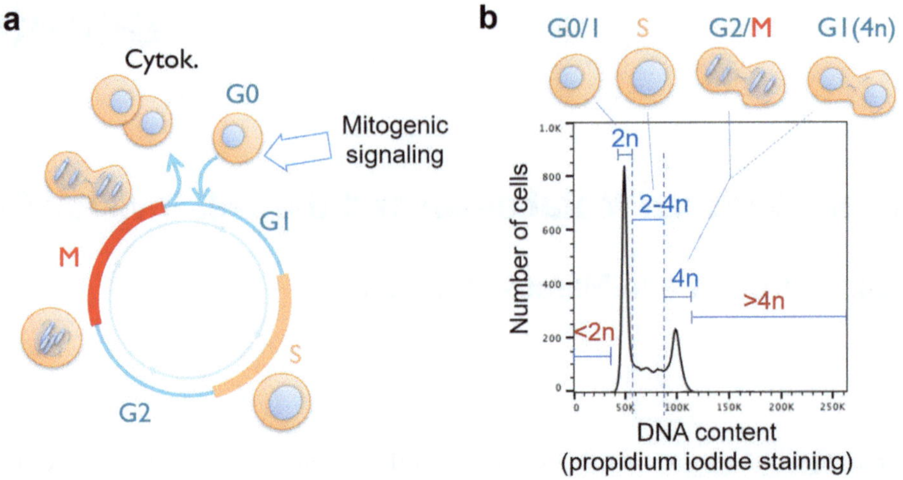

Fig. 1 Schematic representation of cell cycle phases and its identification by basic flow cytometry. (**a**) Basic phases of the cell cycle in which quiescent cells (G0) enter into the cell cycle after receiving mitogenic signals. DNA replication occurs in the S phase, whereas chromosomes are segregated during mitosis (M) and cytoplasms are completely separated after cytokinesis (Cytok.). G1 and G2 represent two gap phases. (**b**) Example of DNA content quantification by flow cytometry using propidium iodide as a colored DNA binder. In a normal cell cycle, 2n, intermediate 2-4n, and 4n populations typically correspond to G0/G1, S-phase, and G2/M cells. Under specific treatments or conditions, 4n cells may represent tetraploid G1 cells (i.e., after cytokinesis failure) and further >2n (fragmented or apoptotic) or > 4n (aneuploid or polyploid) populations may be present

been repaired, thus ensuring genomic stability. The coordination of all these processes requires a major coordinated effort of molecular and cellular pathways. Among the core regulators of this process, cyclin-dependent kinases (CDKs) are known as the engines of the cell cycle, owing to their central role in sensing regulatory inputs and coordinating multiple cytoplasmic and nuclear changes [1].

The connections between the cell division cycle and metabolism have been studied under many different perspectives in the last century. Classical and recent studies suggest that cells accumulate energy, using both oxidative phosphorylation and glycolysis, during the first phases of the cell cycle, whereas chromosome segregation is energetically deficient [2]. Yet, our understanding of these processes mostly comes from a few studies in model organisms or selected mammalian cell lines, with limited number of technologies and physiological conditions. Recent data in yeast suggest that metabolic changes are temporally segregated during the cell cycle with acceleration of glucose-uptake flux in G1 and phase-shifted oscillations of oxygen and carbon dioxide exchanges [3]. Protein synthesis, one of the most demanding cellular processes, specifically peaks in G1 and S/G2/M, whereas lipid and polysaccharide biosynthesis activity peaks during S/G2/M [3]. Different CDK complexes participate in coupling these metabolic changes with the transcriptional and translational waves required for G1 progression and the synthesis of the cell cycle machinery or the massive cellular rearrangements required for DNA replication, nuclear disassembly, and the segregation of chromosomes together with many cellular organelles [2, 4, 5]. Additional biosynthetic/metabolic oscillations that are not primarily generated by cell cycle activity may also be present [3, 6], although the identity of these control mechanisms is mostly unknown.

Importantly, diseased cells such as cancer cells may rewire metabolic pathways and these changes have been proposed to be at the root of tumor formation and growth [2, 7–9]. Genetic and biochemical analyses of tumor cells suggest that cancer metabolism is an attractive target for therapeutic exploitation [10]. This idea not only refers to antimetabolites (e.g., nucleotide analogs) widely used in current cancer therapies, but also to specific targeted therapies focused on the inhibition of critical metabolic enzymes, whose inhibition may cause synthetic lethal effects in the presence of specific tumor-associated mutations in patients [10].

In this chapter, we provide and discuss several protocols to analyze some metabolic properties and activities coupled to a careful consideration of the different phases of the cell cycle. It is very important to consider that most manipulations to synchronize cells in specific cell cycle phases trigger checkpoints and should therefore be considered as checkpoint-arrested rather than unperturbed cells transitioning specific cell cycle phases. In addition, whereas the protocols discussed here are limited to the analysis of a few metabolic properties or activities, they can be easily extended to additional studies (e.g., nuclear magnetic resonance or massive metabolomics), as far as synchronization protocols are properly combined and adjusted.

2 Materials

2.1 Materials for Centrifugal Elutriation

1. Centrifuge compatible with elutriation rotor.
2. JE 5.0 rotor and elutriation chamber (Beckman Coulter Inc., California, United States) (*see* **Note 1**).
3. Variable speed pump.
4. Manometer.
5. 50 mL conical tubes.
6. Ethanol.
7. Double distilled water (ddH_2O).
8. Many of these protocols have been used in our lab to analyze MDA-MB-231 breast cancer cells but are easily extrapolated to any other cultured cell line.

9. Culture medium: DMEM supplemented with 10% fetal bovine serum.
10. Elutriation medium: culture medium but supplemented with 1–10% FBS.
11. Phosphate buffered saline (PBS).
12. Trypsin–EDTA (0.05%).

2.2 Materials for Chemical Synchronization

1. MDA-MB-231 or other cultured cell line.
2. Culture medium: DMEM supplemented with 10% FBS.
3. 100 mm plates.
4. Phosphate buffered saline (PBS).
5. Trypsin–EDTA (0.05%).
6. Lovastatin, a fungal polyketide employed as a cholesterol-lowering drug that inhibits (3S)-hydroxy-3-methylglutaryl–coenzyme A (HMG-CoA) reductase.
7. Thymidine (1-(2-deoxy-β-D-ribofuranosyl)-5-methyluracil).
8. CDKl inhibitor RO-3306.
9. Nocodazole, a reversible agent that interferes with the polymerization of microtubules.

2.3 Materials for Validation of Synchronization by Flow Cytometry

1. Culture medium: DMEM supplemented with 10% FBS.
2. Phosphate buffered saline (PBS).
3. Trypsin–EDTA (0.05%).
4. Propidium iodide (PI).
5. RNase A.
6. Triton ×-100.
7. Any flow cytometer.

2.4 Materials for Measurement of Changes in Oxygen Concentration and pH by Seahorse

1. Seahorse XF Analyzer.
2. Seahorse XFe24 FluxPak.
3. Non-CO_2 incubator.
4. pH meter.
5. Cell counter.
6. Water bath.
7. MDA-MB-231 or other culture cell line.
8. DMEM supplemented with 10% FBS.
9. Assay medium: unbuffered DMEM or Seahorse XF (extracellular flux) DMEM (or unbuffered RPMI or Seahorse XF RPMI).
10. D-glucose.

11. L-glutamine solution.
12. Pyruvate solution.
13. The ATPase inhibitor oligomycin.
14. Carbonyl cyanide 4-(trifluoromethoxy)phenylhydrazone (FCCP), a potent uncoupler of mitochondrial oxidative phosphorylation.
15. Rotenone, a naturally occurring toxin that inhibits complex I of the mitochondrial electron transport chain.
16. Antimycin, an inhibitor of electron transport from cytochrome b to cytochrome complex III (cytochrome c reductase).
17. 2-Deoxyglucose.
18. Cell-Tak (tissue adhesive, Corning).
19. Centrifuge.
20. Trypsin–EDTA (0.05%).

2.5 Materials for the Analysis of Nutrients Uptake

1. Culture medium: DMEM supplemented with 10% FBS.
2. Phosphate buffered saline (PBS).
3. Trypsin–EDTA (0.05%).
4. Diisononyl phthalate (DINP). Used to modulate the viscosity in oil–water emulsions.
5. Poly(dimethylsiloxane-co-methylphenylsiloxane) (PDMS). Used to modulate the viscosity in oil–water emulsions.
6. ^{3}H or ^{14}C Radiolabeled glucose and/or amino acid.
7. DMEM-free glucose.
8. Hanks' balanced salt solution (HBSS).
9. Eppendorf tubes.
10. Six-well plates.
11. NaOH 1 M.
12. Scintillation counter and vials.
13. Scintillant suitable for aqueous solutions.

3 Methods

3.1 Synchronization by Centrifugal Elutriation

Centrifugal elutriation is based on the sedimentation properties of cells, and it may be the preferred method when minimal perturbation of cellular functions is desired. Since this method is based on mechanical (size, density) rather than molecular properties, it may provide limited resolution of separation for specific assays.

1. Culture MDA-MB-231 cells (or any other cell line) at 37 °C, 5% CO_2 with DMEM supplemented with 10% FBS,

maintaining low density (no more than 50–60%), to allow logarithmic growth.

2. On the day before the experiment, if needed, sterilize the elutriator system filling the reservoir system with 70% ethanol at 20 mL/min.
3. Check that there are no leaks and centrifugate at 1000 rpm to remove the bubbles.
4. On the day of the experiment, turn the pump on at 25–30 mL/min and replace 70% ethanol first with sterile water and later with elutriation medium (*see* **Note 2**).
5. Replace the reservoir with a mix of water and ethanol for a new one filled with elutriation medium.
6. Trypsinize cells and resuspend in 20 mL of ice-cold medium.
7. Filtrate the solution through a cell strainer to remove cell aggregates.
8. Load the cells into the system and put centrifugation speed and flow rate at starting conditions which will allow to retain the cells in the elutriation chamber. Reserve a small fraction of cell solution as non-elutriated control (*see* **Note 3**).
9. Increase the pump speed slowly (1 mL/min) and start to collect elutriation fractions (50–100 mL each).
10. Take some ml of each fraction to measure cell cycle phase and cell size.
11. Use remaining fraction to perform the elected metabolic assays (*see* **Note 4**).

3.2 Chemical Synchronization

Synchronization of cells can be achieved by imposing molecular or metabolic blocks that prevent cell cycle progression at specific phases. In some cases, the arrest is reversible allowing cells to resume cell cycle progression in a synchronized manner. It is very important to consider that most if not all chemical synchronization methods work by either perturbing metabolic pathways or by triggering specific cell cycle checkpoints, and this needs to be carefully analyzed and discussed when presenting the results from assays using these methods. Some classical handbooks to introduce basic cell cycle synchronization methods have been published previously [11, 12].

3.2.1 G_0/G_1 Arrest by Serum Deprivation

1. Maintain MDA-MB-231 cells or other mammalian cells in the appropriate medium supplemented with 10% of FBS at 37 °C with 5% CO_2.
2. Seed 2.5×10^6 cells in 100 mm plates with 10 mL of culture medium (40% of confluence).
3. Next day, wash the cells twice very gently with 5 mL of PBS.

4. Add 10 mL of culture medium without any FBS.
5. Next 24–48 h cells will be arrested in early G1.

3.2.2 G_1 Arrest by Lovastatin

1. Maintain MDA-MB-231 cells or other mammalian cells in the appropriate medium supplemented with 10% of FBS at 37 °C with 5% CO_2.
2. Seed 2.5 × 10^6 cells in 100 mm plates with 10 mL of culture medium (40% of confluence).
3. Next day change the media for 10 mL of culture medium with 10% of FBS plus 20 μM of lovastatin (*see* **Note 5**).
4. Next day cells will be arrested in G1. To reverse this block, wash the cells with PBS and culture them in growth media with at least 2 mM mevalonic acid (*see* **Note 6**).

3.2.3 S-phase Arrest by Double Thymidine Block

1. Maintain MDA-MB-231 cells or other mammalian cells in the appropriate medium supplemented with 10% of FBS at 37 °C with 5% CO_2.
2. Seed 2.5 × 10^6 cells in 100 mm plates with 10 mL of culture medium (40% of confluence).
3. Next day, change the media for 10 mL of culture medium with 10% of FBS plus 2 mM of thymidine.
4. After 16 h, cells will be blocked in S phase, release them from the block washing two times with 5 mL of phosphate buffered saline (PBS), and incubate for 8 h with complete medium.
5. After 8 h, incubate the cells for 16 h with the second thymidine block adding between 2 μM and 2 mM thymidine to the medium (*see* **Note** 7).
6. Next day, wash the cells twice very gently with 5 mL of phosphate buffered saline (PBS) and add 10 mL of complete media. During at least one cycle, cells will be cycling synchronized (*see* **Note 8**).

3.2.4 G_2 Arrest by RO-3306

1. Maintain MDA-MB-231 cells or other mammalian cells in the appropriate medium supplemented with 10% of FBS at 37 °C with 5% CO_2.
2. Seed 2.5 × 10^6 cells in 100 mm plates with 10 mL of culture medium (40% of confluence).
3. Next day, change the media for 10 mL of culture medium with 10% of FBS plus 10 μM of RO-3306.
4. After 12 h, treatment cells will be arrested in G_2. To release them, wash them with PBS and culture with growth medium.

3.2.5 Prometaphase Arrest by Nocodazole

1. Maintain MDA-MB-231 cells or other mammalian cells in the appropriate medium supplemented with 10% of FBS at 37 °C with 5% CO_2.
2. Seed 2.5 × 10^6 cells in 100 mm plates with 10 mL of culture medium (40% of confluence).
3. Next day, change the media for 10 mL of culture medium with 10% of FBS plus 0.8 μM nocodazole for 16 h.
4. Next day, cells will be arrested in prometaphase.

3.3 Validation of Synchronization by Flow Cytometry

1. Trypsinize cells and add double amount of growth medium to block trypsin–EDTA.
2. Centrifuge cells 300 g for 5 min and resuspend the pellet in 1 mL of PBS.
3. Repeat **step 2**.
4. Mix cell suspension with 2.5 mL of absolute ethanol (final concentration approximately 70%) using vortex low speed while adding the ethanol. This will prevent clustering of cells during the fixation (*see* **Note 9**).
5. Incubate on ice for 15 min or overnight at −20 °C (cells can be stored in this step for a week).
6. Pellet the cells at 300 g for 5 min.
7. Resuspend the pellet in 500 μL of PBS with 50 μg/mL of PI, 0.1 mg/mL RNase A, and 0.05% Triton ×-100 (*see* **Note 7**).
8. Incubate for 40 min at 37 °C.
9. Add 3 mL of PBS and centrifuge cells 300 g for 5 min.
10. Resuspend the pellet with 500 μL of PBS.
11. Analyze by flow cytometry, using DNA content to differentiate between 2n, 2n-to-4n, and 4n cells (typically corresponding to G0/G1, S-phase, and G2/M cells, respectively) (Fig. 1b; *see* **Notes 10** and **11**).

3.4 Measurement of Changes in Oxygen Concentration and pH by Seahorse

On the day before the experiment:

1. Switch on Seahorse XF Analyzer to warm up overnight. Manufacturer recommends at least 5 h in advance.
2. Put 1 mL per well of Seahorse XF Calibrant on the utility plate of a sensor XF24 cartridge. Put the Hydro Booster on the top of the utility plate and later the sensor cartridge, being sure that sensors are well-submerged. Incubate overnight 37 °C in a non-CO_2 incubator.
3. Prepare a XF24 cell culture microplate with Cell-Tak solution (*see* **Note 12**): mix 22.8 μL of Cell-Tak, 11.4 μL of 1 M NaOH, and 1.465 mL of 0.1 M NaHCO3 pH 8.0. This solution needs to be used within 10 min, adding 50 μL/well. Leave it for at

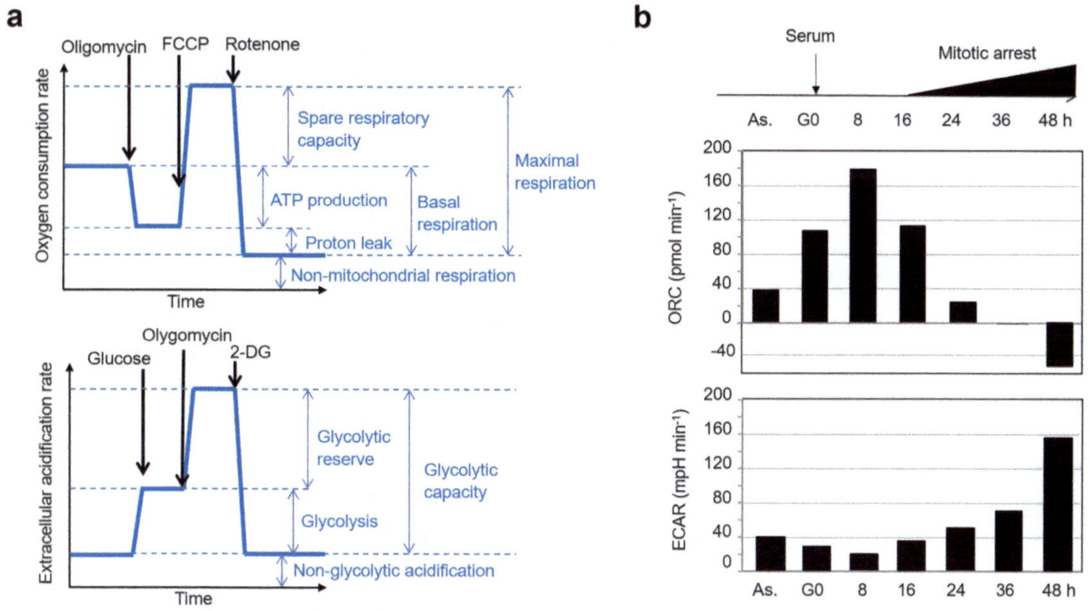

Fig. 2 Testing mitochondrial and glycolytic activities using the seahorse apparatus. (**a**) Typical outputs of oxygen consumption (top) and glycolytic activity (bottom). Top: key parameters of mitochondrial function are basal respiration, ATP-linked respiration, spare (reserve) respiratory capacity, proton leak, maximal respiration, and non-mitochondrial respiration in the absence or presence of the indicated treatments. Bottom: addition of glucose or the glucose analog 2-DG in the presence or absence of mitochondrial function (oligomycin) provides measurements of glycolytic activity, maximum glycolytic reserve and total glycolytic capacity as indicated. (**b**) Example of changes in oxygen consumption rate (OCR) and extracellular acidification rate (ECAR) in asynchronous (As.) or serum-starved breast cancer cells, and the same cells after mitotic arrest, showing exhaustion of mitochondrial respiration and increased in the glycolytic capacity during prolonged mitotic arrest. (Data from Ref. Domenech et al. [14])

least 30 min at room temperature (RT). Wash twice with sterile H2O, and dry at RT before storing at 4 °C overnight.

4. Alternative to **step 3**, if Cell-Tak solution is not used, trypsinize your cells and seed them in a XF24 cell culture microplate, using only 100 μL of growth medium. Do not seed cells in background correction wells (A1, B4, C3, D6); only leave the medium. Leave cells to adhere in the biosafety cabinet for 1–5 h and add 150 μL of growth medium very gently. If cells are being chemical arrested, it could be done at this point (*see* **Note 13**).

The day of the assay:

1. Warm assay medium in 37 °C water bath (*see* **Note 14**).
2. For cells seeded the day before, remove media except 50 μL (to avoid cells dry) and add 1 mL of assay medium. Leave them to rest for 1 h in a non-CO_2 incubator at 37 °C and later replace the medium with 500 μL of assay medium.

3. For cells that are going to be seed in a Cell-Tak-covered microplate just the day of the assay, prepare a cell suspension with enough cells for 25 wells. Centrifuge at 300 g for 5 min and resuspend them in 2500 μL of assay medium. Add 100 μL of the cell suspension per well, but background wells that only add medium. Centrifuge the microplate at 300 g for 1 min without brake. Place the microplate into a non-CO_2 incubator at 37 °C for 30 min. Add 400 μL of assay medium per well and leave the cells rest for another 30 min.
4. Prepare compound solutions to perform Cell Mito Stress Test (10 μM oligomycin, 10 μM FCCP, 10 μM rotenone, and 10 μM antimycin A, all in assay medium) or Glycolytic Rate Assay (10 μM rotenone, 10 μM antimycin A, and 0.5 M 2-deoxyglucose in assay medium) (*see* **Note 15**).
5. Load solutions into the ports of sensor cartridge.
6. Cell Mito Stress Test (Port A: 56 μL of 10 μM oligomycin; Port B: 62 μL of 10 μM FCCP; Port C: 34.5 μL of 10 μM rotenone plus 34.5 μL of 10 μM antimycin A).
7. Glycolytic Rate Assay (Port A: 28 μL of 10 μM rotenone plus 28 μL of 10 μM antimycin A; Port B: 62 μL of 0.5 M 2-deoxyglucose).
8. Load the template onto the Seahorse XF Analyzer and run the assay (Fig. 2a, b).

3.5 Radiolabeled Nutrient uptakes

3.5.1 For Suspension Cells

1. Mix in 1:1 proportion diisononyl phthalate with poly (dimethylsiloxane-co-methylphenylsiloxane) in a 50 mL conical tube. Vortex oil mix and remove bubbles centrifugating at 300 g for 3 min.
2. Put 500 μL of oil mix per Eppendorf. Every condition should be tested in triplicate.
3. Prepare the glucose or amino acid tracer buffer, combining 0.2 μCi per sample of radiolabeled glucose or amino acid (glutamine, leucine, phenylalanine) and 200 μL of glucose free DMEM media (for glucose uptake) or HBSS (for amino acid uptake).

   This amount needs to be multiplied by sample number considering that every condition will be tested by triplicate.
4. Put 200 μL of tracer mix per Eppendorf. This will form like a bubble into the oil mix. Centrifuge the Eppendorf if there are more than one bubble.
5. Wash 1.2×10^6 cells with glucose free DMEM medium or HBSS (0.9×10^6 cells to perform triplicates plus 0.3×10^6 extra) (*see* **Note 16**).
6. Remove the supernatant leaving tubes upside down to not leave any liquid.

7. Resuspend cells in 800 μL of glucose free media or HBSS.
8. Put 200 μL of cells per Eppendorf and be sure that cells go into the bubble.
9. Wait 4 min and centrifugate 3 min 8.000 rpm (*see* **Note 17**).
10. Remove the oil mix and the tracer buffer without touching cell pellet.
11. Wash 2 times every Eppendorf with Milli-Q H_2O.
12. Lyse the pellet with 200 μL 1 M NaOH.
13. Put the lysis in scintillation vials with 3 mL of OptiPhase HiSafe 3 and count.

3.5.2 For Adherent Cells

1. Seed the cells in 6-well plates at 40–50% of confluence. Every condition should be tested in triplicate.

   Next day, perform treatments for cell cycle synchronization and wait the necessary time to be in the desired cell cycle phase. Prepare the glucose or amino acid tracer buffer, by combining 1 μCi per sample of radiolabeled glucose or amino acid (glutamine, leucine, phenylalanine) and 1 mL of glucose-free DMEM media (for glucose uptake) or HBSS (for amino acid uptake). This amount needs to be multiplied by sample number considering that every condition will be tested by triplicate.
2. Wash the cells with PBS twice and add the tracer mix into the well.
3. Wait 5 min and wash the cells with PBS twice (*see* **Note 17**).
4. Lyse the cells with 200 μL 1 M NaOH.
5. Put the lysis in scintillation vials with 3 mL scintillant for aqueous solutions and count it.

4 Notes

1. There are three sizes of elutriation chambers: 4 mL chamber, a 40 mL chamber, and a 5.5 mL Sanderson chamber (with also a different shape). The 4 ml chamber is good for fractioning 2×10^7–10^9 cells, 40 mL chamber for 2×10^8–10^{10}, and the Sanderson for 2×10^6–10^9.
2. The elutriation medium is the normal growth medium with a 10% FBS. For economical reason, lower FBS percentage could be used (1–10%) but it may have metabolic consequences if cells are used straight forward for metabolic assays.
3. Increased flow rate and decreased centrifugation speed have similar effects allowing high resolution of synchronization. However, to keep a low fixed centrifugal speed, increasing the pump speed will avoid cell damage and will enhance

reproducibility between experiments. The best flow rate and centrifugal speed need to be tested in advance to obtain and good resolution, but some suggestions about how to calculate these parameters as well as other suggestions on these techniques have been already published [12, 13].

4. Before performing the elected metabolic assays, cells will need to recover from the fractioning considering that it will be synchronized for at least the next cycle.
5. Lovastatin treatment could be ranged from 5 to 60 μM of lovastatin.
6. Mevalonic acid treatment should be at least 100 times greater than lovastatin. Recovery from lovastatine-induced G1 arrest could prolong the time in this phase; therefore, this time needs to be considered if the experiment will be performed in the next phases.
7. A single treatment with thymidine may be sufficient for good levels of synchronization in many cell types. In some cells, the second block of thymidine could be less concentrated reducing some negative effects of the treatment.
8. The time of each phase will depend on the cells used. Therefore, to know the duration of each phase, tests must be done in advance.
9. Alternatively, cells could be fixed with 4% paraformaldehyde for 15 min.
10. Most publications mention to G1, S, or G2/M cells when referring to 2n, intermediate 2n-4n, or 4n populations. However, this may be wrong in many cases. For instance, 4n cells are frequently 4n (tetraploid) G1 cells that existed the cell cycle without segregation or without cytokinesis. This is a very frequent mistake especially when using specific cell types or when using treatments that perturb normal cell cycle progression.
11. Whereas the DNA content analysis using PI may be appropriate for most cases as a control of synchronization, some specific assays may require a more informative method in which DNA content is combined with a specific marker as a second color. Two frequently used examples are DNA content + a reporter of DNA replication (e.g., BrdU or EdU; see, for instance, Click-iT™ EdU; Thermo Fisher cat. No. C10337) or DNA content + a mitotic marker such as phospho S10-histone H3 or MPM2 antigens recognized by specific antibodies. The expression of specific cyclins (i.e., cyclin E) or additional cell cycle regulators (i.e., TPX2, Aurora B, Geminin) could also be used for understanding the efficacy of the synchronization protocol and the stage of the cell. As indicated in the Introduction, it is

important to consider the effect that several synchronization protocols may have in specific cell cycle checkpoints.

12. Cell-Tak needs to be used when cells are nonadherent, or you want to attach the cells to the plate after centrifuge elutriation or chemical synchronization without waiting cells attach by themselves. If Cell-Tak is used, to normalize Seahorse XF data, use cell number or protein amount, because fluorescence detection of DNA could be affected.
13. For ECAR and OCR measurement, a 50–90% cell confluence is typically recommended. However, cell division may be limited at high density, and 50–70% of confluence will be more appropriate. Typical breast cancer cells numbers will be ranged between 1×10^4 and 4×10^4 cells per well.
14. Manufacturers recommend supplementing the assay media with 1 mM pyruvate, 2 mM glutamine, and 10 mM of glucose, but this composition could be changed according with the cell type or the assays condition that want to be tested.
15. Compound solutions should be prepared at higher concentration than suggested ones in DMSO. This will allow appropriate storage for a long period. On the day of the assay, compound solutions need to be diluted in assay medium into the suggested concentrations. These concentrations could be ranged between 5 μM and 25 μM for oligomycin and 1.25 μM to 20 μM for FCCP. Consider that final concentration (in the well) will be 10 times diluted.
16. This number of cells is appropriate for most breast cancer cells but, for other cell types it will be necessary to work out sensitivity over background. This can be performed doing uptakes on a serial dilution of cells.
17. To calculate optimal nutrient uptake time, an initial time-course analysis using 1–20 min may be required. The selected time needs to be in the linear range.

Acknowledgments

D.V-C. received support from the Fundación Amigos CNIO. This work was funded by the Ministry of Science and Innovation (MICINN/AEI/FEDER, PID2021-128726, and PDC2022-133408-I00) and Comunidad de Madrid (Y2020/BIO-6519 and S2022/BMD-7437).

References

1. Morgan DO (2006) The cell cycle: principles of control (primers in biology). Sinauer Associates, Inc
2. Salazar-Roa M, Malumbres M (2017) Fueling the cell division cycle. Trends Cell Biol 27(1): 69–81. https://doi.org/10.1016/j.tcb.2016.08.009
3. Takhaveev V, Ozsezen S, Smith EN, Zylstra A, Chaillet ML, Chen H, Papagiannakis A, Milias-Argeitis A, Heinemann M (2023) Temporal segregation of biosynthetic processes is responsible for metabolic oscillations during the budding yeast cell cycle. Nat Metab 5(2):294–313. https://doi.org/10.1038/s42255-023-00741-x
4. Fajas L (2013) Re-thinking cell cycle regulators: the cross-talk with metabolism. Front Oncol 3:4. https://doi.org/10.3389/fonc.2013.00004
5. Solaki M, Ewald JC (2018) Fueling the cycle: CDKs in carbon and energy metabolism. Front Cell Dev Biol 6:93. https://doi.org/10.3389/fcell.2018.00093
6. Papagiannakis A, Niebel B, Wit EC, Heinemann M (2017) Autonomous metabolic oscillations robustly gate the early and late cell cycle. Mol Cell 65(2):285–295. https://doi.org/10.1016/j.molcel.2016.11.018
7. Warburg O (1956) On the origin of cancer cells. Science 123(3191):309–314. https://doi.org/10.1126/science.123.3191.309
8. Leal-Esteban LC, Fajas L (2020) Cell cycle regulators in cancer cell metabolism. Biochim Biophys Acta Mol basis Dis 1866(5):165715. https://doi.org/10.1016/j.bbadis.2020.165715
9. Gatenby RA, Gillies RJ (2004) Why do cancers have high aerobic glycolysis? Nat Rev Cancer 4(11):891–899. https://doi.org/10.1038/nrc1478
10. Luengo A, Gui DY, Vander Heiden MG (2017) Targeting metabolism for cancer therapy. Cell Chem Biol 24(9):1161–1180. https://doi.org/10.1016/j.chembiol.2017.08.028
11. Pagano M (1996) Cell cycle: materials and methods. Springer Lab Manuals, Berlin/Heidelberg. https://doi.org/10.1007/978-3-642-57783-3
12. Banfalvi G (2017) Overview of cell synchronization. Methods Mol Biol 1524:3–27. https://doi.org/10.1007/978-1-4939-6603-5_1
13. Marraccino RL, Keng PC (1996) Centrifugal elutriation. In: Pagano M (ed) Cell cycle: materials and methods. Springer Lab Manual, Berlin, Heidelberg. https://doi.org/10.1007/978-3-642-57783-3_5
14. Domenech E, Maestre C, Esteban-Martinez L, Partida D, Pascual R, Fernandez-Miranda G, Seco E, Campos-Olivas R, Perez M, Megias D, Allen K, Lopez M, Saha AK, Velasco G, Rial E, Mendez R, Boya P, Salazar-Roa M, Malumbres M (2015) AMPK and PFKFB3 mediate glycolysis and survival in response to mitophagy during mitotic arrest. Nat Cell Biol 17(10): 1304–1316. https://doi.org/10.1038/ncb3231

Chapter 10

Give and Take: The Reciprocal Control of Metabolism and Cell Cycle

Romain Riscal, Blanche Riquier-Morcant, Gilles Gadea, and Laetitia K Linares

Abstract

Cell cycle is an ordered sequence of events that occur in a cell preparing for cell division. The cell cycle is a four-stage process in which the cell increases in size, copies its DNA , prepares to divide, and divides. All these stages require a coordination of signaling pathways as well as adequate levels of energy and building blocks. These specific signaling and metabolic switches are tightly orchestrated in order for the cell cycle to occur properly. In this book chapter, we will provide information on the basis of metabolism and cell cycle interplay, and we will finish by an unexhaustive list of metabolomics approaches available to study the reciprocal control of metabolism and cell cycle.

Key words Cell cycle, Cell division, Metabolism, Energy, Metabolites, Metabolomics approaches, Metabolic profiles

1 Introduction

Cell division is a metabolically demanding event where large amounts of protein, lipids, and nucleotides are required to double cell mass. Not surprisingly, cells monitor the internal and external availability of nutrients to ensure that conditions are optimal before initiating division. The concept that different metabolic environments require specific adaptation is also supported by the recent findings elucidating the metabolic requirements of cell cycle [1]. Works over the last decade have shown that metabolic regulation is intimately linked with many aspects of cell biology, including cell cycle. Here, we will describe the recent understanding of how metabolic regulation is tuned to match the physiological state and enable progression through the cell cycle phases. We will also discuss how cell cycle regulators contribute to lipid synthesis, glucose production, and glycolytic metabolism.

Anna Castro and Benjamin Lacroix (eds.), *Cell Cycle Control: Methods and Protocols*, Methods in Molecular Biology, vol. 2740,
https://doi.org/10.1007/978-1-0716-3557-5_10,

2 Cellular Metabolism at a Glance

The importance of metabolism resides in pathways generating sufficient amounts of energy to conduct biological functions. Those pathways generate intermediates for biosynthesis of macromolecules such as proteins, lipids, nucleic acids, and carbohydrates. In contrast, these macromolecules can also be catabolized and give simpler structures – fatty acid, amino acids, sugar, and nucleotides. Catabolic and anabolic reactions can generate or consume respectively energy and cofactors to sustain chemical reactions of living organisms and dictate divergent biological outcomes. Initial thoughts were that proteins, kinases, and transcription factors initiate all the important decisions, giving metabolic networks the only task of providing building blocks or energetic resources in response to cell behavior and fate. However, metabolism studies of the last decade clearly show that metabolism is a central piece of the decision-making apparatus of the cell and participate to signaling pathways, gene expression, cell division, cell death, and adaptation to different types or levels of stress.

2.1 Signaling Pathways and Their Role in Metabolism

Mammalian cells aim to maintain anabolic and catabolic ratio acceptable to sustain survival. Metabolic-dependent signaling pathways drive this equilibrium by promoting nutrient uptake and responding to changes in nutrients and growth factors. As an example, during period of nutrients and growth factors abundance, the mammalian target of rapamycin (mTOR) promotes protein and lipid synthesis and also stimulates metabolic pathways such as glycolysis and mitochondrial metabolism important to sustain cell proliferation. In contrast, when cells are energetically stressed, mTORC1 is inhibited to rewire the flux from anabolic to catabolic pathway such as autophagy to maintain survival. PI3K/AKT and AMPK signaling are also involved in this metabolism regulation of cellular fate [2]. Moreover, intracellular metabolites can control signaling through protein covalent modifications inducing change in activity, localization, or stability of target proteins. A key metabolite involved in all of these processes is acetyl coenzyme A (Acetyl-CoA), known to be a precursor of lipids, but CoA is also a substrate for protein and/or histones acetylation resulting in chromatin opening and gene expression. Acetyl-CoA availability drives cell proliferation and differentiation, linking metabolism and induction of genes involved in cell cycle to make sure that the cells have enough biosynthetic and bioenergetic stocks to expand.

2.2 Noncanonical Functions of Metabolic Enzymes

As mentioned earlier, various metabolic enzymes have "moonlighting" functions outside of their established metabolic role, including the regulation of gene expression with metabolic enzymes observed in the nucleus in transcriptional complex such as fructose

1,6-bisphosphatase 1 (FBP1), suppressing hypoxia-inducible factors (HIFs) target genes by colocalizing with HIF at hypoxia responsive elements (HRE) [3]. Epigenic modifications are also dependent on metabolic status of the cells, because acetyl, methyl, or cofactors such as nicotinamide adenine dinucleotide (NAD+) are required substrates and generated by metabolic reactions. Aside from controlling gene expression, metabolic enzymes also participate in DNA damage repair, apoptosis, or cell cycle regulation. Over the past 10 years, many of us have started to study those beyond metabolism reactions, and we are now beginning to understand that metabolic enzymes play a role in a broad range of cellular processes including cell cycle homeostasis. Because cell cycle deregulation leads to aberrant proliferation characteristic, the control and regulation of the cell cycle phases by metabolic enzymes have been of interest in field such as cancer.

3 Cell Cycle at a Glance

The cell cycle refers to all mechanisms a cell undergoes from its formation to its division into two new daughter cells. Many proteins and pathways are involved in this process, which could be divided in two main phases: interphase and mitotic phase. The interphase encompasses three steps: cell growth (G1 phase), DNA replication (S (synthesis) phase), and DNA damages' self-check plus division planning (G2 phase). The mitotic phase (M phase) corresponds to cell division (mitosis and cytokinesis). Segregation of cellular content during mitosis is subdivided into four stages : prophase (DNA condensation), metaphase (chromosome alignment), anaphase (chromatid separation), and telophase (nuclei formation), while cytokinesis (cellular content separation) overlap those phases and starts in anaphase or telophase [4]. In the adult body, most cells are in a so-called G0 phase, also known as "quiescence" or resting state, which characterized cells that exited the G1 phase to enter a non-replicative state.

3.1 Regulatory Mechanisms and Checkpoints

Cyclins and cyclin-dependent kinases (CDKs) are the main actors of cell cycle regulation. Cyclins are regulatory subunits of CDKs that undergo sequential modifications during cell cycle progression allowing CDKs' periodical activation. Throughout cell cycle, protein degradation inhibition and CDKs-mediated transcription lead to specific cyclin accumulation, enabling cell to bypass cell cycle checkpoints. CDKs are serine/threonine enzymes, activated through phosphorylation by cyclin in order to activate and/or inhibit cell cycle checkpoint. Therefore, CDK inhibitors (CKIs) also play a part in cell cycle regulation. Cell cycle checkpoints are crucial to prevent DNA damage and ROS accumulation and propagation that might lead to different pathologies, including/

especially cancer. The first checkpoint, called restriction point, once overcome leads to an irreversible entry into cell cycle (transition G1/S) [5]. At this stage, growth factors are, in part, responsible for the cyclin D regulation by different mechanisms [6]. Specific checkpoints are responsible for slowing down or arresting the cell cycle in response to DNA damage (G1 or G2 phase), damages in the replicative fork (S phase), or incomplete spindle assembly (mitosis).

3.2 Signaling Pathways and Their Role in Cell Cycle Regulation

Connection between metabolism and cell cycle essentially occurs in G1/S phase. The PIK3/Akt pathway is responsible for exiting the resting stage and transitioning from G1 to S phase [7]. Growth factors and other metabolites activate phosphoinositide 3-kinases (PIK3) and thereby phosphorylate its lipid substrate [8]. Protein kinase (PKB), also known as AKT, phosphorylates several partners leading to mTORC1 phosphorylation and activation. mTOR is downregulated by AMP-activated protein kinase (AMPK) mediated pathway [9]. AMPK act as an energy sensor activated by metabolic stress. AMPK activation leads to regulation of AMP/ATP ratio by inhibition of ATP consummation and stimulation of AMP production. These pathways act on the cell cycle headway or delay.

CDK1 regulates the transition between S phase and mitotic phase. If cells are in a replicative stress, signaled by an accumulation of single DNA strands, CHK1 phosphorylates CDC25 leading to its degradation and phosphorylates Wee1 inducing CDK1 phosphorylation and inhibition [10]. This mechanism prevents replicative stress which is a major DNA damage inductor.

After replication, chromatids need to be partitioned in the two daughter cells by mitotic spindles. Disruption of this phenomena could lead to gains or losses in chromosome number; therefore, spindle assembly checkpoint (SAC) is highly regulated. SAC proteins recognize every kinetochore unattached or incorrectly attached to mitotic spindle giving rise to mitotic checkpoint complex (MCC) protein accumulation which acts as an anaphase-promoting complex/cyclosome (APC/C) inhibitor [11].

4 Interplay Between Metabolism and Cell Cycle

The different phases of the eukaryotic cell cycle are extremely well-preserved phenomena. DNA decompaction, RNA and protein synthesis followed by DNA replication, and lipid synthesis occur after resting cells are committed to proliferate. To achieve such organization, cell cycle is closely interrelated with metabolism. Both processes are tightly regulated and the interplay between metabolism and the cell cycle is complex and essential for proper cell function [12]. Disruptions in this interplay can lead to alterations in cell cycle regulation and metabolism, contributing to the development of cancer and other diseases. Several metabolic pathways can influence cell cycle progression and regulate cell division. Metabolic pathways, such as glycolysis and oxidative phosphorylation, produce the energy and building blocks required for the cell to divide and replicate its genetic material [4]. Other pathways, such as the PI3K/Akt pathway, can regulate cell cycle progression by promoting cell division and inhibiting cell cycle arrest [13]. Conversely, cell cycle regulates metabolic pathways by controlling the activity of metabolic enzymes and the availability of metabolic intermediates to ensure that the cell has sufficient energy and building blocks for cell division [14].

4.1 Metabolic Pathways That Control Cell Cycle

Glycolysis produces ATP, the energy currency of the cell, which is essential for cell division. In the G1 phase particularly, aerobic glycolysis actively supports RNA and protein synthesis that are required to increase cell biomass before mitosis [12]. Between G1 and S phases, there is a metabolic checkpoint that determines whether cells undergo G1 arrest or proliferate [15, 16]. AMP-activated protein kinase (AMPK) serves as a sensor for low energy, and its activation inhibits cell growth through mechanisms involving inhibition of mTOR [9]. G1/S checkpoint controls the existence of all conditions (nutrients and enzymes) required for DNA synthesis, including energy as G1/S transition requires an "energy boost." Oxidative phosphorylation generates ATP through the oxidation of metabolic intermediates and provides the energy required for the cell to replicate its DNA and divide.

The pentose phosphate pathway directs glucose flux to its oxidative branch and produces a reduced form of nicotinamide adenine dinucleotide phosphate (NADPH), an essential reductant in anabolic processes. It has become clear that the pentose phosphate pathway plays a critical role in regulating cell growth by supplying cells with ribose-5-phosphate, which is required for the synthesis of nucleotides, the building blocks of DNA. The amino acid metabolism also provides the building blocks required for the synthesis of proteins, which are essential for cell division and DNA replication. But this is not the only contribution of amino acid metabolism in cell cycle. Amino acid metabolism is also involved in nucleotide synthesis and gene regulation as exemplified by serine metabolism. *De novo* serine synthesis contributes to cell cycle progression by providing one-carbon units to generate purine nucleobases and to regenerate methionine from homocysteine, a key component of RNA and DNA methylation [17].

4.2 Cell Cycle Regulators That Control Metabolic Pathways

The cell cycle regulates the uptake of nutrients, such as cholesterol, to ensure that the cell has enough energy and building blocks for cell division. For instance, the sterol response element binding protein (SREBP) pathway surveys cholesterol levels in endoplasmic reticulum membranes. When cellular cholesterol levels decrease, SREBP undergoes proteolytic cleavage and is released from the membrane. Subsequently, SREBP goes to the nucleus and induces expression of lipoprotein receptors to increase cholesterol acquisition from extracellular sources and expression of lipogenic enzymes to enhance *de novo* biosynthesis [18].

But the cell cycle also regulates the expression and the activity of metabolic enzymes, such as those involved in glycolysis and oxidative phosphorylation, to ensure that the cell has enough energy for cell division. Several core cell cycle regulators and components, notably cyclin D, CDKs, pRB, and E2Fs, specifically target metabolic enzymes and pathways and thereby control metabolism and growth [19]. These metabolic targets operate in different pathways, including glucose metabolism, oxidative metabolism [20], amino acid (AA) metabolism, and nucleotide and lipid metabolism [21], and are involved in the delivery of metabolic precursors for cell growth.

4.3 Cross Talk Between Metabolic Pathways and Cell Cycle Regulators Is Essential to Coordinate Both Cell Division and Metabolism

CDKs, which are key cell cycle regulators, are regulated by metabolic pathways, including the availability of metabolic substrates and the activity of metabolic pathways. One example of how metabolic pathways regulate CDK activity is through the regulation of cyclin synthesis [1]. The level of cyclin expression is regulated by a number of factors, including metabolic signals. For example, changes in the level of glucose or other metabolic substrates can alter the expression of cyclins, which in turn affects CDK activity. Another example is the regulation of CDK activity by the mTOR pathway, which is a metabolic signaling pathway that controls cell growth and division [22]. The TOR pathway senses the availability of nutrients, such as amino acids, and regulates cell growth accordingly. When nutrients are abundant, the TOR pathway activates CDKs, promoting cell cycle progression. However, when nutrients are scarce, the TOR pathway suppresses CDK activity, slowing down cell cycle progression. In addition, other metabolic pathways, such as the AMP-activated protein kinase (AMPK) pathway, can also regulate CDK activity [23]. The AMPK pathway senses changes in the energy status of the cell and adjusts CDK activity accordingly. When the energy levels of the cell are low, AMPK activation leads to the inhibition of CDK activity, slowing down cell cycle progression. Metabolic pathways play a crucial role in regulating CDK activity, which in turn controls cell cycle progression. These pathways allow cells to coordinate cell cycle progression with the availability of metabolic substrates and the energy status of the cell.

The regulation of retinoblastoma protein (Rb) by metabolic pathways is complex and involves the interplay between multiple metabolic pathways. These pathways can either promote or inhibit Rb phosphorylation and influence the regulation of the cell cycle. For example, the Warburg effect, which is a metabolic shift toward glycolysis in cancer cells, has been shown to promote Rb phosphorylation [24]. This is thought to occur through the activation of the metabolic kinase hexokinase, which leads to the activation of CDKs and the phosphorylation of Rb. The AMPK is also a metabolic sensor that is activated by cellular energy stress, such as low ATP levels. Activation of AMPK has been shown to lead to the phosphorylation of Rb and the inhibition of CDK activity. This suggests that metabolic pathways can regulate Rb phosphorylation through the activation of AMPK, which leads to the inhibition of CDKs and the inhibition of Rb phosphorylation. The regulation of Rb by mTOR occurs through the phosphorylation of Rb by the mTORC1 complex following activation of the PI3K/Akt signaling pathway by growth factors and nutrients, such as insulin and amino acids. The phosphorylation of Rb results in the release of the transcription factor E2F from Rb–E2F complexes. The released E2F transcription factor can then activate target genes involved in cell growth and division [25].

Energy availability can control CDKs through the regulation of metabolic pathways that are involved in the production and utilization of cellular energy [26]. The AMPK is activated by low ATP levels and acts to restore energy balance in the cell. In addition to AMPK, the regulation of CDKs by energy availability can also occur through the regulation of metabolic pathways that produce ATP. For example, the glycolytic pathway, which is responsible for the production of ATP from glucose, can regulate CDK activity. Increased glucose uptake and glycolysis have been shown to promote CDK activity and cell cycle progression, while inhibition of glycolysis has been shown to inhibit CDK activity and cell cycle progression [26].

Oxidative phosphorylation (OXPHOS) is the main process by which cells generate ATP, and plays a critical role in regulating the DNA damage response and cell cycle progression. The regulation of these processes by OXPHOS occurs through several mechanisms. DNA damage triggers the activation of cellular sensors, such as ATR and Chk1, which initiate the DNA damage response. These sensors lead to the activation of a cascade of signaling pathways that coordinate the repair of DNA damage and the arrest of the cell cycle [17]. The DNA damage response requires a significant amount of ATP, and OXPHOS provides the necessary energy for this process [27]. This energy is necessary for the activation of the mTOR pathway and the progression of the cell cycle. OXPHOS is also involved in the regulation of cellular metabolism through the control of oxidative stress. DNA damage can cause oxidative stress, which can further disrupt cellular metabolism and lead to additional DNA damage. OXPHOS provides the energy necessary for the maintenance of cellular metabolism and the reduction of oxidative stress, which helps to prevent further DNA damage.

5 Dysregulation of Metabolism and Cell Cycle

Dysregulation of metabolism and cell cycle can result in a variety of cellular problems and diseases, including cancer, neurodegenerative diseases, aging, or infertility [4, 28–32]. Cancer cells exhibit abnormal metabolism and cell cycle regulation, often involving increased glucose metabolism and increased cell division. Mutations in genes involved in metabolism and cell cycle regulation can lead to abnormal metabolism and cell cycle regulation, contributing to the development of cancer and other diseases. Environmental factors, such as oxidative stress, can disrupt metabolism and cell cycle regulation, leading to cellular problems and disease. Chronic diseases, such as diabetes, can result from dysregulation of metabolic pathways, which can also affect cell cycle regulation and contribute to cellular problems. Indeed, dysregulation of metabolism and cell cycle can have serious consequences for cellular function and health, contributing to the development of cancer and other diseases. Therefore, maintaining proper regulation of metabolism and cell cycle is critical for cellular health and the prevention of disease.

These pathologies highlight the importance of maintaining proper regulation of metabolism and cell cycle in order to prevent cellular problems and disease. They also highlight the need of strategies to target metabolism and cell cycle in order to release effective therapies. Inhibiting key metabolic enzymes, such as hexokinase and lactate dehydrogenase, or blocking metabolic pathways, such as the Warburg effect, can disrupt cancer cell metabolism and growth [33–35]. Inhibiting cell cycle regulators, such as cyclin-dependent kinases (CDKs) and cyclins, can block cell cycle progression and replication of DNA in cancer cells. Drugs that modify cellular metabolism, such as metformin, can target cancer cell metabolism and influence cell cycle regulation [36]. Radiotherapy and chemotherapy target rapidly dividing cancer cells by damaging DNA and inducing cell death [37]. Even immune therapy, such as checkpoint inhibitors, can target immune checkpoints and enhance the immune response against cancer cells. These therapeutic strategies demonstrate the potential for targeting metabolism and cell cycle in the treatment of cancer and other diseases. Further research into the mechanisms underlying these processes will be critical for developing new and more effective treatments.

There are several current research trends and developments in the fields of metabolism and cell cycle. Research is focused on understanding the mechanisms underlying metabolic reprogramming in cancer cells and developing drugs that target metabolic pathways in cancer cells. Currently, there are no drugs that are specifically designed to reprogram cells. Reprogramming of cells typically involves introducing specific genes into cells using techniques such as viral transduction or CRISPR-mediated genome editing. Some drugs may be used to facilitate this process, such as small-molecule inhibitors that can modulate signaling pathways and support the transition of cells into a desired state [38]. However, these drugs are not directly reprogramming the cells themselves. It is worth noting that the field of cellular reprogramming is rapidly advancing, and new methods and technologies are being developed all the time. In the future, it is possible that drugs specifically designed to reprogram cells may become available. Research is also focused on better understanding the mechanisms of cell cycle regulation and identifying new cell cycle regulators which might be of importance in the interplay between metabolism and cell cycle regulation.

Despite the progress made in the fields of metabolism and cell cycle, there are still limitations and gaps in our understanding of these processes. Metabolism and cell cycle are complex processes that involve multiple pathways, regulators, and interactions. Our understanding of these processes is still limited, and further research is needed to fully comprehend their complexity. Metabolism and cell cycle regulation can vary significantly between individuals and across different diseases, making it challenging to develop generalized therapeutic strategies. Metabolism and cell cycle regulation are closely intertwined with other biological processes, such as the immune system and epigenetic regulation, and the interplay between these processes is not well-understood. While treatments targeting metabolism and cell cycle have been developed, there is still a need for more effective and targeted therapies. These limitations and gaps in our understanding are critical to define the future research directions in the fields of metabolism and cell cycle, including systems biology, metabolic reprogramming, or even personalized medicine. Cross talk between metabolic pathways and cell cycle regulators is essential for proper cell division and replication of DNA. Disruptions in this cross talk can lead to alterations in cell cycle regulation and metabolism, contributing to the development of cancer and other diseases.

6 Metabolomics Approaches of the Cell Cycle

Metabolomics is a field of study that aims to characterize the complete set of small-molecule metabolites in a cell or tissue. Studying the cell cycle with metabolomics involves measuring changes in the metabolic profile of cells as they progress through the cell cycle. This can provide valuable insights into the metabolic changes that occur during cell division and can help to identify potential regulatory pathways and metabolic switches that control the cell cycle.

There are several different approaches that can be used to study the cell cycle with metabolomics, including the four main strategies described hereafter (Fig. 1).

6.1 Time-Course Metabolomics

In this approach, cells are synchronized in the cell cycle and then harvested at regular intervals to capture the metabolic changes that occur over time. Time-course metabolomics is a research approach that involves measuring changes in the metabolic profile of cells, tissues, or organisms over time. The goal of time-course metabolomics is to capture the dynamic changes in the metabolic state of a system as a function of time and to use this information to understand the underlying biological processes [39]. In time-course metabolomics experiments, samples are taken at regular intervals over a period of time and analyzed for their metabolic content. The resulting time-series data can then be used to study the temporal evolution of the metabolic state and to identify changes in the levels of specific metabolites that are associated with specific biological events or processes. For example, time-course metabolomics can be used to study the metabolic changes that occur in cells as they progress through the cell cycle in response to environmental stimuli or genetic modifications [40–42]. The data obtained from time-course metabolomics studies can be used to identify metabolic pathways that are altered in response to these changes and to better understand the relationships between cell cycle and metabolic changes. Overall, time-course metabolomics is a powerful tool for investigating complex biological systems as cell cycle and understanding the metabolic changes that occur over time in response to various environmental stimuli or genetic modifications.

6.2 Stable Isotope Labeling

In this approach, cells are labeled with stable isotopes, such as 13C, and the resulting changes in the metabolic profile are analyzed as the cells progress through the cell cycle. Stable isotope labeling is a technique used in metabolomics to track the flux of specific metabolites within a biological system [43]. The technique involves introducing isotopically labeled precursors into the system and then measuring the incorporation of the labeled precursors into the metabolic network. In stable isotope labeling, specific metabolites or precursors are labeled with stable isotopes, such as 13C, 15N, or 2H, which do not decay over time and can be easily distinguished from their non-labeled counterparts using mass spectrometry or other analytical techniques [44]. By incorporating the labeled precursors into the metabolic network, changes in the metabolic profile can be tracked in real time, allowing for the study of the dynamic metabolic changes that occur in response to various inputs and stimuli. Stable isotope labeling is widely used in

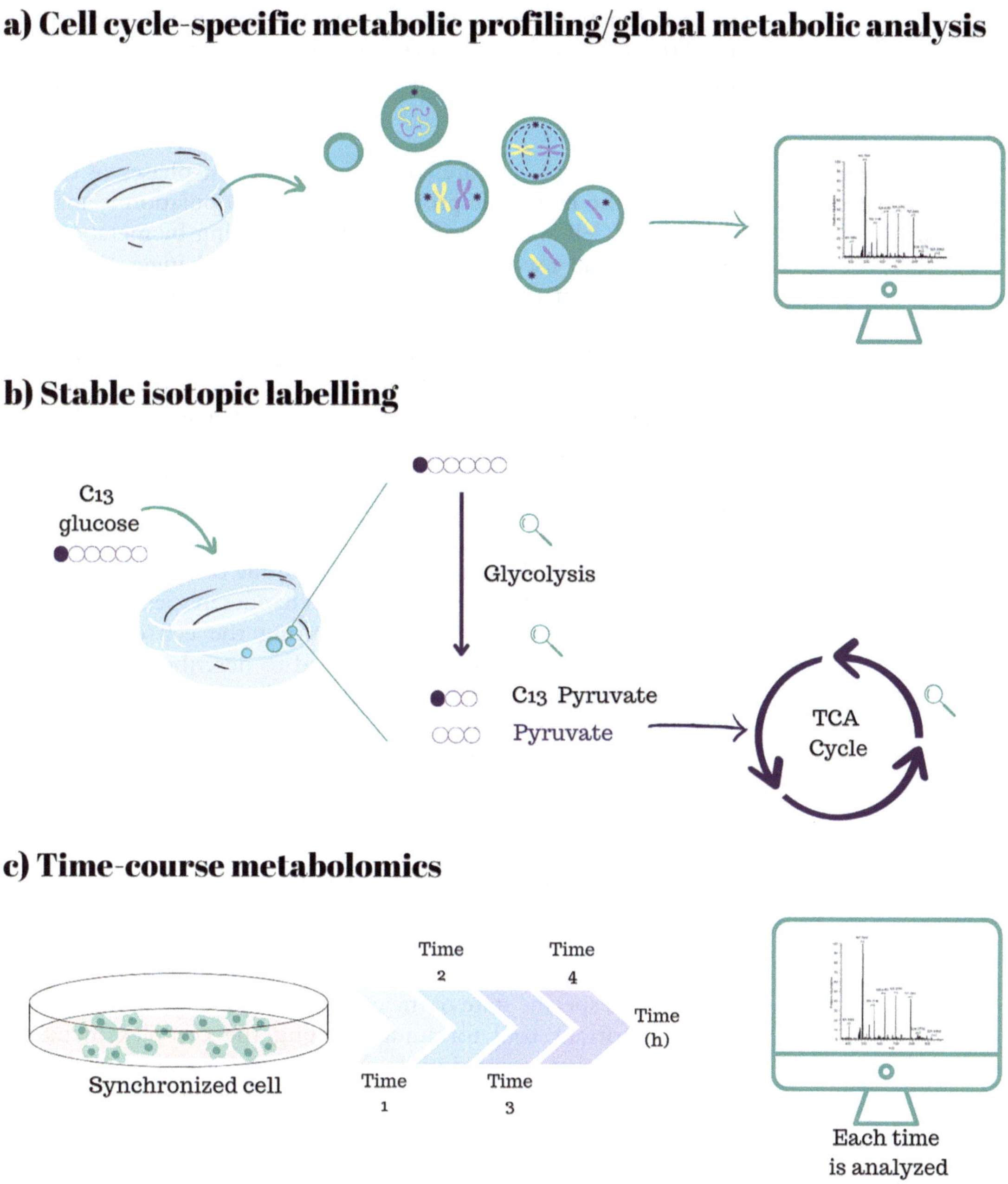

Fig. 1 Metabolic approaches of the cell cycle. (**a**) Cell cycle metabolic profiling allows analysis of cellular metabolite involved in cell cycle phases. Metabolites are analyzed using mass spectrometry (MS) or nuclear magnetic resonance (NMR). Global metabolic analysis gives an overall picture of metabolic content using the same techniques. (**b**) In a stable isotopic labelling analysis, molecules are labeled with stable isotopes (like 13C, 15N) allowing metabolic flux to be tracked. (**c**) Time course metabolomics. Cells are synchronized and analyzed by MS or NMR

metabolic research as a tool for studying the dynamics of metabolic pathways, for characterizing the metabolic fluxes within a cell, and for understanding the relationships between metabolism and cellular function. For example, stable isotope labeling can also be used to study the dynamic metabolic changes that occur in cell cycle progression in response to environmental stress or genetic modifications.

6.3 Cell Cycle-Specific Metabolic Profiling

In this approach, cells are sorted into different phases of the cell cycle and the metabolic profile of each cell population is analyzed separately. Cell cycle-specific metabolic profiling by metabolomics is a technique used to study the metabolic changes that occur during different phases of the cell cycle [45]. Cell cycle-specific metabolic profiling is usually performed to quantify changes in the levels of hundreds to thousands of metabolites. By comparing the metabolic profiles of cells in different phases of the cell cycle, researchers can identify metabolic pathways that are upregulated or downregulated. This information can be used to gain a better understanding of the metabolic regulation of the cell cycle and how it is involved in the control of cell growth and division.

6.4 Global Metabolic Analysis

In this approach, the entire metabolic profile of cells is analyzed at once, allowing for the identification of changes in specific metabolic pathways or networks that occur during the cell cycle. Global metabolic analysis is a broad term used to describe the systematic and comprehensive study of the metabolic composition of a cell [46]. This analysis aims to identify and quantify all the small-molecule metabolites present in a sample, including sugars, amino acids, nucleotides, fatty acids, and other metabolites. Global metabolic analysis is typically performed using high-throughput analytical technologies such as mass spectrometry (MS) or nuclear magnetic resonance (NMR) spectroscopy [47]. These techniques enable the simultaneous measurement of hundreds to thousands of metabolites in a single sample. The data obtained from these analyses can provide a comprehensive overview of the metabolic state of a cell at a specific state of the cell cycle. Global metabolic analysis is an important tool for understanding the complex metabolic interactions that occur during cell cycle and for discovering new metabolic pathways and biomarkers for various diseases. The information obtained from these analyses can be used to develop new strategies for diagnosing and treating diseases, as well as to optimize metabolic processes in biotechnology and industrial applications. Regardless of the approach used, the data obtained from cell cycle-focused metabolomics studies is typically analyzed using a combination of statistical and bioinformatics tools, such as multivariate analysis, pathway analysis, and network analysis, to identify meaningful patterns and connections in the data.

7 Conclusion

We have discussed here findings and technics showing the interconnection between metabolism and cell cycle. This dual regulation is a key cellular process for the successful viability and proliferation of cells. Understanding of the reciprocal control of metabolism and cell cycle should help to prevent disease, such as cancer.

References

1. Kalucka J, Missiaen R, Georgiadou M, Schoors S, Lange C, De Bock K, Dewerchin M, Carmeliet P (2015) Metabolic control of the cell cycle. Cell Cycle 14:3379–3388
2. Dalle Pezze P, Ruf S, Sonntag AG, Langelaar-Makkinje M, Hall P, Heberle AM, Razquin Navas P, Van Eunen K, Tölle RC, Schwarz JJ, Wiese H, Warscheid B, Deitersen J, Stork B, Fäßler E, Schäuble S, Hahn U, Horvatovich P, Shanley DP, Thedieck K (2016) A systems study reveals concurrent activation of AMPK and mTOR by amino acids. Nat Commun 7: 13254
3. Pan C, Li B, Simon MC (2021) Moonlighting functions of metabolic enzymes and metabolites in cancer. Mol Cell 81:3760–3774
4. Matthews HK, Bertoli C, de Bruin RAM (2022) Cell cycle control in cancer. Nat Rev Mol Cell Biol 23:74–88
5. Blagosklonny MV, Pardee AB (2013) The restriction point of the cell cycle. Landes Bioscience. https://www.ncbi.nlm.nih.gov/books/NBK6318/
6. Pardee AB (1974) A restriction point for control of normal animal cell proliferation. Proc Natl Acad Sci USA 71:1286–1290
7. Roche S, Koegl M, Courtneidge SA (1994) The phosphatidylinositol 3-kinase ar is required for DNA synthesis induced by some, but not all, growth factors. Proc Natl Acad Sci U S A 91(19):9185–9189
8. Auger KR, Serunian LA, Soltoff SP, Libby P, Cantley LC (1989) PDGF-dependent tyrosine phosphorylation stimulates production of novel polyphosphoinositides in intact cells. Cell 57:167–175
9. Gwinn DM, Shackelford DB, Egan DF, Mihaylova MM, Mery A, Vasquez DS, Turk BE, Shaw RJ (2008) AMPK phosphorylation of raptor mediates a metabolic checkpoint. Mol Cell 30:214–226
10. Lee J, Kumagai A, Dunphy WG, Hunt T (2001) Positive regulation of Weel by Chkl and 14-3-3 proteins. MBoC 12:551–563
11. Afonso O, Castellani CM, Cheeseman LP, Ferreira JG, Orr B, Ferreira LT, Chambers JJ, Morais-de-Sá E, Maresca TJ, Maiato H (2019) Spatiotemporal control of mitotic exit during anaphase by an aurora B-Cdkl cross-talk. elife 8:e47646
12. Lunt SY, Van der Heiden MG (2011) Aerobic glycolysis: meeting the metabolic requirements of cell proliferation. Annu Rev Cell Dev Biol 27:441–464
13. Chang F, Lee JT, Navolanic PM, Steelman LS, Shelton JG, Blalock WL, Franklin RA, McCubrey JA (2003) Involvement of PI3K/Akt pathway in cell cycle progression, apoptosis, and neoplastic transformation: a target for cancer chemotherapy. Leukemia 17:590–603
14. Fajas L (2013) Re-thinking cell cycle regulators: the cross-talk with metabolism. Front Oncol 3. https://doi.org/10.3389/fonc.2013.00004
15. Jones RG, Plas DR, Kubek S, Buzzai M, Mu J, Xu Y, Birnbaum MJ, Thompson CB (2005) AMP-activated protein kinase induces a p53-dependent metabolic checkpoint. Mol Cell 18:283–293
16. Shaw RJ (2006) Glucose metabolism and cancer. Curr Opin Cell Biol 18:598–608
17. Tajan M, Hennequart M, Cheung EC, Zani F, Hock AK, Legrave N, Maddocks ODK, Ridgway RA, Athineos D, Suárez-Bonnet A, Ludwig RL, Novellasdemunt L, Angelis N, Li VSW, Vlachogiannis G, Valeri N, Mainolfi N, Suri V, Friedman A, Manfredi M, Blyth K, Sansom OJ, Vousden KH (2021) Serine synthesis pathway inhibition cooperates with dietary serine and glycine limitation for cancer therapy. Nat Commun 12:366
18. Brown MS, Goldstein JL (1997) The SREBP pathway: regulation of cholesterol metabolism by proteolysis of a membrane-bound transcription factor. Cell 89:331–340
19. Kumari R, Jat P (2021) Mechanisms of cellular senescence: cell cycle arrest and senescence associated secretory phenotype. Front Cell Dev Biol 9:645593

20. Lopez-Mejia IC, Fajas L (2015) Cell cycle regulation of mitochondrial function. Curr Opin Cell Biol 33:19–25
21. Epstein RJ (2013) The unpluggable in pursuit of the undruggable: tackling the dark matter of the cancer therapeutics universe. Front Oncol 3. https://doi.org/10.3389/fonc.2013.00304
22. Gonzalez S, Rallis C (2017) The TOR signaling pathway in spatial and temporal control of cell size and growth. Front Cell Dev Biol 5:61
23. Mihaylova MM, Shaw RJ (2011) The AMPK signalling pathway coordinates cell growth, autophagy and metabolism. Nat Cell Biol 13:1016–1023
24. Liberti MV, Locasale JW (2016) The Warburg effect: how does it benefit cancer cells? Trends Biochem Sci 41:211–218
25. Bracken AP, Ciro M, Cocito A, Helin K (2004) E2F target genes: unraveling the biology. Trends Biochem Sci 29:409–417
26. Solaki M, Ewald JC (2018) Fueling the cycle: CDKs in carbon and energy metabolism. Front Cell Dev Biol 6:93
27. Wilson DF (2017) Oxidative phosphorylation: regulation and role in cellular and tissue metabolism. J Physiol 595:7023–7038
28. Martínez-Reyes I, Chandel NS (2021) Cancer metabolism: looking forward. Nat Rev Cancer 21:669–680
29. Procaccini C, Santopaolo M, Faicchia D, Colamatteo A, Formisano L, De Candia P, Galgani M, De Rosa V, Matarese G (2016) Role of metabolism in neurodegenerative disorders. Metabolism 65:1376–1390
30. Palmer AK, Jensen MD (2022) Metabolic changes in aging humans: current evidence and therapeutic strategies. J Clin Investig 132:e158451
31. Torre SD, Benedusi V, Fontana R, Maggi A (2014) Energy metabolism and fertility – a balance preserved for female health. Nat Rev Endocrinol 10:13–23
32. Joseph C, Mangani AS, Gupta V, Chitranshi N, Shen T, Dheer Y, Kb D, Mirzaei M, You Y, Graham SL, Gupta V (2020) Cell cycle deficits in neurodegenerative disorders: uncovering molecular mechanisms to drive innovative therapeutic development. Aging Dis 11:946
33. Zheng M, Wu C, Yang K, Yang Y, Liu Y, Gao S, Wang Q, Li C, Chen L, Li H (2021) Novel selective hexokinase 2 inhibitor Benitrobenrazide blocks cancer cells growth by targeting glycolysis. Pharmacol Res 164:105367
34. Miholjcic TBS, Halse H, Bonvalet M, Bigorgne A, Rouanne M, Dercle L, Shankar V, Marabelle A (2023) Rationale for LDH-targeted cancer immunotherapy. Eur J Cancer 181:166–178
35. Kozal K, Jóźwiak P, Krześlak A (2021) Contemporary perspectives on the Warburg effect inhibition in cancer therapy. Cancer Control 28:107327482110412
36. Andrzejewski S, Siegel PM, St-Pierre J (2018) Metabolic profiles associated with metformin efficacy in cancer. Front Endocrinol 9:372
37. Kery M, Papandreou I (2020) Emerging strategies to target cancer metabolism and improve radiation therapy outcomes. BJR 93:20200067
38. Pranzini E, Pardella E, Paoli P, Fendt S-M, Taddei ML (2021) Metabolic reprogramming in anticancer drug resistance: a focus on amino acids. Trends Cancer 7:682–699
39. Jendoubi T, Ebbels TMD (2020) Integrative analysis of time course metabolic data and biomarker discovery. BMC Bioinf 21:11
40. Yatsuoka W, Ueno T, Miyano K, Enomoto A, Ota S, Sugimoto M, Uezono Y (2021) Time-course of salivary metabolomic profiles during radiation therapy for head and neck cancer. JCM 10:2631
41. Buchweitz LF, Yurkovich JT, Blessing C, Kohler V, Schwarzkopf F, King ZA, Yang L, Jóhannsson F, Sigurjónsson ÓE, Rolfsson Ó, Heinrich J, Dräger A (2020) Visualizing metabolic network dynamics through time-series metabolomic data. BMC Bioinf 21:130
42. Müller N, Warwick T, Noack K, Malacarne PF, Cooper AJL, Weissmann N, Schröder K, Brandes RP, Rezende F (2022) Reactive oxygen species differentially modulate the metabolic and transcriptomic response of endothelial cells. Antioxidants 11:434
43. Chokkathukalam A, Kim D-H, Barrett MP, Breitling R, Creek DJ (2014) Stable isotope-labeling studies in metabolomics: new insights into structure and dynamics of metabolic networks. Bioanalysis 6:511–524
44. Jang C, Chen L, Rabinowitz JD (2018) Metabolomics and isotope tracing. Cell 173:822–837
45. Roci I, Watrous JD, Lagerborg KA, Jain M, Nilsson R (2020) Mapping metabolic oscillations during cell cycle progression. Cell Cycle 19:2676–2684
46. Bunning BJ, Contrepois K, Lee-McMullen B, Dhondalay GKR, Zhang W, Tupa D, Raeber O, Desai M, Nadeau KC, Snyder MP, Andorf S (2020) Global metabolic profiling to model biological processes of aging in twins. Aging Cell 19. https://doi.org/10.1111/acel.13073
47. Doerr A (2017) Global metabolomics. Nat Methods 14:32–32

Chapter 11

Preparation of *Xenopus borealis* and *Xenopus tropicalis* Egg Extracts for Comparative Cell Biology and Evolutionary Studies

Maiko Kitaoka, Gabriel Guilloux, Rebecca Heald, and Romain Gibeaux

Abstract

Cytoplasmic extracts prepared from eggs of the African clawed frog *Xenopus laevis* are extensively used to study various cellular events including the cell cycle, cytoskeleton dynamics, and cytoplasm organization, as well as the biology of membranous organelles and phase-separated non-membrane-bound structures. Recent development of extracts from eggs of other *Xenopus* allows interspecies comparisons that provide new insights into morphological and biological size variations and underlying mechanisms across evolution. Here, we describe methods to prepare cytoplasmic extracts from eggs of the allotetraploid Marsabit clawed frog, *Xenopus borealis*, and the diploid Western clawed frog, *Xenopus tropicalis*. We detail mixing and "hybrid" experiments that take advantage of the physiological but highly accessible nature of extracts to reveal the evolutionary relationships across species. These new developments create a robust and versatile toolbox to elucidate molecular, cell biological, and evolutionary questions in essential cellular processes.

Key words *Xenopus*, *Xenopus borealis*, *Xenopus tropicalis*, Egg extracts, Mixed extracts, Hybrids

1 Introduction

Cell-free extracts from *Xenopus* eggs were first prepared by Lohka and Maller in 1985, who collected and crushed the eggs to fractionate their components and harvest undiluted cytoplasm [1, 2]. The authors took advantage of the well-established allotetraploid African clawed frog, *Xenopus laevis*, whose eggs are arrested in metaphase II of meiosis and stockpiled with maternal RNAs and proteins to support rapid early embryonic cell cycles and embryogenesis. Thus, once isolated, cytoplasmic extracts contain all the necessary biomolecules and machineries necessary to recapitulate the cell cycle in a test tube, including bipolar spindle assembly when a DNA source, such as sperm nuclei or

Maiko Kitaoka and Gabriel Guilloux have equally contributed to this chapter.

Anna Castro and Benjamin Lacroix (eds.), *Cell Cycle Control: Methods and Protocols*, Methods in Molecular Biology, vol. 2740, https://doi.org/10.1007/978-1-0716-3557-5_11,

DNA-coated beads, is added [1, 3]. Subsequent optimization of extract preparation by Andrew Murray and colleagues allowed induction of interphase and transit through multiple cell cycles in vitro, demonstrating the power and tractability of egg extracts to investigate central cell biological questions [2, 4, 5]. Notably, metaphase spindles, mitotic chromosomes, and interphase nuclei can all be formed in vitro, and a number of fundamental cellular processes, including DNA replication, chromosome condensation, and kinetochore formation, are fully supported in extracts [6, 7]. The extract system undergoes a simple cell cycle, mimicking the timing and functions of the early embryonic cleavages, lacking gap phases and oscillating rapidly between S and M phases. Moreover, the extract's accessible cell-free nature allows for unparalleled control over the cell cycle, as it can be arrested in either metaphase or interphase, and is amenable to biochemical and pharmacological manipulations, such as protein immunodepletions and small-molecule inhibitor treatments. Thus far, *X. laevis* egg extracts have been key to our understanding of the fundamental principles, including DNA replication [8], cyclin degradation [9], as well as microtubule dynamics and spindle assembly [3, 10–12].

While *X. laevis* egg extracts remain a system of choice to study a variety of cellular events in vitro, comparisons across multiple *Xenopus* species have provided a unique evolutionary window to examine conservation and divergence of these mechanisms. In vitro egg extract systems have now been developed and applied to a variety of species with different egg and genome sizes. Beginning in 2007, Brown and colleagues developed the egg extracts of the diploid Western clawed frog, *Xenopus tropicalis* [13]. Remarkably, spindles formed in egg extracts of this smaller frog were shorter in length compared to *X. laevis* due to cytoplasmic composition differences between species. Indeed, mixing *X. laevis* and *X. tropicalis* extracts produced spindles of intermediate size and morphological characteristics in a dose-dependent manner, regardless of the source of chromosomes, demonstrating that cytoplasmic factors are sufficient to scale subcellular structures and build the associated morphological architecture [13, 14]. Comparisons between these two species' egg extracts led to the discovery of numerous mechanisms of nuclear and spindle-size scaling [14–17]. More recently, we developed egg extracts from a third *Xenopus* species, the allotetraploid Marsabit clawed frog, *X. borealis* [18]. Intermediate in size, but more recently diverged from *X. laevis* than *X. tropicalis*, *X. borealis* thus occupies a unique place in evolutionary history that allows us to address the conservation of spindle morphology, assembly, and scaling mechanisms as well as chromosome biology across evolution [18–20].

These three *Xenopus* species egg extracts make it possible to question how various central cellular processes have evolved and adapted. In this protocol, we focus on the preparation of egg extracts from eggs of *X. borealis* and *X. tropicalis*, as *X. laevis* protocols have been extensively detailed elsewhere [21, 22]. In addition, we provide specific advice regarding frog care for *Xenopus* species. We also highlight two experimental schemes that have been key to our understanding of scaling and evolutionary divergence in cell cycle events: mixing cytoplasmic extracts and "hybrid" reactions.

2 Materials

2.1 Xenopus *Frogs*

1. Marsabit clawed frog, *Xenopus borealis*, females.
2. Western clawed frog, *Xenopus tropicalis* females.

X. borealis and *X. tropicalis* frogs can be obtained from the National *Xenopus* Resource (NXR), Woods Hole, USA. In Europe, *X. tropicalis* frogs can be obtained from the European *Xenopus* Resource Centre (EXRC), Portsmouth, UK. Females are used for egg collection to make extracts, while males are used to prepare demembranated sperm chromatin [23].

2.2 *Hormones*

1. Pregnant mare serum gonadotropin (PMSG): 200 IU/mL PMSG in sterile MilliQ H_2O. Store at 4 °C.
2. Human chorionic gonadotropin hormone (HCG): 1000 IU/mL HCG in sterile MilliQ H_2O (e.g., HCG from Sigma lyophilized with buffer) or provided diluent (e.g., Chorulon HCG from MSD). Store at 4 °C for the time recommended by the provider.

2.3 *Extract Preparation Buffers and Chemicals*

1. 20× Marc's Modified Ringer solution (MMR): 2 M NaCl, 40 mM KCl, 40 mM $CaCl_2$, 20 mM $MgCl_2$, 100 mM HEPES (free acid), 2 mM EDTA, NaOH pH 7.8. Prepare 1 L in MilliQ H_2O and filter-sterilize.
2. 20× XB salts: 2 M KCl, 20 mM $MgCl_2$, 2 mM $CaCl_2$. Prepare 1 L in MilliQ H_2O, and filter-sterilize. Store at 4 °C.
3. Dejellying solution for *X. borealis* eggs: 1× XB salts and 3% L-cysteine in MilliQ H_2O. Adjust to pH 7.8 with 10 N NaOH. Prepare 500 mL fresh just before use.
4. Dejellying solution for *X. tropicalis* eggs: 3% L-cysteine in MilliQ H_2O. Adjust to pH 7.8 with 10 N NaOH. Prepare 500 mL fresh just before use.
5. CSF-XB: 1× XB salts, 50 mM sucrose, 10 mM K-HEPES pH 7.8, 2 mM $MgCl_2$, 10 mM K-EGTA pH 7.8 in MilliQ H_2O. Adjust to pH 7.8 with 10 N KOH. Prepare 1 L fresh just before use.

6. LPC protease inhibitor cocktail: 10 mg/mL each of leupeptin, pepstatin, and chymostatin in DMSO. Aliquot and store at − 20 °C.
7. Cytochalasin B or D: 10 mg/mL in DMSO. Aliquot and store at −20 °C.
8. 50× energy mix: 190 mM creatine phosphate, 25 mM adenosine triphosphate, 25 mM MgC12, 2.5 mM K-EGTA pH 7.7 in MilliQ H_2O. Aliquot and store at −20 °C.
9. CSF-XB+: CSF-XB supplemented with 10 μg/mL each of LPC. Prepare 50 mL fresh just before use by adding 50 μL 10 mg/mL LPC stock to 50 mL CSF-XB.

2.4 Extract Preparation Equipment

1. Square plastic containers for individual storage of *X. borealis* frogs (about 22 × 22 × 12 cm, 4–6 L capacity) with tight lids allowing air exchange.
2. Square glass containers for individual storage of *X. tropicalis* frogs (about 20 × 20 × 25 cm, 8–10 L capacity) with lids allowing air exchange.
3. Room or non-airtight incubator set to 16 °C for *X. borealis.*
4. Transfer pipettes, including wide bore and narrow stem transfer pipettes.
5. Thin-wall centrifuge tubes, 5 mL (SW-55 ultraclear thin-wall; Beckman).
6. Adaptor tubes, 13 mL (Sarstedt).
7. Clinical centrifuge set to 16 °C equipped with a swinging-bucket rotor.
8. High-speed floor centrifuge set to 16 °C equipped with a swinging-bucket rotor (e.g., Sorvall HB-6 or Beckman JS13.1) and corresponding rubber adaptors.

2.5 Extract Reactions

1. Demembranated sperm nuclei, prepared as detailed elsewhere [23].
2. 10× calcium solution, for interphase induction: 4 mM $CaCl_2$, 100 mM KCl, 1 mM $MgCl_2$.
3. Spindle fix: 1× MMR, 48% (w/w) glycerol, 11% (w/v) formaldehyde, 5 μg/mL Hoechst DNA dye.
4. Glass slides and square coverslips (18 × 18 or 22 × 22 mM) for fluorescence microscopy.

3 Methods

All *Xenopus* egg extracts described so far follow the same basic protocol flow. However, key differences exist in hormone doses, buffer recipes, and species-specific nuances that have been made for maximal frog health and experimental output [6, 13, 18, 21, 22, 24]. Please refer to Table 1 for a full comparison between *X. laevis*, *X. tropicalis*, and *X. borealis* egg extract protocols.

3.1 Frog Care Considerations

Xenopus frogs can be prone to stress during ovulation for egg extracts, particularly *X. tropicalis*. Here, we detail several considerations and signs to look out for during any frog handling and ovulation. Information about frog housing has been detailed elsewhere [25–27].

1. All primed frogs should be fasted after hormone injections to reduce the possibility of fecal matter contaminating the eggs.
2. If *X. tropicalis* frogs are raised throughout their life in buffered 0.1× MMR, keeping them in this solution and at husbandry temperature throughout the experiment, including ovulation, does not impair egg and extract quality while reducing animal stress.
3. *X. tropicalis* frogs are sensitive to the container plastic quality. While some plastic tubs are used without any problem by some labs, we opted for glass containers here, for extra safety.
4. If frogs lay eggs or their water/buffer is cloudy from lysed eggs after the priming (first) hormone dose, check to see if the frog is still actively laying before deciding whether or not to proceed with a second boosting dose. Reduce the priming dose if this happens consistently.
5. Frogs can suffer from hyper-ovulation after hormone injections, leading to death within a few days post-ovulation. This is particularly true for *X. tropicalis*, and less of a concern with *X. laevis* and *X. borealis*. Occasionally, this is indicative of a "sensitive" batch of frogs, but if the problem is persistent, the boosting dose should be adjusted.
6. Frogs will continue to lay for 12–24 h after use and may suffer from post-experiment problems such as internal ovulation. Be sure to store them isolated and under observation overnight to a few days before returning them to their permanent housing systems. This prevents eggs and unhealthy or dead animals from clogging and contaminating the housing system water and flow.
7. It is recommended that female frogs are rested for specific intervals between ovulations to maintain the highest egg quality over time (*see* Table 1).
8. Depending on whether frog colonies are bred in-house or shipped from a supplier, care must be taken to acclimate frogs and adapt them slowly to their housing environment. For instance, *X. tropicalis* and *X. borealis* can be more sensitive

Table 1
Comparison between 3 *Xenopus* species' egg extract protocols

	X. laevis	*X. tropicalis*	*X. borealis*
Priming	100 IU PMSG, at least 48 h and up to 2 weeks before boost; or 15 IU hCG, 72 to 30 h before boost	10 IU hCG, 20 h before boost	60 IU PMSG, at least 48 h and up to 1 week before boost
Boosting	500 IU hCG, or up to 750 IU if primed with hCG, the evening before the experiment, 16–18 h	200–250 IU hCG, the morning of the experiment, ~3–4 h before	300 IU hCG, the evening before the experiment, 16–18 h
Egg laying conditions	1× MMR, 16 °C Overnight	ddH2O or buffered 0.1× MMR, 23 °C Frogs will start laying ~3 h after boosting. Can be helped/sped along by gently holding females to promote kicking 3–4 times at 15-min intervals	0.5× MMR, 16 °C Overnight
Egg collection	Pour off excess 1× MMR, until all eggs can be transferred to a clean beaker	Eggs will stick to the tub (glass or plastic). Use a gloved hand to gently sweep and detach the eggs. Collect them in one corner of the tub while carefully pouring excess water off. Collect them into a clean beaker	Pour off excess 0.5× MMR, until all eggs can be transferred to a clean beaker
CSF-XB	100 mM KCl 0.1 mM $CaCl_2$ *2 mM $MgCl_2$* 50 mM sucrose *5 mM K-EGTA* 10 mM K-HEPES pH 7.8	100 mM KCl 0.1 mM $CaCl_2$ *3 mM $MgCl_2$* 50 mM sucrose *10 mM K-EGTA* 10 mM K-HEPES pH 7.8	100 mM KCl 0.1 mM $CaCl_2$ *3 mM $MgCl_2$* 50 mM sucrose *10 mM K-EGTA* 10 mM K-HEPES pH 7.8
Dejelly	2% L-cysteine, 1× XB salts, in ddH_2O, pH 7.7–7.8 with NaOH	3% L-cysteine, in ddH_2O, pH 7.7–7.8 with NaOH	3% L-cysteine, 1× XB salts, in ddH_2O, pH 7.7–7.8 with NaOH
Approx. extract yield per frog	1 mL	200 μL	600 μL
Egg extract storage	4 °C, on ice	16 °C	16 °C
Extract reaction timing	CSF spindles in 60 min Cycled nuclei in 45 min	CSF spindles in 30 min Cycled nuclei in 30 min	CSF spindles in 30–45 min Cycled nuclei in 45 min

(continued)

Table 1 (continued)

	X. laevis	*X. tropicalis*	*X. borealis*
Resting interval between ovulations	6 months	3 months	4 months

Differences between *X. laevis*, *X. tropicalis*, and *X. borealis* egg extract protocols are often subtle but important for frog health and egg/extract quality

than *X. laevis* to changes in conductivity and salt content in water. In addition, *X. tropicalis* frogs are particularly sensitive to *Batrachochytrium dendrobatidis* (*Bd*) chytrid fungal infections (Kitaoka and Heald, unpublished observations) and thus should be tested prior to introducing them into permanent housing systems.

9. Each frog species should have dedicated spaces in the lab (i.e., benches) and housing rooms. Care should be taken to ensure that *X. tropicalis* are separated from other species, but *X. laevis* and *X. borealis* can be housed together within the same recirculating system, though separate tanks should be used. Cross-contamination between species can be particularly dangerous as *X. laevis* and *X. borealis* are more tolerant of *Bd* chytrid fungus, while *X. tropicalis* are highly susceptible.

3.2 Preparation of X. borealis Egg Extracts

1. Prime *X. borealis* females (usually 3–4) by injecting 0.3 mL (60 IU) of PMSG subcutaneously into the dorsal lymph sac using a 1 mL syringe and a 30-gauge needle, at least 48 h before boosting (*see* **Note 1**).
2. Boost *X. borealis* females on the evening preceding the experiment by injecting 0.3 mL (300 IU) of HCG to induce ovulation, 16–18 h before eggs are needed. Store each frog individually in 2 L of freshly prepared 0.5× MMR in their plastic container overnight at 16 °C (*see* **Note 2**).
3. The next morning, make sure that both the clinical and the high-speed centrifuges are set to 16 °C.
4. Just before collecting the eggs, prepare the dejellying solution and the CSF-XB. Set aside a 50 mL aliquot of CSF-XB (to later prepare the CSF-XB+), and, if possible, store all solutions at 16 °C until use. Allow frozen aliquots of Cytochalasin, LPC cocktail, and energy mix to thaw at room temperature.
5. Remove the females from their containers and analyze the egg quality to select batches of eggs to be used. Avoid batches that contain a significant amount of misshaped, lysed, or stringy eggs, and remove these from the chosen batches with a wide-

bore transfer pipette (*see* **Note 3**). Retrieve the selected batches of eggs by gently pouring off the majority of 0.5× MMR from the tub and collecting eggs into a 500 mL glass beaker.

6. For the dejellying and washing steps **(steps 7–8)**, take extra care to avoid exposing the eggs to air and swirling them excessively. Remove misshaped and lysing eggs throughout these steps using a transfer pipette (*see* **Note 4**).
7. To dejelly the eggs, pour off as much 0.5× MMR as possible and replace with the *X. borealis* dejellying solution (about 1/3 of the prepared amount). Let sit for a few minutes, gently swirling a couple of times before exchanging. Repeat twice to use all the solution (500 mL). In about 6 min, the eggs should settle and pack with gravity without any space left in between. In addition, eggs will rotate so that the animal pole (pigmented hemisphere) will face upward once it has dejellied, indicating when the process is complete (*see* **Note 5**).
8. As soon as the eggs are dejellied, pour off the dejellying solution and start washing with the CSF-XB solution. Pour a generous amount of CSF-XB buffer into the beaker of eggs and allow the eggs to swirl and settle before proceeding to the next wash. Wash 3–4 times until all the solution (~1 L) is used. During this step, prepare the CSF-XB+ solution. Then, take out 1 mL of CSF-XB+ to fill each thin-wall centrifuge tubes and supplement with 10 μL of 10 mg/mL Cytochalasin (100 μg/mL final).
9. Returning to the eggs, pour off the CSF-XB solution and replace with the CSF-XB+ solution, taking care to ensure that all eggs remain submerged. Often, the beaker should remain tilted to keep all eggs submerged.
10. Cut a transfer pipette at the beginning of its larger diameter, and use it to carefully transfer the eggs to the thin-wall centrifuge tubes containing the Cytochalasin-supplemented CSF-XB + without exposing the eggs to air or adding too much liquid to the tube.
11. Add about 200 μL of CSF-XB+ to the adaptor 13 mL tube, and without touching the eggs, use forceps to place the thin-wall centrifuge tubes containing the eggs inside it.
12. To pack the eggs, centrifuge the tubes in the clinical centrifuge at 500 g for 1 min at 16 °C, and then remove the excess buffer on top of the packed eggs using a narrow stem transfer pipette (*see* **Note 6**).
13. To crush the eggs, immediately transfer the tubes to the high-speed centrifuge and spin for 15 min at 17,000 × g at 16 °C (*see* **Note 7**).

14. After the spin, take the thin-wall centrifuge tubes out of the adaptor tubes. Using a 1 mL syringe and an 18-gauge needle, perforate the tube at the bottom of the cytoplasmic fraction (the second one out of three) and aspirate the isolated cytoplasm. Take care that only the cytoplasmic layer is aspirated, not the surrounding yolk or lipid layers as this will affect extract activity and quality. Remove the syringe from the needle and transfer the cytoplasm to a tube that was precooled at 16 °C.
15. Supplement the extract with the LPC cocktail (1:1000 dilution of stock), Cytochalasin (1:500 dilution), and energy mix (1:50 dilution). Mix by inverting the tube or flicking gently. Keep the extract at 16 °C (*see* **Note 8**). Extracts should be used freshly, but good quality extracts will last 4–8 h.

3.3 Preparation of X. tropicalis Egg Extracts

1. Prime *X. tropicalis* females (usually 5) by injecting 0.1 mL (10 IU) of HCG (HCG stock solution diluted 1:10 in MilliQ H_2O or diluent) subcutaneously into the dorsal lymph sac using a 1 mL syringe and a 30-gauge needle, about 20 h before boosting (*see* **Note 1**). Store the frogs in 30 L sodium bicarbonate-buffered 0.1× MMR in a glass tank overnight at husbandry temperature (*see* **Note 9**).
2. The next morning, boost *X. tropicalis* females by injecting 0.2 mL (200 IU) of HCG to induce ovulation, ~3–4 h before eggs are needed. Store each frog individually in 4 L of buffered 0.1× MMR in their glass container at husbandry temperature (*see* **Notes 2** and **10**).
3. Frogs should start laying about 3 h after boosting. Let the females lay for about 2 h, frequently watching for decreases in egg quality. When the needed amount of eggs is laid or at the first sign of quality decreases, start collecting the eggs (*see* **Note 11**).
4. During **step 3**, make sure that both the clinical and the high-speed centrifuges are set to 16 °C; prepare the dejellying solution and CSF-XB. Set aside a 50 mL aliquot of CSF-XB (to later prepare the CSF-XB+), and, if possible, store all solutions at 16 °C until use. Allow frozen aliquots of Cytochalasin, LPC cocktail, and energy mix to thaw at room temperature.
5. Remove the females from the containers and analyze egg quality to select batches of eggs to be used. Avoid batches that contain a significant amount of misshaped or lysed eggs (*see* **Note 3**). Retrieve the selected batches of eggs in a 400 mL glass beaker (*see* **Note 12**).
6. For the dejellying and washing steps (**steps 7–8**), take extra care to avoid exposing the eggs to air and swirling them excessively (*see* **Note 4**). Remove misshaped and lysing eggs throughout using a transfer pipette.

7. To dejelly the eggs, pour off as much buffered 0.1× MMR as possible and replace with the *X. tropicalis* dejellying solution (about 1/3 of it). Let sit for a few minutes, gently swirling a couple of times before exchanging. Repeat twice to use all the solution (500 mL). In about 8 min, the eggs should settle and pack with gravity without any space left in between. In addition, eggs will rotate so that the animal pole (pigmented hemisphere) will face upward once it has dejellied, indicating when the process is complete (*see* **Note 5**).
8. As soon as the eggs are dejellied, pour off the dejellying solution and start washing with the CSF-XB solution. Pour a generous amount of CSF-XB buffer into the beaker of eggs, and allow the eggs to swirl and settle before proceeding to the next wash. Wash 3–4 times until all the solution (~1 L) is used. Prepare the CSF-XB+ solution during this step. Then, take out 1 mL of CSF-XB+ to fill each thin-wall centrifuge tubes and supplement with 10 μL of 10 mg/mL Cytochalasin (100 μg/mL final).
9. Pour off the CSF-XB solution and replace with the CSF-XB+ solution, making sure to keep all eggs submerged by keeping the beaker angled if necessary.
10. Cut a transfer pipette at the beginning of its larger diameter, and use it to carefully transfer the eggs to the thin-wall centrifuge tubes containing the Cytochalasin-supplemented CSF-XB + without exposing the eggs to air or adding too much liquid to the tube.
11. Add about 200 μL of CSF-XB+ to the adaptor 13 mL tube, and, without touching the eggs, use forceps to place the thin-wall centrifuge tubes containing the eggs inside it.
12. To pack the eggs, centrifuge the tubes in the clinical centrifuge at 500 g for 1 min at 16 °C, and then remove the excess of buffer on top of the packed eggs using a narrow stem transfer pipette (*see* **Note 6**).
13. To crush the eggs, immediately transfer the tubes to the high-speed centrifuge and spin for 15 min at 17,000 × g at 16 °C (*see* **Note** 7).
14. After the spin, take the thin-wall centrifuge tubes out of the adaptor tubes. Using a 1 mL syringe and an 18-gauge needle, perforate the tube at the bottom of the cytoplasmic fraction (the second one out of three) and aspirate the isolated cytoplasm. Take care that only the cytoplasmic layer is aspirated, not the surrounding yolk or lipid layers as this will affect the extract activity and quality. Separate the syringe from the needle, and transfer the cytoplasm to a tube that was precooled at 16 °C.

15. Supplement the extract with the LPC cocktail (1:1000 dilution of stock), Cytochalasin (1:500 dilution), and energy mix (1:50 dilution). Mix by inverting the tube or flicking gently. Keep the extract at 16 °C (*see* **Note 13**). Extracts should be used freshly, but good quality extracts will last 4–8 h.

3.4 Mixing *Xenopus* Egg Extracts

Xenopus egg extracts have been instrumental in understanding the mechanisms of spindle assembly and scaling. Importantly, the cell-free nature of extract cytoplasm uniquely allowed mixing experiments to demonstrate that cytoplasmic contents control size scaling and morphology [13, 14, 18]. Here, we provide the example of mixing *X. borealis* and *X. tropicalis* extracts and studying the consequence on spindle assembly and size (Fig. 1), but this experiment can apply to any two frog extracts or other cellular structures or organelles. All reactions are at room temperature.

1. Prepare *X. borealis* and *X. tropicalis* egg extracts as described above (*see* Subheadings 3.2 and 3.3). It is best to use both extracts as freshly as possible to ensure that both remain functional throughout the experiment.
2. Supplement extracts with 40 ng/μL of rhodamine-labelled tubulin to visualize microtubules.
3. Prepare mixing reactions in 1.5 mL microcentrifuge tubes. The total volume of extract reactions should be 20–50 μL.
4. First, pipette the appropriate amount of *X. borealis* extract using wide-bore P200 tips (Fig. 1a).
5. Add *X. tropicalis* extract in the appropriate proportions on top of *X. borealis* extract, pipetting with wide-bore P200 tips.
6. Mix by flicking the tube gently to ensure that both cytoplasmic extracts are fully merged.
7. Controls are internal to this experiment, as one sample should be 100% *X. borealis* extract and another should be 100% *X. tropicalis* extract (Fig. 1b).
8. Add 500–1000 *X. tropicalis* sperm nuclei per μL to each reaction (*see* **Notes 14** and **15**). Flick the tube gently to mix, as this will help the sperm disperse throughout the cytoplasm and prevent clumping of structures.
9. Follow each reaction with fixed "squashes" as desired every 15 min, taking care to flick the tube at this time as well.
10. To "squash" samples, pipette 1 μL of spindle fix onto a glass slide, then pipette 1 μL of extract into the spindle fix. Carefully place a square coverslip on top – the extract should spread immediately across the coverslip. DNA and microtubules should be readily visible by fluorescence microscopy, allowing observation of the cell cycle progression as desired.

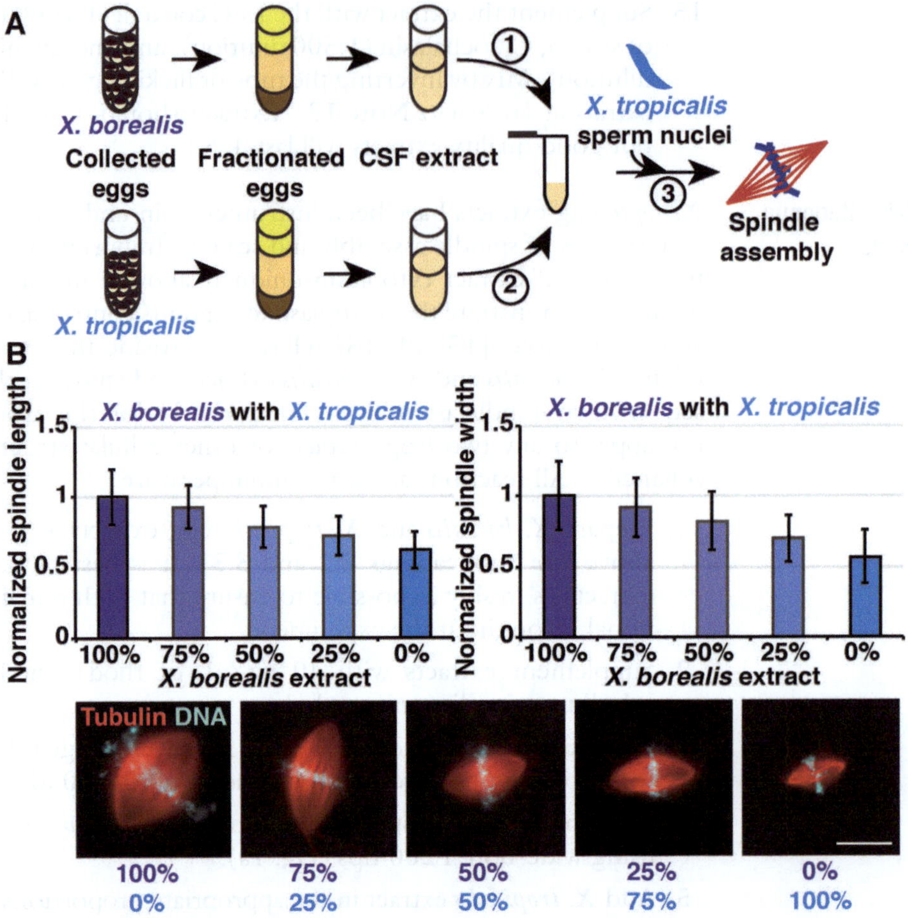

Fig. 1 Mixing extracts uniquely demonstrates size control via cytoplasmic origin. (**a**) Experimental schematic of a mixing experiment. (**b**) Mixing *X. borealis* and *X. tropicalis* extract shrinks spindle size as more *X. tropicalis* extract is added. Scale is 20 μm. (Reproduced from Kitaoka et al. with permission from John Wiley and Sons [18])

11. Conclude the experiment by spinning down the reactions onto coverslips if, for example, immunofluorescence analysis is desired, as detailed elsewhere [22] (*see* **Note 15**).

3.5 "Hybrid" Extract Reactions

Xenopus frogs provide a phenomenal, vertebrate model system to study speciation barriers and mechanisms of hybrid inviability that lead to embryonic lethality [20, 28]. While other model systems, particularly *Drosophila*, provide genetic insights to hybrid incompatibility [29, 30], the egg extract system opens the door for detailed molecular, cell biological, and evolutionary investigation of maternal vs. paternal contributions to hybrid embryos. Here, we add a twist to classical extract reactions by mix-and-matching sperm chromosomes and egg cytoplasmic extracts from different species. This mimics the early embryonic cell divisions in hybrids, which are

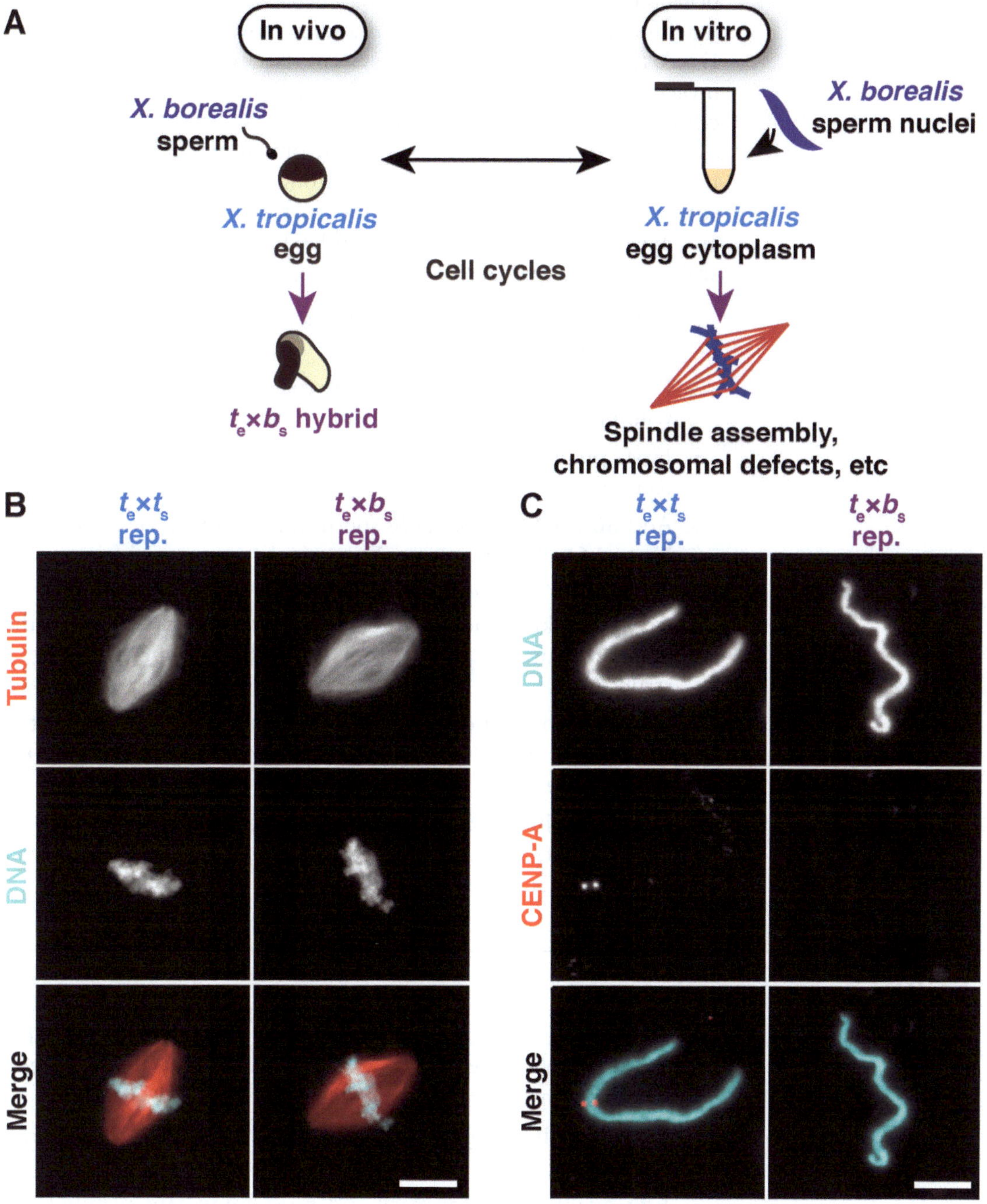

Fig. 2 Using "hybrid" reactions to understand maternal vs. paternal contributions to hybrid inviability. (**a**) Experimental schematic of "hybrid" reactions, demonstrating how the extract can mimic the early cell cycles in hybrid embryos. (**b**) *X. tropicalis* (left) and *X. borealis* (right) sperm DNA can both induce spindle assembly in *X. tropicalis* egg extract. (**c**) "Hybrid" reactions can reveal conflicts between species, such as loss of the centromeric histone CENP-A from replicated *X. borealis* chromosomes in *X. tropicalis* extract (right). Scale is 10 μm. Nomenclature denotes maternal egg or egg extract as subscript e and paternal sperm chromosomes as subscript s. (Reproduced from Kitaoka et al. with permission from Cell Press [20])

powered exclusively by maternal cytoplasmic components and machinery but also need to handle "foreign" paternal sperm DNA (Fig. 2a–c). Thus, this system can reveal mismatches and molecular conflicts in hybrids. In this protocol, we investigate combining *X. tropicalis* egg extract with *X. borealis* sperm DNA, as described recently to investigate chromosomal defects in inviable hybrids [20]. However, other combinations of sperm chromosomes and egg extracts are bound to reveal new insights as well and have already been used to understand spindle assembly and scaling mechanisms across frog species [13, 14, 31].

1. Prepare *X. tropicalis* egg extract as described above (*see* Subheading 3.3).
2. Prepare reactions in 1.5 mL microcentrifuge tubes. The total volume of extract reactions should be 20–50 μL.
3. First, pipette the appropriate amount of *X. tropicalis* extract using wide-bore P200 tips.
4. Add 500–1000 *X. borealis* sperm nuclei per μL of reaction (*see* **Note 16**). Be sure to use *X. tropicalis* sperm nuclei in control reactions (Fig. 2a).
5. Mix by flicking the tube gently. This will help the sperm disperse throughout the cytoplasm and prevent clumping of structures.
6. Follow each reaction with fixed "squashes" every 15 min, taking care to flick the tube at this time as well (see **step 10** in Subheading 3.4).
7. Conclude the experiment by spinning down the reactions onto coverslips to analyze either spindles or individualized chromosomes, as detailed elsewhere [6].

4 Notes

1. Priming the female frogs improves egg yield and quality. While they may be usable for 2 weeks like *X. laevis*, we use *X. borealis* frogs within 1 week of priming.
2. Individual housing prevents mixing egg batches of different quality.
3. Egg quality is the primary and most important parameter as this will determine the extract quality. Careful analysis of eggs is thus highly recommended. Yet, egg strings can be used for *X. tropicalis*, if they do not contain any lysed eggs.
4. Exposure to air, especially once eggs are dejellied, will induce egg lysis and corrupt extract quality. Excessive swirling

increases the risk of releasing the metaphase arrest and obtaining an interphasic or non-functional extract.

5. The dejellying step should not take more than 10 min. If eggs are not dejellied after 10 min, quality may not be good. Also, overexposure to the dejellying solution will lead to egg lysis.
6. The tube can be quickly tipped to 45° to remove as much buffer as possible. Do not hesitate to sacrifice a few eggs.
7. An extra minute can be added to make sure the eggs are centrifuged at full speed for 15 min.
8. This temperature was empirically determined as it was known to allow the conservation of both *X. laevis* and *X. tropicalis* egg extracts. Sensitivity of *X. borealis* extracts to cold was not determined. *X. borealis* extracts are stable at this temperature for at least 4 h.
9. *X. tropicalis* frogs can be primed and boosted successfully in ddH_2O, which does not impair egg and extract quality.
10. The boosting dose for *X. tropicalis* can be increased up to 250 IU HCG, in our hands.
11. If high yield of extracts is an objective, one can accelerate egg release by simply holding the frog gently, allowing her to wiggle and kick her legs. Repeat 3–4 times at 15-min intervals. Be aware that this may cause animal stress, though it should not be detrimental if done gently.
12. *X. tropicalis* eggs stick to the container. Wear nitrile gloves and use your finger to gently sweep and detach the eggs. Collect them in one corner of the tub while carefully pouring excess water off. The eggs will settle and stick again, so keep them in suspension in water when collecting them into a clean beaker.
13. *X. tropicalis* extracts cannot be kept on ice. This will impair extract activity.
14. Recent work has demonstrated that hybrids from *X. borealis* eggs and *X. tropicalis* sperm are viable and have no obvious cell cycle defects, while the reverse hybrid is inviable and has chromosomal defects [20, 28]. Here, we chose to add *X. tropicalis* sperm nuclei to assay spindle assembly without complicating our results with additional hybrid incompatibility concerns.
15. Centrifugation of extract reactions onto coverslips and their fixation are required for further processing as detailed elsewhere [6, 22] but will shrink the size of subcellular structures. This is not an issue if all experiments compared are spun down. Conversely, squashes do not change the size of structures, providing more accurate size measurements, but are not compatible with immunofluorescence.

16. If induction into interphase is desired, add 1× calcium solution to release the extract from its metaphase arrest and flick the tube gently. Incubate the reaction for 5–6 min, and then add sperm nuclei.

Acknowledgments

We thank members of the Gibeaux and the Heald labs for their support. We are grateful to Celine Callens for the critical reading of the manuscript. M.K. was supported by a National Science Foundation (NSF) GRFP fellowship and G.G. by a PhD fellowship (ARED) from the Region Bretagne. R.H. was supported by NIH MIRA grant R35 GM118183 and the Flora Lamson Hewlett Chair and R.G. by a Human Frontier Science Program grant CDA00019/2019-C.

References

1. Lohka MI, Maller JL (1985) Induction of nuclear envelope breakdown, chromosome condensation, and spindle formation in cell-free extracts. J Cell Biol 101:518–523. https://doi.org/10.1083/jcb.101.2.518
2. Murray AW (1991) Cell cycle extracts. Methods Cell Biol 36:581–605. https://doi.org/10.1016/S0091-679X(08)60298-8
3. Heald R, Tournebize R, Blank T, Sandaltzopoulos R, Becker P, Hyman A, Karsenti E (1996) Self-organization of microtubules into bipolar spindles around artificial chromosomes in Xenopus egg extracts. Nature 382:420–425
4. Murray AW, Solomon MJ, Kirschner MW (1989) The role of cyclin synthesis and degradation in the control of maturation promoting factor activity. Nature 339:280–286. https://doi.org/10.1038/339280a0
5. Murray AW, Kirschner MW (1989) Cyclin synthesis drives the early embryonic cell cycle. Nature 339:275–280
6. Maresca TJ, Heald R (2006) Methods for studying spindle assembly and chromosome condensation in Xenopus egg extracts. Methods Mol Biol (Clifton, NJ) 322:459–474. https://doi.org/10.1007/978-1-59745-000-3_33
7. French BT, Straight AF (2017) The power of Xenopus egg extract for reconstitution of centromeres and kinetochore function. Prog Mol Subcell Biol 56:59–84. https://doi.org/10.1007/978-3-319-58592-5
8. Blow JJ, Laskey RA (1986) Initiation of DNA replication in nuclei and purified DNA by a cell-free extract of Xenopus eggs. Cell 47: 577–587. https://doi.org/10.1016/0092-8674(86)90622-7
9. Glotzer M, Murray AW, Kirschner MW (1991) Cyclin is degraded by the ubiquitin pathway. Nature 349:132–138. https://doi.org/10.1038/349132a0
10. Belmont LD, Hyman A, Sawin KE, Mitchison T (1990) Real-time visualization of cell cycle-dependent changes in microtubule dynamics in cytoplasmic extracts. Cell 62:579–589. https://doi.org/10.1016/0092-8674(90)90022-7
11. Verde F, Labbé JC, Dorée M, Karsenti E (1990) Regulation of microtubule dynamics by cdc2 protein kinase in cell-free extracts of Xenopus eggs. Nature 343:233–238. https://doi.org/10.1038/343233a0
12. Sawin KE, Mitchison TJ (1991) Mitotic spindle assembly by two different pathways in vitro. J Cell Biol 112:925–940
13. Brown KS, Blower MD, Maresca TJ, Grammer TC, Harland RM, Heald R (2007) Xenopus tropicalis egg extracts provide insight into scaling of the mitotic spindle. J Cell Biol 176:765–770. https://doi.org/10.1083/jcb.200610043
14. Helmke KJ, Heald R (2014) TPX2 levels modulate meiotic spindle size and architecture in Xenopus egg extracts. J Cell Biol 206:385–393. https://doi.org/10.1083/jcb.201401014

15. Levy DL, Heald R (2010) Nuclear size is regulated by importin a and Ntf2 in Xenopus. Cell 143:288–298. https://doi.org/10.1016/j.cell.2010.09.012
16. Loughlin R, Heald R, Nédélec F (2010) A computational model predicts Xenopus meiotic spindle organization. J Cell Biol 191: 1239–1249. https://doi.org/10.1083/jcb.201006076
17. Loughlin R, Wilbur JD, McNally FJ, Nédélec FJ, Heald R (2011) Katanin contributes to interspecies spindle length scaling in xenopus. Cell 147:1397–1407. https://doi.org/10.1016/j.cell.2011.11.014
18. Kitaoka M, Heald R, Gibeaux R (2018) Spindle assembly in egg extracts of the Marsabit clawed frog, Xenopus borealis. Cytoskeleton 75:244–257. https://doi.org/10.1002/cm.21444
19. Session AM, Uno Y, Kwon T, Chapman JA, Toyoda A, Takahashi S, Fukui A, Hikosaka A, Suzuki A, Kondo M, Van Heeringen SJ, Quigley I, Heinz S, Ogino H, Ochi H, Hellsten U, Lyons JB, Simakov O, Putnam N, Stites J, Kuroki Y, Tanaka T, Michiue T, Watanabe M, Bogdanovic O, Lister R, Georgiou G, Paranjpe SS, Van Kruijsbergen I, Shu S, Carlson J, Kinoshita T, Ohta Y, Mawaribuchi S, Jenkins J, Grimwood J, Schmutz J, Mitros T, Mozaffari SV, Suzuki Y, Haramoto Y, Yamamoto TS, Takagi C, Heald R, Miller K, Haudenschild C, Kitzman J, Nakayama T, Izutsu Y, Robert J, Fortriede J, Burns K, Lotay V, Karimi K, Yasuoka Y, Dichmann DS, Flajnik MF, Houston DW, Shendure J, Dupasquier L, Vize PD, Zorn AM, Ito M, Marcotte EM, Wallingford JB, Ito Y, Asashima M, Ueno N, Matsuda Y, Veenstra GJC, Fujiyama A, Harland RM, Taira M, Rokhsar DS (2016) Genome evolution in the allotetraploid frog Xenopus laevis. Nature 538:336–343. https://doi.org/10.1038/nature19840
20. Kitaoka M, Smith OK, Straight AF, Heald R (2022) Molecular conflicts disrupting centromere assembly contribute to Xenopus hybrid inviability. Curr Biol 32:1–13. https://doi.org/10.1016/j.cub.2022.07.037
21. Good MC, Heald R (2018) Preparation of cellular extracts from Xenopus eggs and embryos. Cold Spring Harb Protoc 1–12. https://doi.org/10.1101/pdb.prot097055
22. Hannak E, Heald R (2006) Investigating mitotic spindle assembly and function in vitro using Xenopus laevis egg extracts. Nat Protoc 1:2305–2314. https://doi.org/10.1038/nprot.2006.396
23. Hazel JW, Gatlin JC (2018) Isolation and demembranation of xenopus sperm nuclei. Cold Spring Harb Protoc 2018:449–454. https://doi.org/10.1101/pdb.prot099044
24. Sharp JA, Blower MD (2013) Production of xenopus tropicalis egg extracts to identify microtubule-associated rnas. J Vis Exp. https://doi.org/10.3791/50434
25. McNamara S, Wlizla M, Horb ME (2018) Husbandry, general care, and transportation of Xenopus laevis and Xenopus tropicalis. Methods Mol Biol 1865. https://doi.org/10.1007/978-1-4939-8784-9_1
26. Shaidani NI, McNamara S, Wlizla M, Horb ME (2020) Animal maintenance systems: Xenopus laevis. Cold Spring Harb Protoc 2020:433–437. https://doi.org/10.1101/pdb.prot106138
27. Shaidani N-I, McNamara S, Wlizla M, Horb ME (2020) Animal maintenance systems: Xenopus tropicalis. Cold Spring Harb Protoc 1–6. https://doi.org/10.1101/pdb.prot106146
28. Gibeaux R, Acker R, Kitaoka M, Georgiou G, Van Kruijsbergen I, Ford B, Marcotte EM, Nomura DK, Kwon T, Veenstra GJC, Heald R (2018) Paternal chromosome loss and metabolic crisis contribute to hybrid inviability in Xenopus. Nature 553:337–341. https://doi.org/10.1038/nature25188
29. Presgraves DC, Balagopalan L, Abmayr SM, Orr HA (2003) Adaptive evolution drives divergence of a hybrid inviability gene between two species of drosophila. Nature 423:715–719. https://doi.org/10.1038/nature01679
30. Phadnis N, Baker EP, Cooper JC, Frizzell KA, Hsieh E, de la Cruz AFA, Shendure J, Kitzman JO, Malik HS (2015) An essential cell cycle regulation gene causes hybrid inviability in drosophila. Science 350(80):1552–1555. https://doi.org/10.1126/science.aac7504
31. Miller KE, Session AM, Heald R (2019) Kif2a scales meiotic spindle size in Hymenochirus boettgeri. Curr Biol 29:3720–3727.e5. https://doi.org/10.1016/j.cub.2019.08.073

Chapter 12

Measuring Mitotic Spindle and Microtubule Dynamics in Marine Embryos and Non-model Organisms

Janet Chenevert, Morgane L. V. Robert, Jérémy Sallé, Sébastien Cacchia, Thierry Lorca, Anna Castro, Alex McDougall, Nicolas Minc, Stefania Castagnetti, Julien Dumont, and Benjamin Lacroix

Abstract

During eukaryotic cell division a microtubule-based structure, the mitotic spindle, aligns and segregates chromosomes between daughter cells. Understanding how this cellular structure is assembled and coordinated in space and in time requires measuring microtubule dynamics and visualizing spindle assembly with high temporal and spatial resolution. Visualization is often achieved by the introduction and the detection of molecular probes and fluorescence microscopy. Microtubules and mitotic spindles are highly conserved across eukaryotes; however, several technical limitations have restricted these investigations to only a few species. The ability to monitor microtubule and chromosome choreography in a wide range of species is fundamental to reveal conserved mechanisms or unravel unconventional strategies that certain forms of life have developed to ensure faithful partitioning of chromosomes during cell division. Here, we describe a technique based on injection of purified proteins that enables the visualization of microtubules and chromosomes with a high contrast in several divergent marine embryos. We also provide analysis methods and tools to extract microtubule dynamics and monitor spindle assembly. These techniques can be adapted to a wide variety of species in order to measure microtubule dynamics and spindle assembly kinetics when genetic tools are not available or in parallel to the development of such techniques in non-model organisms.

Key words Mitotic spindle, Microtubule dynamics, Non-model organisms, Marine embryos, Tubulin labelling, Histone purification, Confocal live imaging

1 Introduction

Mitotic spindles are self-organized macromolecular structures that partition chromosomes during cell division. At mitotic entry, the interphasic microtubule network is disassembled and reorganizes around the nucleus in radial arrays called asters made of short and highly dynamic microtubules. In most species and cell types, the asters are nucleated around two centrosomes which duplicate during interphase and contribute to the nucleation of a great portion of

Anna Castro and Benjamin Lacroix (eds.), *Cell Cycle Control: Methods and Protocols*, Methods in Molecular Biology, vol. 2740, https://doi.org/10.1007/978-1-0716-3557-5_12,

spindle microtubules. In the early phase of mitosis (prophase), the two microtubule asters contribute to the separation of the centrosomes that migrate to opposite sides of the nucleus and provide bipolarity to the mitotic structure. Following nuclear envelope breakdown (NEBD) microtubules grow toward condensed mitotic chromosomes, align them on a metaphasic plate, and then segregate the two sets of chromosomes between the two daughter cells. Numerous microtubule-associated proteins including molecular motors are involved in aster movement, chromosome capture, alignment, and segregation, but microtubules and their dynamic properties play a central role in these processes. Indeed, the assembly of the mitotic spindle relies on the dynamic properties of microtubule filaments, hollow cylindrical tubes composed of polymerized alpha−/beta-tubulin heterodimers. This assembly is reversible such that, both in vivo and in vitro, microtubules stochastically oscillate between phases of growth and shrinkage in a process called dynamic instability [1]. Phases of growth and shrinkage are interspersed with transition events known as catastrophe and rescue. A catastrophe corresponds to a rapid depolymerization event occurring after a growth excursion, while rescue corresponds to a regrowth event interrupting a depolymerization phase. The characterization of microtubule dynamics requires the measurement of growth and shrinkage velocities, also called polymerization and depolymerization rates, and of the frequencies of catastrophe and rescue [2–4]. These four parameters define the dynamic status of individual microtubules but can also be used to predict average microtubule length or collective behavior of microtubule networks and microtubule-based structures [5].

Mitotic spindles and microtubule components are remarkably conserved across eukaryotes. However, depending on cell types or species, the mitotic spindle exhibits high variability of sizes, different nucleation pathways, and kinetics of assembly. Despite the variety of forms of life, studies of spindle assembly mechanisms have focused on very few model organisms. This is mostly due to technical limitations because visualizing microtubule dynamics and spindle assembly requires high spatial and temporal resolution, live imaging, and the use of bright fluorescent probes. In genetically tractable organisms, such conditions are met by the use of transgenic strains or cell lines stably expressing fluorescently tagged proteins to visualize microtubules (tubulins or microtubule binding proteins) and chromosomes (typically histones). Exploring microtubule and mitotic spindle assembly in many different species is necessary to understand how distinct and divergent paths could have emerged throughout evolution while resulting in the same outcome. Conversely, identifying conserved mechanisms among a large variety of species will likely reveal the most essential pathways.

To apprehend the diversity of microtubule and spindle assembly dynamics, we need standardized tools and approaches to enable microtubule and spindle visualization in species which lack genetic tools. Prior to the development of genetic tools and gene editing strategies, in the 1980s, visualization of cellular processes in live cells and embryos was performed using microinjection of chemically labelled proteins [6–10]. These approaches have been remarkably useful to reveal the dynamic nature of microtubules and spindles in vivo during cell division and to document cell lineage and fate mapping during embryogenesis. However poor resolution, fluorescent background due to out-of-focus light, and high phototoxicity of light sources were major limitations to the application of this methodology. Today, with the enormous advance in confocal microscopy and high-resolution live imaging, this approach can be adapted to overcome the lack of genetic tools or the inability of some species to translate injected mRNAs during the early stages of embryonic development.

Here, we describe an updated protocol for analyzing microtubule dynamics based on protein purification and injection in marine embryos. We present the use of purified and chemically labelled tubulin and affinity purification of recombinant fluorescent histone coupled to a high-resolution imaging to follow cell division, monitor mitotic spindle assembly, and quantify microtubule dynamics. We also provide a standardized methodology for extraction and interpretation of microtubule dynamics parameters as well as an estimation of the mitotic spindle assembly kinetics. We are confident that these techniques could be used in a wide variety of species provided that their eggs are transparent enough to be compatible with light microscopy and can withstand microinjection.

2 Materials

2.1 Tissues for Tubulin Extraction

Visualization of microtubules and spindles in embryos by fluorescence microscopy is facilitated by the injection of "homemade" fluorescent tubulin. The tubulin used for the methodology described here is prepared from freshly collected pig brains (*see* **Notes 1** and **2**). Purification and labelling of tubulin are performed as previously described by Castoldi and Popov, and Hyman A. A. [11, 12] and require the following materials:

1. Fresh brain tissues, preferentially pig brains (*see* **Note 2**) collected from a slaughterhouse. Time between slaughtering of the animal, brain collection, and protein extraction should be minimized (less than 2 h) to increase purification yield.
2. Cooler filled with ice and plastic food bags to reduce air and avoid direct contact with ice.

2.2 Tubulin Purification and Labelling

1. A robust blender to crush brain tissues without heating them.
2. Gauze swabs (not necessarily sterile).
3. Glass Dounce or Potter-Elvehjem homogenizers with a glass or PTFE pestle of different volumes. Typically, 50 mL for the first steps, 10 mL and 2 mL for the last steps of tubulin labelling.
4. Refrigerated ultracentrifuge with corresponding tubes for large volumes (250–500 mL) with a centrifugal force of at least 25,000 × g.
5. Ultracentrifuge for volumes between 50 and 100 mL and micro-ultracentrifuge with volume < 1 mL able to reach a centrifugal force of at least 50,000 × g for sedimentation of polymerized microtubules.
6. N-hydroxysuccinimide (NHS) ester fluorescent dye (*see* **Note 3**). About 5 mg is required to label 50–100 mg of tubulin. We usually use 5 mg of N-hydroxysuccinimide (NHS) ester fluorescent dye to label 80 mg of tubulin.
7. Prior to use, NHS dyes are diluted to 50 mM in water-free DMSO.
8. Depolymerization Buffer (DB): 50 mM MES pH 6.6 with KOH and 1 mM $CaCl_2$ (kept at 4 °C).
9. Glycerol 100%; kept warm (from hereafter warm means 37 °C).
10. High-molarity PIPES buffer (HMPB): 1 M PIPES pH 6.8 with KOH, 10 mM $MgCl_2$, and 20 mM EGTA; kept at 37 °C.
11. 5xBRB: 400 mM PIPES pH 6.8 (KOH), 5 mM $MgCl_2$, and 5 mM EGTA.
12. 1xBRB (Brinkley BR buffer 80 or BRB80): 80 mM PIPES pH 6.8 with KOH, 1 mM $MgCl_2$, and 1 mM EGTA; kept at 4 °C. 1xBRB is freshly prepared by diluting 5xBRB with pure water.
13. 150 mM ATP solution. It is recommended to adjust the pH of the solution to pH ~7 using pH paper strips. ATP can be resuspended in 5–10 mM HEPES and then buffered with KOH. Aliquots can be stored at −20 °C.
14. 100 or 200 mM GTP solution. The solution can be adjusted to pH ~7 similar to the ATP solution. Aliquots can be stored at −20 °C or −80 °C.
15. High-pH cushion: 0.1 M HEPES pH 8.6 (NaOH), 1 mM $MgCl_2$, 1 mM EGTA, and 60% (volume/volume) glycerol.
16. Low-pH cushion: 60% (v/v) glycerol in 1xBRB.
17. Labeling buffer: 0.1 M HEPES pH 8.6 (NaOH), 1 mM $MgCl_2$, 1 mM EGTA, and 40% (v/v) glycerol.
18. Glutamate depolymerization buffer (GDB): 50 mM glutamate pH 7.0 (KOH) and 0.5 mM $MgCl_2$.

19. 1 M $MgCl_2$ solution in pure water.
20. Centrifuge tubes compatible with DMSO (*see* **Note 4**).
21. Ultracentrifuge must be able to function at 4 °C and at 30–37 °C for tubulin polymerization-depolymerization cycles. To maximize purification yield, it is recommended to have both warm and cold ultracentrifuges with their respective rotors to minimize time for warming or cooling down of rotors.

2.3 Purification of Histone-RFP

1. 1 M solution of IPTG (isopropyl β-D-1-thiogalactopyranoside) in water. Store aliquots at −20 °C.
2. BL21(DE3) *E. coli*-competent cells.
3. Plasmid pET11-H2B-RFP-6His. A mouse H2B-RFP sequence was obtained from a pCS2 plasmid [13] and subcloned into a pET11 vector for bacterial expression. This plasmid can be provided upon request.
4. LB liquid medium (Luria-Bertani broth). Combine 10 g of tryptone, 5 g of yeast extract, 10 g of NaCl for 1 L of distilled water; adjust pH to 7.0 with 1 N NaOH and autoclave the mixture for 25 min at 120 °C.
5. Ampicillin stock solution at 100 mg/mL in water; stored at −20 °C.
6. Lysozyme solution 100×: 10 mg/mL in water.
7. Ultrasonicator with an appropriate probe.
8. Refrigerate ultracentrifuge and tubes for at least 30,000 × g.
9. TALON Superflow metal resin or any resin coupled to a metal for 6× histine affinity purification.
10. Empty gravity flow columns.
11. Lysis buffer: 1× PBS (137 mM NaCl, 2.7 mM KCl, 10 mM Na_2HPO_4, and 1.8 mM KH_2PO_4) supplemented with 300 mM NaCl and protease inhibitors.
12. Buffer A1: 50 mM Tris-HCl pH 8, 500 mM NaCl, 5% glycerol, and 6 M urea. Store at 4 °C.
13. Buffer A2: 50 mM Tris-HCl pH 8, 1 M NaCl, 5% glycerol, and 6 M urea. Store at 4 °C.
14. Elution buffer: 50 mM Tris-HCl pH 8, 500 mM NaCl, 5% glycerol, 6 M urea, and 20 mM imidazole. Store at 4 °C.
15. Dialysis buffer D1: 50 mM Tris-HCl pH 8, 10 mM DTT, 2 mM EDTA, and 6 M urea, 2 M NaCl. Store at 4 °C.
16. Dialysis buffer D2: 20 mM Tris-HCl pH 8, 5 mM DTT, 1 mM EDTA, 1 mM PMSF, 5% glycerol, an 2 M NaCl. Store at 4 °C.

17. Dialysis buffer D3: 20 mM Tris-HCl pH 8, 5 mM DTT, 1 mM EDTA, 1 mM PMSF, 5% glycerol, and 1 M NaCl. Store at 4 °C.
18. Dialysis buffer D4: 20 mM Tris-HCl pH 8, 5 mM DTT, 1 mM EDTA, 1 mM PMSF, 5% glycerol, and 0.5 M NaCl. Store at 4 °C.
19. Dialysis buffer D5: 20 mM Tris-HCl pH 8, 5 mM DTT, 1 mM EDTA, 1 mM PMSF, 5% glycerol, and 140 mM NaCl. Store at 4 °C.
20. Dialysis tubing, cassette, or device compatible with proteins and with a cutoff of <30 kDa.

2.4 Collection or Ordering of Marine Animals

Animal availability varies throughout the year depending on reproductive seasons and on geographical location (*see* **Note 5**). In Europe, wild animal sampling is often not allowed without specific authorities' approval. However, a large variety of species can be obtained from the European Biological Resource Center (EMBRC https://www.embrc.eu and its national branches such as https://www.embrc-france.fr/fr). These infrastructures provide animals while ensuring a regular survey on wild populations limiting the impact of research on ecosystems. Worldwide, researchers can certainly have access to local marine biological resource suppliers. We have successfully used the protocols described here in two echinoderm species: the sea urchin *Paracentrotus lividus* [14] and the starfish *Hacelia attenuata*, in a cnidarian the jellyfish *Clytia hemisphaerica* and a tunicate the ascidian *Phallusia mammillata*. We hope that having tested this methodology in several animals from different and distant taxa provides sufficient evidence that it can be applied to a large variety of species.

2.5 Embryo Mounting for Injection and Imaging

This chapter aims at the visualization of microtubules and chromatin and the quantification of microtubule dynamics in a wide variety of species. Our methods do not depend on the specific injection procedure, which varies with the species and is highly user dependent. However, we find that a modification of the horizontal injection configuration described previously [15] represents a versatile setup which is compatible with many types of eggs. We describe here the materials required to build an injection chamber and an imaging chamber for eggs of different sizes.

1. Plastic or metallic customized chamber holder (Fig. 1a, b) (*see* **Note 6**).
2. Glass-cutting diamond pen.
3. Artificial or natural filtered seawater.
4. Glass-bottom dishes of 35 mm diameter.
5. Dow Corning high vacuum grease.

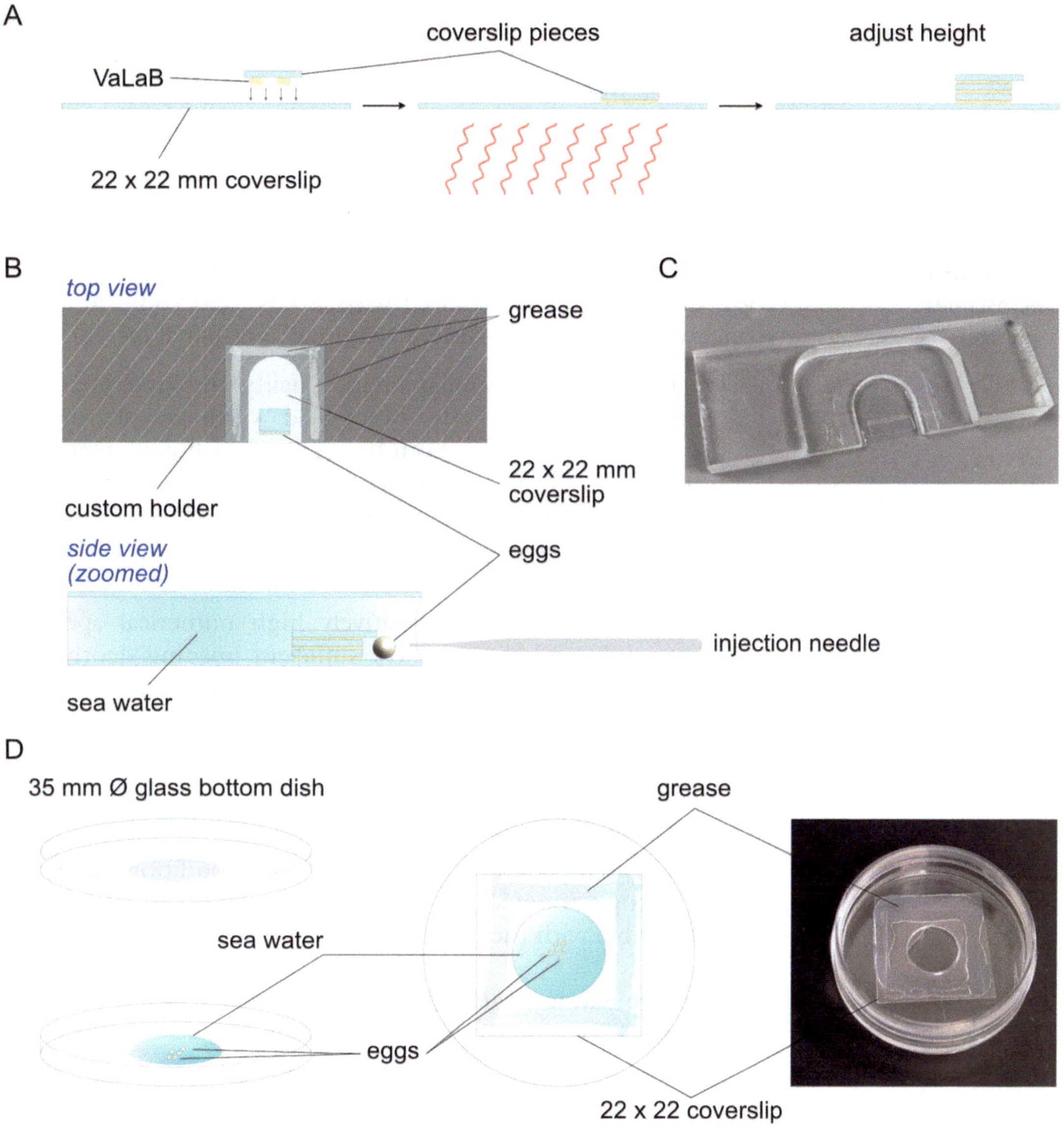

Fig. 1 Chambers for injection and live imaging (**a**) Schematic representation (side view) of how to prepare a glass ledge to align eggs for horizontal injection (inspired by Jaffe and Terasaki [18]). Small pieces of glass are stacked and glued with wax (VaLaB) onto a coverslip to create a ledge of the height of the egg to be injected. The wax melts with the addition of heat (red wavy lines) and solidifies in place when returned to room temperature. (**b**) Diagram depicting creation of the injection chamber. Using Dow Corning grease, the coverslip containing the ledge is sealed to the bottom, and another coverslip is sealed on top of the U-shaped opening in the plastic holder. Side view shows an egg immobilized by compression between the ledge and the needle. (**c**) Picture of an injection chamber. (**d**) Live imaging chamber used for inverted confocal microscopy. A drop (~100 μL) of seawater is placed on the glass bottom of a 35 mm diameter dish. Fertilized eggs are placed into the seawater drop. The chamber is closed with a 22 × 22 mm coverslip sealed with Dow Corning grease at the edges to prevent evaporation and contact with air. These chambers allow for up to 24 h of normal development

6. 10 mL syringe filled with Dow Corning high vacuum grease fitted with a cutoff plastic pipette tip.
7. 22 × 22 mm coverslips.
8. VaLaB: Lipid mixture composed of Vaseline, lanolin, and beeswax at an equal 1:1:1 weight proportion.
9. Heating plate which heats to at least 50 °C.

2.6 High-Resolution Microscopes

To allow tracking of individual microtubules by fluorescent microscopy, movies must be acquired with an inverted confocal microscope, which provides sufficient temporal and spatial resolution.

1. We have obtained similar image quality and outcomes with spinning disk [14] or swept-field confocal microscopes [16, 17]. Similar images can be achieved with new-generation scanning confocal microscopes, especially if equipped with photomultiplier tube of greater quantum efficiency such as gallium arsenide phosphide (GaAsP) PMT or hybrid detector HyD.
2. Objectives should have relatively high numerical apertures (above 1.1) and should allow sufficient imaging depth especially for large embryos. 30× to 60× magnification water or silicon immersion objectives are particularly appropriate for such imaging (*see* **Note 7**).
3. Acquisition devices (camera or PMT) must allow a temporal resolution of less than 1 s.
4. A stage temperature controller or air conditioning in the microscopy room can be required to maintain temperature compatible with the organism during imaging (*see* **Note 8**).

2.7 Image Analysis

1. Data are extracted using the ImageJ software https://imagej.nih.gov/ij/download.html.
2. Macros to create kymographs can be downloaded from https://github.com/benlacroix/ImageJ_Macros.
3. Datasheet to extract dynamics data can be found on https://github.com/benlacroix/MTDynamX_Data_Xtraction.

3 Methods

3.1 Purification of Tubulin

Tubulin is purified following a protocol described by Castoldi and Popov [11] based on temperature-induced cycles of polymerization and depolymerization and the use of a high-molarity buffer to remove microtubule-associated proteins.

1. Fresh brains obtained at the slaughterhouse should be kept on ice during transport.

2. Back at the lab and in a cold room: Peel the exterior membranes off of the brains, and remove blood vessels and residual pieces of bones and cartilage.
3. Weigh brains and add 1 L of DB per 1 kg of brain.
4. Blend for 30 s at a low speed, pause for 30 s to cool down samples, and then 30 s at a high speed.
5. Filter through a double layer of gauze.
6. Centrifuge at 29,000 × g at 4 °C for 1 h.
7. Filter the supernatant through the gauze directly from the tubes into a graduated cylinder.
8. Measure volume. This corresponds to the "initial volume" that we will refer to for further steps of this protocol.
9. Add 1 volume of warm HMPB (37 °C) and 1 volume of warm (37 °C) glycerol and mix by inverting.
10. Add ATP to 1.5 mM and GTP to 0.5 mM final concentrations.
11. Mix by inverting and transfer to prewarmed (37 °C) ultracentrifuge tubes.
12. Incubate for 1 h at 37 °C preferably in the rotor inside the ultracentrifuge. If the volume exceeds that of your centrifuging capacities, keep the remaining volume in a cylinder at 37 °C in a water bath and repeat the following centrifugation step once the first batch has been pelleted (manipulate this microtubule solution gently when transferring to tubes).
13. Pellet polymerized microtubules at 100,000 × g for 30 min at 37 °C.
14. Discard supernatant and resuspend pellet (polymerized microtubules) in a smaller volume of ice-cold DB. Typically a 10 mL per 70 mL of initial volume.
15. Homogenize with a Dounce/Potter on ice (20 strokes up and down).
16. Incubate for 15 min on ice to depolymerize microtubules.
17. Spin this suspension at 4 °C, 120,000 × g for 30 min.
18. Keep the supernatant and add 1 volume of warm HMPB, 1 volume of warm glycerol, 1.5 mM ATP (final), and 0.75 mM GTP (final).
19. Incubate for 1 h at 37 °C (incubation can be done directly in centrifuge tubes).
20. Sediment polymerized microtubules by centrifugation at 150,000 × g at 37 °C for 30 min.
21. Resuspend the pellet in a minimal volume of ice-cold 1× BRB (use 10 mL for 500–600 mL of the initial volume to get a highly concentrated tubulin).

22. Homogenize with a Dounce/Potter on ice and incubate for 15 min to depolymerize microtubules.
23. Spin at 100,000 × g at 4 °C for 30 min.
24. The supernatant contains pure tubulin.
25. Measure absorbance at 280 nm of a 1/10 dilution (in 1× BRB) and calculate the protein concentration, given an extinction coefficient of tubulin at 280 nm of 115,000 $M^{-1}.cm^{-1}$ and a molecular weight of 110,000 g/mol for tubulin dimers.

$$C_{(mol/L)} = \text{dilution factor} \cdot A_{280\,nM}/115,000$$
$$C_{(mg/mL)} = C_{(mol/L)} \cdot 110,000$$

26. Aliquot (1–2 μL) in microtubes and snap freeze in liquid nitrogen.
27. Store at −80 °C.
28. Run 10 μg on a 10% acrylamide SDS-PAGE gel and stain with Coomassie blue. A single strong band of around 50 kDa should appear (*see* **Note 9**).

3.2 Labelling of Tubulin

Labelling is performed using NHS (N-hydroxysuccinimide) ester-coupled dyes that will react with protein amino groups and create a stable covalent bond. The reaction of NHS esters with amines is pH dependent, because amines need to be deprotonated. The reaction is thus optimal at "high" pH (here, pH 8.6). The reaction is then stopped when the pH is lowered to 6.5–6.8 (typical pH for tubulin buffers) as amines get protonated. A subsequent microtubule polymerization/depolymerization cycle selects for tubulin that is still able to polymerize after labelling and removes excess of free labelling dye.

1. Thaw an aliquot of concentrated tubulin as quickly as possible, and place it on ice as soon as it starts melting.
2. For each labelling reaction, prepare 10 mL of 8 mg/mL tubulin in 1× BRB in a 50 mL Falcon tube by diluting the concentrated tubulin stock. Keep on ice.
3. Bring the $MgCl_2$ concentration to 3.5 mM using a 1 M stock solution (remember that 1× BRB already contains 1 mM $MgCl_2$, i.e., add 25 μL of 1 M $MgCl_2$ per 10 mL of tubulin solution). Bring the GTP concentration to 1 mM using a 100 mM stock solution. Store on ice for 5 min.
4. Transfer to 37 °C and add 5 mL of prewarmed glycerol to promote polymerization (*see* **Note 10**).
5. Let polymerize for 40 min at 37 °C.

6. In an ultracentrifugation tube, layer polymerized tubulin onto a 9 mL warm (37 °C) high-pH cushion.
7. Pellet microtubules at 150,000 × g for 45 min at 35 °C (*see* **Note 11**). The rotor should accelerate and brake slowly, and the balance tube must contain a 30% glycerol solution on a 60% glycerol cushion to balance properly.
8. Aspirate the supernatant above the cushion and discard it. Rinse the supernatant/cushion interface twice with warm labelling buffer to remove unpolymerized tubulin.
9. Remove the cushion and resuspend the pellet in 1.9 mL of warm labelling buffer using a cut pipette tip with a wide opening (*see* **Note 12**). Keep the tubulin warm during resuspension and continue resuspending (pipetting up and down) until no aggregates are visible (this takes a while).
10. Resuspend the succinimidyl ester-coupled fluorophore in water-free DMSO to make a 50 mM solution (10×).
11. Add the fluorophore (dissolved in DMSO) in two steps. Add the first half to reach a 2.5 mM final concentration, and mix by gentle pipetting. This step must be done in a DMSO compatible tube (*see* **Note 4**).
12. Incubate for 20 min at 37 °C.
13. Add the second half of the fluorophore solution to reach the final concentration of 5 mM (*see* **Note 13**). Always pipette into the microtubule solution while slowly rotating the tube by hand. Avoid air bubbles.
14. Keep at 37 °C for 20 min (40 min total incubation with fluorophore), and vortex gently every 5 min. Following this step, always keep the sample covered with aluminum foil to protect it from the light.
15. In two tubes (polyallomer PA) of appropriate volume, layer the labelling reaction onto 1 mL of prewarmed low-pH cushion.
16. Spin at 120,000 × g for 40 min at 35 °C. Accelerate and brake slowly, typically 30–50% of maximal acceleration/deceleration capacity of your ultracentrifuge.
17. Aspirate the supernatant above the cushion and discard it. Rinse the supernatant/cushion interface twice with warm 1× BRB. Remove the cushion, wash the pellet briefly with warm 1× BRB, and resuspend each pellet in 400 μL of ice-cold glutamate depolymerization buffer (GDB). Homogenize on ice using a Dounce homogenizer. Leave for 30 min on ice. Use another 200 μL of cold GDB to wash tubes and the Dounce homogenizer and pool with the initial 400 μL.
18. Spin the depolymerized tubulin at 200,000 × g for 10 min at 2 °C.

19. Recover the supernatant from the cold spin, add BRB to 1× final concentration using a 5× stock solution, $MgCl_2$ to 4 mM using a 1 M stock, and GTP to 1 mM using a 100 mM stock. Incubate on ice for 3 min.
20. Warm up to 37 °C for 2 min, add 0.5 volume of pure glycerol, and mix well.
21. Polymerize at 37 °C for 30 min. In a 2–4 mL ultracentrifuge tube, layer the polymerization reaction onto a prewarmed 1 mL low-pH cushion and pellet the microtubules at 150,000 × g for 30 min at 37 °C. Accelerate and brake slowly, typically 30% of maximal acceleration and deceleration.
22. Aspirate the supernatant above the cushion and discard it. Rinse the supernatant cushion interface twice with warm 1× BRB. Remove the cushion and briefly rinse the pellet twice with 1 mL warm 1× BRB. Resuspend the pellet using a cut pipette tip in a 0.2 mL of ice-cold 1× BRB.
23. Incubate on ice for 20 min.
24. At this step, it is helpful to use a small-volume glass 2 mL Dounce/Potter to aid in resuspension and increase recovery.
25. Spin the depolymerized tubulin in a microvolume ultracentrifuge at 200,000–250,000 × g for 10 min at 2 °C. Recover the supernatant.
26. Quickly estimate the tubulin concentration (see Subheading 3.1), ideally adjust to a final concentration of 20 mg/mL with 1× BRB without GTP (*see* **Note 14**) and freeze in 1–2 μL aliquots in liquid nitrogen.
27. Store at −70 °C or in liquid nitrogen.

3.3 Purification of Histone-RFP

1. Transform BL21(DE3)-competent bacteria with the pET11-H2B-RFP-6His plasmid, and select colonies on ampicillin plate.
2. Pick a single colony and grow overnight in LB medium with 100 mg/L ampicillin.
3. Dilute the culture 1/100 in fresh LB containing 100 mg/L ampicillin and leave shaking at 37 °C until it reaches 0.4–0.6 optical density at 600 nm.
4. Transfer the culture to 20 °C and induce protein expression by adding IPTG to the final concentration of 0.1 mM.
5. Incubate overnight at 20 °C under shaking.
6. Sediment the bacteria at 4000 × g for 20 min at 4 °C.
7. Resuspend the bacterial pellet in lysis buffer containing 0.1 mg/mL lysozyme. At this step, the bacterial suspension can be stored at −80 °C.

8. Incubate the suspension for 30 min at 4 °C on a rotating wheel.
9. Place the suspension in a 50 mL plastic centrifugation tube in a plastic beaker filled with ice and water. Clean the ultrasonicator probe with water and alcohol. Dip the probe into the tube (3/4 depth). Sonicate at a 40% amplitude, with pulses of 1 s on/3 s off, for a total of 1 min or until the solution becomes less viscous.
10. Ultracentrifuge at 30,000 × g for 30 min at 4 °C. Insoluble proteins will be found in the pellet. Discard the supernatant.
11. Resuspend the pellet with buffer A1 and homogenize using a Dounce/Potter.
12. Ultracentrifuge at 30,000 × g for 30 min at 4 °C.
13. Collect the supernatant containing the soluble proteins.
14. Centrifuge the TALON Superflow metal resin at 500 × g for 2 min at 4 °C, and wash it three times with buffer A1.
15. Incubate the resin with the supernatant from **step 13** overnight at 4 °C on a rotating wheel.
16. Wash the resin three times with 10 mL of buffer A1.
17. Wash the resin three times with 10 mL of buffer A2.
18. Wash once more with buffer A1.
19. Place the resin in a gravity flow column, and wash with a solution containing 2% buffer B1 and 98% buffer A1 to remove nonspecific proteins.
20. Elute the H2B-RFP His-tagged protein using a small volume of elution buffer, collecting 0.5–1 mL fractions until fractions become near colorless.
21. Run proteins on a 10% SDS-polyacrylamide gel, and check protein purity by Coomassie staining.
22. Quantify protein concentration using the Bradford assay or an equivalent method.
23. Pool the most concentrated fractions containing pure protein and transfer the solution to a dialysis device (tube or cassette).
24. Dialyze the protein with dialysis buffer D1 overnight at 4 °C to remove imidazole.
25. Then dialyze the sample with dialysis buffer D2 for 4 h to overnight at 4 °C to remove urea.
26. Remove aggregates by centrifuging the sample at 30,000 × g for 20 min at 4 °C.
27. Dialyze the supernatant with dialysis buffer D3 for 4 h to overnight at 4 °C.
28. Dialyze with dialysis buffer D4 for 4 h to overnight at 4 °C.

29. Dialyze with dialysis buffer D5 overnight at 4 °C.
30. Centrifuge proteins at 30,000 × g for 20 min at 4 °C to remove aggregates.
31. Make small aliquots (2–4 μL) from the supernatant.
32. Freeze in liquid nitrogen and store at −80 °C.

3.4 Assembly of Injection Chambers

We use an injection chamber adapted from Jaffe and Terasaki [18] which can be adjusted to embryos of distinct sizes (Fig. 1a–c).

1. Using the diamond glass cutter pen, cut small glass pieces (about 5 × 10 mm) from a coverslip. This will serve as the "ledge" against which the eggs will align.
2. With a pipette tip or glass capillary, apply a small amount of VaLaB to one side of the ledge. Place it on a 22 × 22 mm coverslip, about 2 mm from the edge.
3. Place the coverslip + ledge on a warm heating plate (50 °C) for a few minutes.
4. Remove the coverslip + ledge and let it cool down.
5. If the eggs to be injected are larger than 150 μm in diameter, add another overlapping glass piece with VaLaB, so the ledge will be taller than the egg.
6. Take the customized chamber holder, and using the Dow Corning grease-filled syringe, apply a thin line of grease to both sides of the U-shaped opening.
7. Attach the glass coverslip + ledge to the chamber holder on the objective side (bottom). Press the grease to seal it flat.
8. Seal the second 22 × 22 mm glass coverslip on the other (top) side of the holder to complete the chamber (Fig. 1a, b).
9. The injection chamber can now be filled with seawater (or other liquid media for live embryos).
10. Pipette the eggs into the injection chamber, near to the open side but at a safe distance from the opening and exposure to air.
11. Tilt the holder slightly so the eggs will line up along the ledge by gravity, and then place it on the microscope stage for injection.

3.5 Preparation of Proteins for Injection

1. Protein aliquots are thawed quickly by holding them in hand or placing them into room-temperature water.
2. As soon as liquid appears on tube walls, place the tubes on ice.
3. Tubulin is diluted in a calcium-free 1xPBS to a concentration of 2–4 mg/mL.
4. H2B-RFP protein is used at a concentration of 4 mg/mL diluted in a calcium-free 1xPBS.

5. When mixed together, tubulin is diluted directly in the histone solution.
6. To remove aggregates, centrifuge the mixture at 10,000 × g for 5 min at 4 °C.

3.6 Injection

We recommend trying different dilutions and injection volumes. Injection of a large excess of tubulin can cause premature pole fragmentation in anaphase, formation of cortical arrays in meiotic eggs or during mitosis, asymmetric cleavage, or loss of cell cycle synchrony [14].

1. Injected volume is estimated to be between 1% and 10% of the volume of the egg as previously described [8].
2. This volume can be calibrated by measuring the diameter of an aqueous droplet injected into an oil droplet using similar injection parameters. In some eggs, we see a transient round clearing in the cytoplasm upon injection which gives a direct estimate of the injected volume.

3.7 Mounting of Injected Embryos for Imaging

1. Place a drop of seawater onto the glass of a glass-bottom dish, 100–120 μL for a 35 mm diameter dish (Fig. 1d).
2. Pipette the injected and fertilized eggs into the drop.
3. Using a syringe filled with Dow Corning grease, apply a thin outline of grease around but not touching the drop.
4. Place a 22 × 22 mm glass coverslip on top of the drop taking care not to trap air bubbles near the eggs.
5. Press gently on the coverslip to stabilize its contact with the drop and the grease.

3.8 Acquisition Settings on Confocal Microscopes

When mitotic spindles assemble perpendicular to the light path (parallel to the slide), the observer can easily visualize long microtubule trajectories within different subcellular populations of microtubules. In this case, 2D spatial imaging (x-y-t) will be sufficient to observe and track microtubules with high temporal resolution. The key requirement to measure microtubule dynamics, in addition to a high signal-to-noise ratio, is a frame rate between 0.5 and 2 images per second [3, 14, 16]. If microtubules grow at a speed of 0.4 μm/s and the frame rate is 1 image per second, a pixel size of 200 nm would be appropriate, and the objective should be chosen accordingly. This pixel size is typically achieved with 60× oil immersion objectives on most cameras. Water or silicon immersion objectives can also be used with thicker specimens, which is often the case for marine embryos.

3.9 Measurement of Microtubule Dynamics from Movies

1. To better visualize microtubule tracks, create a maximal intensity projection of the time frames corresponding to the period over which you want to extract dynamic parameters (*Image>Stack>Z Project...*) (*see* **Note 15**).
2. Microtubule tracks are drawn on the max intensity projections or directly on the stack using the "Segmented line tool" in ImageJ.
3. Every track position is recorded in the region of interest (ROI) manager (*Edit>Selection>Add to Manager*) or by pressing "T" on the keyboard.
4. Select all the tracks and save them as an RoiSet.zip file (*More>Save*). This RoiSet.zip can be renamed and saved with the corresponding image sequence. It can be re-opened anytime to reuse, modify, or add any ROI.
5. Kymographs can then be created for each ROI using our macro "*Multi_ROI_kymo.ijm*" (available on https://github.com/benlacroix/ImageJ_Macros). This macro will create a kymograph for each ROI and save them as individual images in a single folder (*see* **Note 16**).
6. Kymographs in ImageJ represent the signal in gray value of the ROI (each ROI is a line corresponding to a microtubule track) with the horizontal and vertical axes corresponding to distance and time, respectively. Each horizontal pixel corresponds to a distance equal to pixel size in the acquired image and each vertical pixel represents a single frame of the image sequence.
7. Dynamics data is extracted manually on each kymograph by measuring the slope (straight lines command) of polymerization and depolymerization events (Fig. 2a). To do so, draw a line that matches each event of interest and press the M key (*Analyse>Measure*).
8. ImageJ will open a "Results" window that can be saved as a .csv file and imported into a spreadsheet software. The user can set which measurements are displayed in this window. To extract the dynamic parameters, the "Angle" and the "Length" of each slope are required.

3.10 Analysis and Interpretation of Dynamic Properties

1. Example files to practice data extraction are available on https://github.com/benlacroix/MTDynamX_Data_Xtraction.
2. Angles are measured following a trigonometric circle (counterclockwise) by ImageJ but are displayed in degrees. Slopes of growing microtubules, therefore, generate an angle between 0° and −90°. Depolymerization events generate an angle between −90° and −180° and pause events an angle $\alpha = -90°$ or $= 90°$ (*see* **Notes 17** and **18**).

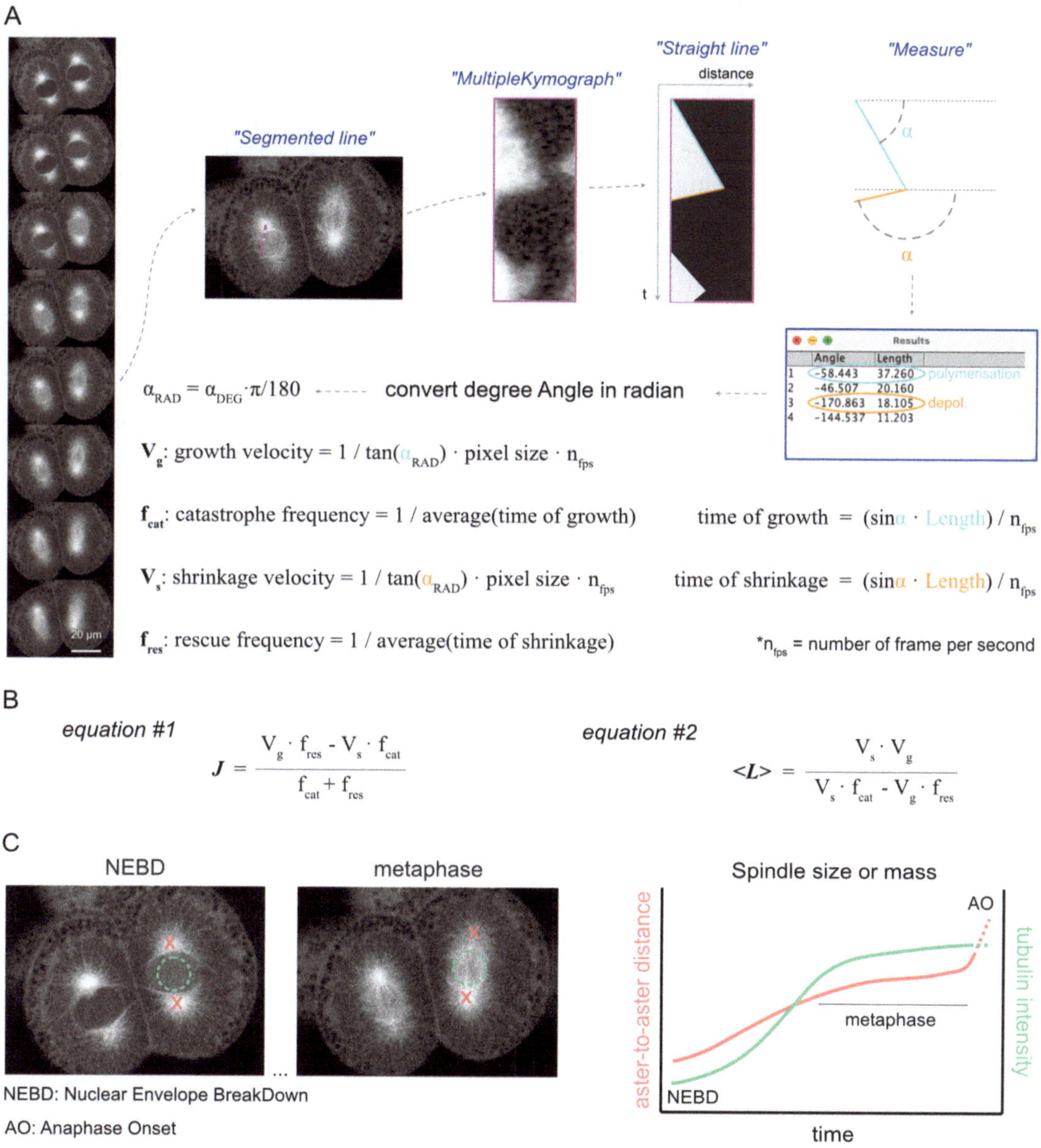

Fig. 2 Extraction of microtubule dynamics and spindle assembly properties. (**a**) Extraction of microtubule dynamics using kymographs. The images show fluorescent tubulin in two cells entering mitosis, from a time-lapse movie of a *P. mammillata* embryo. In quotes are the ImageJ tools and functions used for the analysis. Using a "Segmented line" tool, individual microtubule tracks are drawn. Kymographs are then generated from the movies using the plugin "*MultipleKymograph*" (embedded in the macro "*Multi_ROI_kymo.ijm*," *see* Subheading 3.9, **step 5**). A "Straight line" tool is then used to draw the slopes corresponding to polymerization events (cyan) and depolymerization events (orange). ImageJ can then "Measure" and report the "Angle" and the "Length" for each individual slope. Growing microtubules, therefore, generate an angle between 0° and −90° (cyan). Depolymerization events generate an angle between −90° and −180° (orange). The example shows two cycles, with lines 1 and 3 in the Results box corresponding to the top kymograph and lines 2 and 4 corresponding to the bottom kymograph. Angles are converted to radians, and each parameter is extracted using the formulas shown. (**b**) Eq. 1 is used to estimate the average growth rate of the entire microtubule population. In mitosis, this should be a negative number (bounded regime) [5]. Equation 2 estimates the

3. Polymerization ($0° > \alpha > -90°$) and depolymerization ($-90° > \alpha > -180°$) can be easily sorted by their angle value.
4. Velocities (distance/time) can be extracted from these measurements as they correspond to the inverse of the tangent of the angle measured by ImageJ (Fig. 2a). For this calculation, angles in degrees must be converted into radians.
5. Similarly, growth or shrinkage duration can be extracted using the sine of the angle and the length of the straight line drawn on growth and shrinkage events (*see* Subheading 3.9). Catastrophe and rescue frequencies are calculated, respectively, as the inverse of the average growth duration and the inverse of the average shrinkage duration (*see* **Note 19**).
6. Polymerization and depolymerization distances are extracted using cosine of the angle measured by ImageJ and the corresponding length.

3.11 Determining the Regime of Microtubule Dynamics

In mitosis, microtubule catastrophe is high, and microtubules depolymerize all the way to the spindle poles [5, 19, 20]. The consequence of this behavior is that the average microtubule length is restricted by the microtubule dynamic properties. This dynamic state corresponds to the "bounded" regime [5] as opposed to the "unbounded" regime of interphasic microtubules, which tend to continuously elongate. Verde et al. proposed a mathematical model and a simple equation that describes the dynamic regime of microtubules using the observed values of the four dynamics parameters (Eq. 1, Fig. 2b [5]). The value of microtubule average velocity ("J") becomes negative when microtubules are in the bounded regime and positive in the unbounded regime. In addition, in the bounded regime, it is possible to predict the average microtubule population length ($<L>$) using Eq. 2 (Fig. 2b) proposed by Verde et al. [5]. The mathematical estimation of microtubule average length can be useful if the microtubule length cannot be determined experimentally.

3.12 Assessing Spindle Assembly by Measuring the Spindle Length

Spindle assembly can be followed by measuring the distance between spindle poles or between centrosomes throughout spindle assembly and plotting this distance over time.

1. Based on tubulin fluorescence intensity, draw a straight line between the two points at the centers of the two asters (Fig. 2c).

Fig. 2 (continued) average microtubule length (only if microtubules are in the bounded regime). (**c**) Using time-lapse sequences, spindle length can be extracted over time by measuring aster-to-aster distance (red). Alternatively, tubulin intensity within the spindle area can be analyzed overtime to follow spindle assembly (green curve)

2. Measure the length of this line (*Analyse>Measure*), and repeat this process for every time point.
3. Plot the measured distances relative to time to follow the evolution of spindle length.

3.13 Assessing Spindle Assembly by Measuring the Tubulin Intensity Within the Spindle

Depending on the cell type, the distance between asters or centrosomes may not always be a valid readout of spindle assembly because this distance does not necessarily increase. In addition, this distance does not reflect how the mitotic spindle assembles in 3D. A more valuable readout to estimate spindle assembly could, therefore, be to measure the spindle mass [21, 22] defined by the quantity of microtubules (or tubulin) assembled in the structure. We found that following the evolution of the microtubule density (tubulin signal) within the spindle can serve as a simple and useful proxy to estimate spindle assembly kinetics (Fig. 2c).

1. Make sure that the spindle or the embryo does not move or drift during the time of spindle assembly (*see* **Note 20**).
2. Select an ROI remaining within the spindle from NEBD through to anaphase.
3. Use the command "Measure Stack" (*Image>Stack>Measure Stack*).
4. Plot the integrated density or the mean gray value over time. As shown in Fig. 2c, the intensity will begin to increase at NEBD and will reach a plateau at metaphase.

3.14 Identification of Mitotic Phases Using the Histone Marker

Injection and imaging of recombinant histone-RFP can be performed to better define the stages of cell division in each species (Fig. 3) and compare with the results obtained using tubulin. Mitotic progression is easier to visualize using a chromatin marker as NEBD; metaphase and anaphase onset will be clearly identified (Fig. 3c).

1. NEBD can be characterized by following histone-RFP background intensity within nuclear area. At NEBD, this diffuse histone-RFP signal within the nuclear area drops suddenly as proteins become diluted in the cytoplasm.
2. Cells are considered in metaphase when chromosomes are all aligned on a single plate.
3. We define the anaphase onset as the first timepoint where a clear separation between sister chromatids is observed.

4 Notes

1. Commercially available fluorescent tubulin can also be used. However, in our hands, some batches gave rise to aggregation and poor signal-to-noise ratio impeding visualization of microtubule fibers.

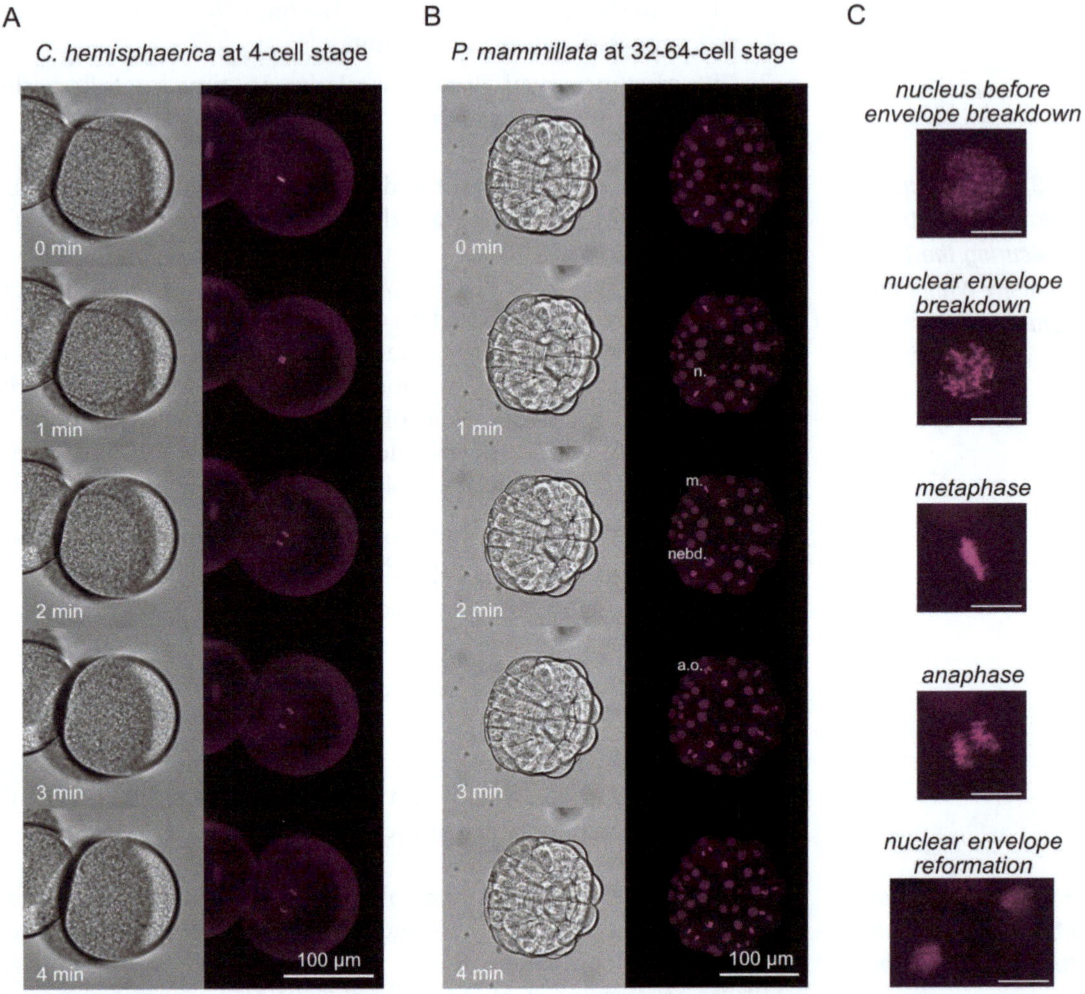

Fig. 3 Mitotic landmarks visualized using histone H2B-RFP as a chromatin marker (**a**) Images from a time-lapse acquisition of a *Clytia hemisphaerica* embryo derived from an egg injected with histone H2B-RFP protein and imaged on a confocal microscope. Transmitted light on the left, RFP emission signal (561 nm) on the right (magenta). (**b**) Same as (**a**) for a *Phallusia mammillata* embryo. On the RFP signal (magenta), mitotic events can be observed: *n.* nucleus prior to nebd, *nebd* nuclear envelope breakdown, *m* metaphase plate, *a. o.* anaphase onset. (**c**) Selected images from a time-lapse sequence showing DNA (H2B-RFP signal, magenta) in a single cell of a developing *Phallusia mammillata* embryo as in **b**. Detectable mitotic landmarks are indicated. Scale bar = 10 μm

2. Pig versus bovine tubulin. We recommend the use of pig brain rather than bovine for two main reasons. Firstly, we observed that in vitro the critical concentration necessary for tubulin polymerization was higher for bovine tubulin. Thus, a higher tubulin concentration is necessary for similar polymerization kinetics. Secondly, the term “bovine” at the slaughterhouse refers to any individual of the species *Bos taurus* and can include animals of very different ages and selected for different

purposes (milk, meat, calving, etc.). "Pig" instead refers to *Sus scrofa* of a relatively narrow range of age (between 6 and 9 months) mostly dedicated to meat products. It is mainly to avoid these age differences and variability in protein quality that we prefer pig brain as tubulin source.

3. We used ATTO 488 and ATTO 565 in our experiments as they have a relatively high fluorescence quantum yield and are resistant to photobleaching.
4. Polysuflone, PVC, or polycarbonate do not withstand high concentration of DMSO. The labelling reaction can be initiated in polypropylene tubes which are DMSO-compatible. Once DMSO concentration has been diluted (to <10%), the solution can be transferred to any type of ultracentrifugation tubes. For tubulin labelling, as the volumes are relatively small, it is recommended to use a microvolume (<1 mL) ultracentrifuge with a high-speed capacity (>50,000 × g).
5. The geographical dependence for animal availability is illustrated by the use of sea urchin species; for example, *Strongylocentrotus purpuratus* is present on the west coast of North America but absent from the east coast and European shores. We know from our previous work and past studies that injection of tubulin is a valuable tool to study spindle assembly in *Lytechinus variegatus* (East coast of North America) and *Paracentrotus lividus* (European Atlantic and Mediterranean coasts) [8, 10, 14].
6. The custom holder can be adjusted for every injection setup but typically has the dimensions of a standard microscopy slide (75 × 25 mm). It should have a U-shaped opening, which allows both good visibility with the objective (using an inverted microscope) and access from the side for a pipette or injection needle. The thickness of the U-shaped part should not be too great (<6 mm) to allow surface tension to maintain seawater in the chamber [15].
7. Egg size can be a limiting factor since the signal will decay with distance from the microscope objective. Higher magnification and numerical aperture objectives usually have a shorter working distance. Using water immersion 40× magnification objective with 1.1 numerical aperture, we have imaged specimens as large as 210 μm diameter using *Clytia hemisphaerica* embryos.
8. As microtubule dynamics parameters are temperature sensitive [23, 24], it is advisable to maintain consistency in temperature during and between acquisitions.
9. Depending on the acrylamide quality and on the bis-acrylamide concentration in the SDS-PAGE gel, alpha- and beta-tubulin can potentially be separated and appear as two separate bands.

10. The solution might become turbid very quickly during addition of glycerol. This does not affect polymerization efficiency.
11. Temperature during centrifugation is set at 35 °C to avoid rotor or centrifuge overheating, which would alter protein quality.
12. This is achieved by cutting off around 5 mm from the narrow opening of a pipette tip as the pellet may otherwise clog the tip.
13. The molecular ratio for labelling should correspond to a 10- to 20-fold excess of fluorophore molecules per tubulin molecule. We estimate the tubulin amount assuming a 70% polymerization efficiency (0.7 × initial tubulin quantity) and a molecular weight of 110,000 g/moL. The final DMSO concentration should be ≤10%.
14. We do not add GTP to the tubulin in order to reduce the risk of polymerization in the injection needle. GTP at a concentration of 0.1 mM is usually used for the long-term storage of tubulin; however, we have injected and imaged labelled tubulin stored without GTP for 8 years at −70 °C without any obvious sign of alteration.
15. Although called "Z-project," this is a timeseries so it corresponds here to a temporal projection.
16. In the macro "*Multi_ROI_kymo.ijm*" as in the "MultipleKymograph" plugin, the linewidth on which the kymograph is made can be adjusted. If $n > 1$, the signal of n pixels will be averaged. Larger linewidths are more likely to capture microtubule events if the fibers move slightly, but the resulting kymograph will become blurry. We usually use $n = 1$ or 3.
17. Microtubule rescue is very rarely detected in mitosis as microtubules usually depolymerize all the way to the organizing center where the tubulin signal is too strong to enable observing such events. Thus, rescue frequency of mitotic microtubules can be considered null [5]. In addition, mathematical modeling of microtubule length in mitosis suggests that the rescue frequency has little impact on microtubule length and behavior and can, therefore, be neglected [5]. It can, therefore, seem counterintuitive to define microtubule rescue as the inverse of the average depolymerization time, which provides a non-null value. However, assessing this parameter can still be relevant since microtubules depolymerize rapidly (high shrinkage rate) and frequently (high catastrophe frequency), maintaining a large pool of soluble tubulin that, together with centrosomal nucleation, ensures that new microtubule growth will compensate for the loss of microtubules.
18. Pause events are very rare in mitosis and were never observed in any of our observations on spindle microtubules.

19. In vivo, it can be hard to observe the total life history of individual microtubules. Walker et al. proposed to estimate catastrophe frequency as the inverse of the average time spent growing [2]. Technically, this frequency is assessed by averaging the growth duration of multiple microtubules observed in a given time period.
20. If the object drifts or moves during imaging, several recalibration or "registration" softwares or ImageJ plugins are available to correct for this movement.

Acknowledgments

We thank all members of Alex McDougall's and Remi Dumollard's lab and Stefania Castagnetti's lab at LBDV, IMEV in Villefranche-sur-mer, France, Julien Dumont's lab in Institut Jacques Monod in Paris and Anna Castro's and Thierry Lorca's lab at CRBM in Montpellier, France. We are grateful to the Clytia team especially Evelyn Houliston and Tsuyoshi Momose at IMEV for their assistance and for providing *Clytia hemisphaerica* gametes. We thank Carsten Janke's lab (Institut Curie, Orsay, France) for providing help with tubulin purification and Lydia Besnardeau (LBDV, IMEV) for cloning the pET11-mouseH2B-RFP-6His plasmid. We are also grateful to Christian Sardet for fruitful discussions and his inspiring comments. This work benefited from access to the Institut de la Mer de Villefranche including the Imaging Platform PIM (member of MICA microscopy platform), an EMBRC-France and EMBRC-ERIC Site. Financial support was provided by ANR-10-INBS-02. The project was initiated thanks to EMBRC FR – AAP2019 – n° 3238 (B. Lacroix). The work was funded by ANR MTDiSco ANR-20-CE13-0033 (B. Lacroix) and the European Research Council ERC-CoG ChromoSOMe 819179 (J. Dumont).

References

1. Mitchison T, Kirschner M (1984) Dynamic instability of microtubule growth. Nature 312:237–242
2. Walker RA, O'Brien ET, Pryer NK et al (1988) Dynamic instability of individual microtubules analyzed by video light microscopy: rate constants and transition frequencies. J Cell Biol 107:1437–1448
3. Shelden E, Wadsworth P (1993) Observation and quantification of individual microtubule behavior in vivo: microtubule dynamics are cell-type specific. J Cell Biol 120:935–945
4. Zwetsloot AJ, Tut G, Straube A (2018) Measuring microtubule dynamics. Essays Biochem 62:725–735. https://doi.org/10.1042/EBC20180035
5. Verde F, Dogterom M, Stelzer E et al (1992) Control of microtubule dynamics and length by cyclin A- and cyclin B-dependent kinases in Xenopus egg extracts. J Cell Biol 118:1097–1108
6. Minden JS, Agard DA, Sedat JW, Alberts BM (1989) Direct cell lineage analysis in Drosophila melanogaster by time-lapse, three-

dimensional optical microscopy of living embryos. J Cell Biol 109:505–516. https://doi.org/10.1083/jcb.109.2.505

7. Kellogg DR, Mitchison TJ, Alberts BM (1988) Behaviour of microtubules and actin filaments in living Drosophila embryos. Development 103:675–686. https://doi.org/10.1242/dev.103.4.675
8. Wadsworth P, Sloboda RD (1983) Microinjection of fluorescent tubulin into dividing sea urchin cells. J Cell Biol 97:1249–1254. https://doi.org/10.1083/jcb.97.4.1249
9. Hamaguchi Y, Toriyama M, Sakai H, Hiramoto Y (1985) Distribution of fluorescently labeled tubulin injected into sand dollar eggs from fertilization through cleavage. J Cell Biol 100:1262–1272. https://doi.org/10.1083/jcb.100.4.1262
10. Salmon ED, Leslie RJ, Saxton WM et al (1984) Spindle microtubule dynamics in sea urchin embryos: analysis using a fluorescein-labeled tubulin and measurements of fluorescence redistribution after laser photobleaching. J Cell Biol 99:2165–2174. https://doi.org/10.1083/jcb.99.6.2165
11. Castoldi M, Popov AV (2003) Purification of brain tubulin through two cycles of polymerization-depolymerization in a high-molarity buffer. Protein Expr Purif 32:83–88. https://doi.org/10.1016/S1046-5928(03)00218-3
12. Hyman AA (1991) Preparation of marked microtubules for the assay of the polarity of microtubule-based motors by fluorescence. J Cell Sci Suppl 14:125–127
13. Prodon F, Chenevert J, Hébras C et al (2010) Dual mechanism controls asymmetric spindle position in ascidian germ cell precursors. Development 137:2011–2021. https://doi.org/10.1242/dev.047845
14. Lacroix B, Letort G, Pitayu L et al (2018) Microtubule dynamics scale with cell size to set spindle length and assembly timing. Dev Cell 45:496–511 e6. https://doi.org/10.1016/j.devcel.2018.04.022
15. Yasuo H, McDougall A (2018) Practical guide for ascidian microinjection: phallusia mammillata. Adv Exp Med Biol 1029:15–24. https://doi.org/10.1007/978-981-10-7545-2_3
16. Lacroix B, Bourdages KG, Dorn JF et al (2014) In situ imaging in C. elegans reveals developmental regulation of microtubule dynamics. Dev Cell 29:203–216. https://doi.org/10.1016/j.devcel.2014.03.007
17. Lacroix B, Maddox AS (2014) Microtubule dynamics followed through cell differentiation and tissue biogenesis in C. elegans. Worm 3: e967611. https://doi.org/10.4161/21624046.2014.967611
18. Jaffe LA, Terasaki M (2004) Quantitative microinjection of oocytes, eggs, and embryos. Methods Cell Biol 74:219–242
19. Belmont LD, Hyman AA, Sawin KE, Mitchison TJ (1990) Real-time visualization of cell cycle-dependent changes in microtubule dynamics in cytoplasmic extracts. Cell 62: 579–589
20. Verde F, Labbe JC, Doree M, Karsenti E (1990) Regulation of microtubule dynamics by cdc2 protein kinase in cell-free extracts of Xenopus eggs. Nature 343:233–238. https://doi.org/10.1038/343233a0
21. Mitchison TJ, Ishihara K, Nguyen P, Wuhr M (2015) Size scaling of microtubule assemblies in early xenopus embryos. Cold Spring Harb Perspect Biol 7:a019182. https://doi.org/10.1101/cshperspect.a019182
22. Rieckhoff EM, Berndt F, Elsner M et al (2020) Spindle scaling is governed by cell boundary regulation of microtubule nucleation. Curr Biol 30:4973–4983.e10. https://doi.org/10.1016/j.cub.2020.10.093
23. Srayko M, Kaya A, Stamford J, Hyman AA (2005) Identification and characterization of factors required for microtubule growth and nucleation in the early C. elegans embryo. Dev Cell 9:223–236. https://doi.org/10.1016/j.devcel.2005.07.003
24. Li G, Moore JK (2020) Microtubule dynamics at low temperature: evidence that tubulin recycling limits assembly. Mol Biol Cell 31:1154–1166. https://doi.org/10.1091/mbc.E19-11-0634

Chapter 13

Whole-Mount Immunofluorescence Staining to Visualize Cell Cycle Progression in Mouse Oocyte Meiosis

Safia El Jailani, Katja Wassmann, and Sandra A. Touati

Abstract

Whole-mount immunofluorescence allows direct visualization of the cellular architecture within cells. Here, we apply this technique to mouse oocytes to visualize spindle morphology and microtubule attachments to kinetochores, using a technique we call "cold treatment," at various phases of the meiotic cell cycle. This method allows the analysis of spindle structures at different meiosis I stages and at metaphase II. An adaptation of the protocol to the cell cycle stage of interest is described.

Key words Mouse oocytes, Spindle, Kinetochore-microtubule attachments, Immunofluorescence

1 Introduction

Mixing of the genetic background and generation of healthy haploid gametes is the fundamental purpose of meiosis. In mammals, and particularly in humans, the number of oocytes is limited, and oocytes undergoing their meiotic divisions are highly prone to chromosome segregation errors. The two successive meiotic divisions of the oocyte require complex cellular machineries such as the bipolar spindle. Spindles are essential for proper tethering of chromosomes, during mitosis and the two meiotic divisions. They are composed of microtubule bundles whose plus (+) ends bind to chromosomes through protein assemblies called kinetochores, located at a chromatin region called the centromere. In oocyte meiosis, spindle appearance and structure are specific to the cell cycle stages and first or second meiotic division. Here, we provide guidelines to comprehensively illustrate the different stages of the cell cycle in terms of kinetochore-microtubule attachments, spindle shape, and chromosome shape. We describe in detail how to depict each step of chromosome segregation by simultaneously immunolabelling chromosomes, centromeres, and spindles from prophase I to metaphase II stages in mouse oocytes. Immunofluorescence of

Anna Castro and Benjamin Lacroix (eds.), *Cell Cycle Control: Methods and Protocols*, Methods in Molecular Biology, vol. 2740, https://doi.org/10.1007/978-1-0716-3557-5_13,

whole-mount oocytes allows the visualization of spindle shape and chromosomal attachments, which is the first step to determine whether progression through meiosis takes place correctly and to spot any potential errors occurring at the various cell cycle stages. The fixation method, immunofluorescence technique, and conventional fluorescence microscopy described here require reagents and equipment accessible in most cell biology laboratories.

Mouse oocytes measure 80–100 μm in diameter and are transparent, providing an advantage to study chromosomes and spindle architecture by microscopical techniques. Oocytes are naturally arrested at the germinal vesicle (GV) stage, equivalent to prophase I, during which they are morphologically characterized by the presence of a nucleus and absence of the spindle. Upon rupture of the nuclear envelope, a stage called germinal vesicle breakdown (GVBD), oocytes resume meiosis and enter prometaphase I. At GV and GVBD stages, chromosomes are poorly condensed. In prometaphase I, chromosomes condense and their cross-shaped structure, due to the presence of chiasmata (sites of recombination between homologous chromosomes), starts to become visible. Throughout meiosis I, each chromosome is attached to microtubules emanating from opposite spindle poles. The network of microtubules begins to organize and nucleate around the chromosomes. Microtubules attach to kinetochores, first by forming lateral interactions, followed by the establishment of stable, end-on attachments that are under tension from the bipolar spindle. These attachments must be strong and oriented to the correct pole of the spindle for subsequent error-free segregation when chromosomes will be pulled apart by the spindle. One exclusive feature in mammalian oocytes is the absence of centrosomes. Instead, numerous small multiple acentriolar microtubule organizing centers (MTOCs) are already present at GVBD and persist during prometaphase I [1–5]. MTOCs are composed of free gamma-tubulin in the cytoplasm, able to generate relatively short and unstable microtubules. A few hours after GVBD, MTOCs begin to cluster to give rise to two poles, conferring bipolarity to the meiotic spindle [6]. Simultaneously, homologous chromosomes begin to line up at the metaphase I plate. For correct chromosome segregation at anaphase I, the spindle must be bipolar and stabilized. To achieve this, proper attachment of kinetochores in metaphase I is essential for bipolar spindle integrity [7]. It was shown that multiple error-correcting attachment cycles between chromosomes and microtubules are necessary to assure stabilization of attachments [8].

Mouse oocytes take around 7–9 h, depending on the mouse strain, to progress from resumption of meiosis with GVBD to anaphase I onset. Around 1 h before anaphase I, spindles are stabilized and chromosomes are aligned along the metaphase I plate. The whole structure migrates to the cortex of the oocyte, with one pole closer to the surface for the very asymmetric cell division that takes place in oocytes [9, 10]. During anaphase I, chromosomes are pulled apart by the spindle through the action of molecular motors that allow the spindle to lengthen and chromosomes to be segregated into opposite directions. Concomitantly, a process called kinetochore individualization occurs with the dissociation of fused sister kinetochores. Indeed, meiosis I homologous chromosomes harbor two sister kinetochores that remain fused together to form one single entity when caught by the spindle microtubules. The subsequent event of individualization in meiosis I has been shown to be essential for sister chromatid segregation in the following meiosis II [11]. Once anaphase I takes place, the genetic material is halved, and one set of chromosomes is extruded into a first small polar body (which degenerates), leaving the other set of chromosomes in the large oocyte. At this stage, chromosomes are in a two-linked-arm structure consisting of two sister chromatids held together in the centromere region. Oocytes enter the second meiotic division without going through an intermediate S-phase. A meiosis II spindle forms, with a morphology that appears distinct from the meiosis I spindle – in meiosis II, the spindle has pointed ends, whereas the meiosis I spindle has a barrel shape due to its poorly focused poles [12]. Throughout the second meiotic division, the metaphase II spindle stays in the subcortical position, parallel to the oocyte surface. The oocyte remains arrested at this cell cycle stage until it is fertilized by a spermatozoid, which induces anaphase II onset, exit from meiosis II, and the formation of a haploid female pronucleus ready to fuse with the male pronucleus [13].

To understand how correct chromosome segregation is mediated, it is often necessary to analyze spindle morphology and chromosome attachments by immunofluorescence techniques. To improve the visualization of kinetochore-microtubule attachments, we apply a technique called "cold treatment" that has originally been developed for mitotic cells in tissue culture [14]. Here, we describe how this method can be adapted to mouse oocytes to destabilize unstable attachments. We provide a detailed description of whole-mount oocyte fixation, cold treatment, and immunofluorescence to properly visualize only stable and, thus, attached microtubule fibers, to determine the quality of chromosome attachments at different cell cycle stages [15, 16].

2 Materials

2.1 M2-BSA Culture Medium Preparation

All solutions are made with water suitable for human embryo transfer and mouse embryos (*see* **Note 1**).

1. Solution A (10×): Dissolve 5.534 g NaCl, 0.36 g KCl, 0.162 g KH_2PO_4, 0.294 g $MgSO_4$, $7H_2O$, 2.6 mL of 60% sodium lactate (solution kept at 4 °C), 1 g D-glucose, and 1 mL of 100× penicillin/streptomycin (0.625 g of penicillin + 0.5 g of streptomycin in a final volume of 10 mL water, kept at −20 °C) in water in a final volume of 100 mL (keep at 4 °C for 3 months) (*see* **Note 1**).
2. Solution B (10×): Dissolve 0.218 g $NaHCO_3$ in water in a final volume of 10 mL (keep at 4 °C for 7 days).
3. Solution C (100×): Dissolve 0.036 g sodium pyruvate in water in a final volume of 10 mL (keep at 4 °C for 7 days).
4. Solution D (100×): Dissolve 0.252 g $CaCl_2$, $2H_2O$ in water in a final volume of 10 mL (keep at 4 °C for 3 months).
5. Solution E (10×): Dissolve 10.334 g HEPES in water in a final volume of 100 mL (keep at 4 °C for 3 months). Adjust pH to 7.4 with 10 M NaOH before adjusting to the final volume.
6. Precision scale, pH meter.
7. 0.22 μm filter cartridges.
8. M2-BSA medium: Mix five stock solutions (A, B, C, D, E). For 100 mL of M2-BSA, mix 10 mL of stock A, 1.6 mL of stock B, 1 mL of stock C, 1 mL of stock D, and 8.4 mL of stock E. Add 400 mg of Bovine Serum Albumin Standard (BSA) and bring volume to 94 mL by adding 78 mL of water. Adjust the pH to 7.4 by adding 3–4 drops of 1 M NaOH. Filter-sterilize using a 0.22 μm filter. Make stock medium in 50 mL sterile tubes and store at 4 °C (not more than 1.5 months).

2.2 Chamber Preparation

1. Flat washers composed of stainless steel.
2. Hotplate with metal top (aluminum/silicon alloy top plate).
3. Paraffin wax.
4. Coverslips, 15 × 15 mm.
5. Disposable scalpel, Paramount size 11.

2.3 Oocyte Harvesting + Culture

1. Dissection scissors.
2. Tweezers.
3. Tissue culture dish 60 × 15 mm.
4. Glass capillary pipettes (*see* **Note 2**).
5. M2-BSA medium previously prepared.
6. Mineral oil, suitable for mouse embryo cell culture (*see* **Note 3**).
7. 0.5 × 16 mm needle, mounted on a 1 mL insulin syringe.
8. Binoculars with a diascopic stand and a heated stage.

9. 9–12-week-old SWISS female mice.
10. 100 μg/mL dibutyryl cyclic adenosine monophosphate (dbcAMP). 50 μL aliquots are stored at −20 °C and 4 °C once thawed.
11. 38 °C incubator and 38 °C water bath.

2.4 Cold Treatment, Fixation, and Permeabilization

1. Tyrodic acid: For 100 mL, mix 0.800 g of NaCl (137 mM), 0.020 g KCl (2.7 mM), 0.0265 g $CaCl_2$, $2H_2O$ (1.8 mM), 0.010 g $MgCl_2$, $6H_2O$ (0.5 mM), 0.100 g D-glucose (5.5 mM), 0.400 g polyvinylpyrrolidone (PVP, 0.09 mM), and complete with water. Adjust pH to 2.3 by adding 1 M HCl. Filter-sterilize using a 0.22 μm filter, aliquot, and store at −20 °C.
2. Phosphate-buffered saline solution (PBS, 10×): Dissolve "buffer PBS" tablets in 100 mL of water. Sterilize by filtration using a 0.22 μm filter. Prior to use on mouse oocytes, prepare 1× solution (1xPBS) by mixing 100 mL of 10× concentrated stock solution with 900 mL of sterile H_2O mQ.
3. Concavaline A: Dissolve 5 mg of concavaline A in 25 mL of 1xPBS. Filter-sterilize using a 0.22 μm filter, aliquot, and store at −20 °C.
4. M2-PVP medium: For 50 mL of M2-PVP, mix the following amounts of the five stock solutions (A, B, C, D, E) – 5 mL of stock A, 800 μL of stock B, 500 μL of stock C, 500 μL of stock D, and 4.2 mL of stock E. Add 200 mg of PVP and add 39 mL of water to adjust the volume to 50 mL. Adjust the pH to 7.4 by adding 3 to 4 drops of 1 M NaOH. Sterilize by filtration using a 0.22 μm filter. Make 0.5 mL aliquots in sterile 1.5 mL tubes, and store at −20 °C.
5. Centrifuge equipped with a swing-out rotor for 50 mL tubes with homemade holders (15 mL tube's lid, inserted in a cell strainer on top of a 50 mL tube).
6. Glass coverslips, 15 × 15 mm.
7. Curved tweezers.
8. Cold treatment solution (5×): Dissolve 6.048 g PIPES (80 mM), 0.051 g $MgCl_2$ (1 mM) in water and bring the volume to 50 mL. Adjust the pH to 7.4 by adding 10 M NaOH. To treat oocytes, prepare 1× solution by diluting 10 mL 5× stock solution with 40 mL water. Keep at 4 °C.
9. 37% Formaldehyde.
10. BRB80 (5×): Dissolve 6.048 g PIPES, 0.051 g $MgCl_2$, and 0.096 g EGTA and bring volume to 50 mL with water. Adjust the pH to 7.4 by adding 10 M NaOH. Prior to fixation of

oocytes, prepare 1× BRB80 solution by diluting 10 mL of 5× stock solution in 40 mL water. Keep at 4 °C.

11. 20% Triton X100 diluted in H_2O mQ.
12. Tissue culture dish, 60 × 15 mm.
13. Fixation solution: Mix 9.350 mL of 1× BRB80, 150 μL of 20% Triton X100 solution, and 500 μL of 37% formaldehyde (*see* **Note 4**).
14. Permeabilization and blocking solution: Mix 10 mL of 1× PBS, 0.3 g of BSA, and 50 μL of 20% Triton X100 solution.

2.5 Immunostaining and Imaging

1. Blocking solution: 3% BSA dissolved in 1× PBS.
2. Glass petri dish, 150 × 25 mm.
3. Petri dish, 145 × 20 mm.
4. Tweezers.
5. Primary antibodies: To stain spindle microtubules, 1:100 monoclonal mouse anti-alpha tubulin (Clone DM1A, Sigma-Aldrich, T6199) in blocking solution (*see* **Note 5**). To stain centromeres, 1:100 human CREST autoimmune serum in blocking solution. This is a human serum from patients with CREST syndrome.
6. Secondary antibodies: 1:200 Alexa Fluor 488 donkey anti-mouse in blocking solution. 1:200 Cy3 donkey anti-human in blocking solution.
7. Wash solution: 1× PBS solution.
8. DNA staining: Vectashield mounting medium supplemented with DAPI (4′,6-diamidino-2-phenylindole) (*see* **Note 6**).
9. Microscope slides, 76 × 26 mm.
10. Spinning disk or confocal microscope. We use a Zeiss spinning disk microscope equipped with appropriate lasers and filter-sets for three-color imaging and with 10×/0.25 Ph1 ACHRO-PLAN and 100×/1.4 Oil DIC PLAN-APOCHROMAT Zeiss objectives.
11. Software for image acquisition (we use Metamorph 6.0 software) and image analysis (Fiji 5.1.0).

3 Methods

Here, we describe how to apply our method to obtain images of immunofluorescent stainings of whole-mount oocytes at different cell cycle stages. We show acquisitions of spindles at GV, GVBD, GVBD+2 h (early prometaphase I), GVBD+4 h (late prometaphase I), GVBD+6 h (metaphase I), GVBD+7 h (anaphase I), and GVBD +16 h (metaphase II). Furthermore, we compare staining with and without cold treatment for specific stages, namely, GVBD+4 h, GVBD+6 h, GVBD+7 h, and GVBD+16 h. At least 2 days are necessary to obtain immunofluorescence images of meiosis I spindles: A day for mouse oocyte harvesting and fixation, overnight permeabilization, and a second day for immunostaining and image acquisition. Count an extra night for oocyte culture to obtain immunofluorescence of meiosis II spindles after in vitro culture.

3.1 Preparation of Chambers

1. Preheat the hotplate to 100 °C.
2. Wash with water and soap the flat washers and scrape off any corrosion deposits.
3. Dispose on the hotplate several flat washers with the bottom side facing upward as shown in Fig. 1a. The steel should quickly warm up.
4. By wrapping wipe paper around tweezers, clean the washers by removing any residual melting paraffin wax.
5. Using a scalpel, pick up a low amount of paraffin wax with the tip of the blade (*see* Fig. 1a).
6. Very carefully, spread the melting paraffin wax by sliding the tip of the blade around the centre (*see* Fig. 1a).
7. Delicately lay down a coverslip on the melted paraffin wax with the help of tweezers. It is important that the coverslip is placed at the center of the chamber, without protruding corners and making sure that the entire coverslip is properly sealed with paraffin (*see* **Note** 7).
8. Using tweezers, pick up the chamber that is now ready, and let it cool down. Repeat the same process for the rest of the washers to convert them into chambers.

3.2 Oocyte Harvesting + Culture

1. Prewarm one 1 mL aliquot of M2-BSA medium supplemented with 1 μL of dbcAMP solution and one 1 mL aliquot of M2-BSA medium in a 38 °C water bath about 10 min before starting. Supplementing dbcAMP allows to maintain oocytes arrested in prophase I of meiosis.
2. Pull out a glass filter pipette using an alcohol burner and connect it to the aspiration tube.
3. Prepare one petri dish containing small 40 μL droplets of M2-BSA + dbcAMP medium, and another with droplets of M2-BSA medium, and place in a 38 °C incubator. Droplets are covered with mineral oil to prevent evaporation.
4. Prepare an 800 μL drop of pre-prepared and heated medium (M2-BSA + dbcAMP) on a dish on the 38 °C heating plate of the binoculars.

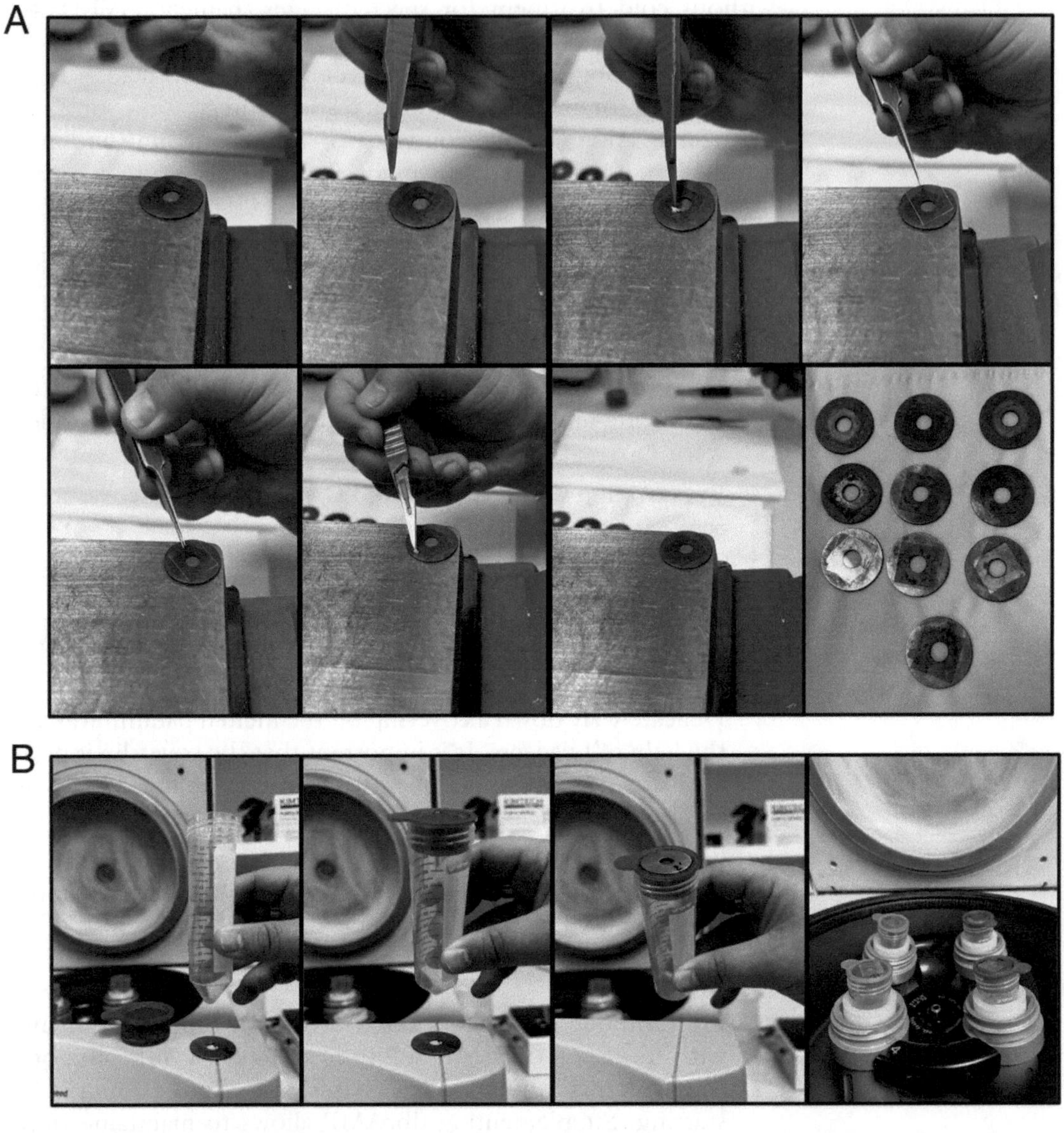

Fig. 1 Illustrations of each step to carry out (**a**) preparation of the chambers and (**b**) placing chambers in the centrifuge. (**a**) As shown, melting paraffin must be equally and delicately spread on the bottom side of the flat washer, using a scalpel. With the help of tweezers, place a coverslip carefully at the center, without protruding corners. If the coverslip is not properly sealed and does not lie flat, add more paraffin by placing the blade next to the "empty" side to fill in the gap by capillarity. (**b**) Before fixation, oocytes must be attached to the concavaline A-treated coverslip at the bottom of the chamber by centrifugation. Appropriate adaptors are needed to hold the chambers steadily during centrifugation. The illustrations show how these homemade holders (15 mL lid inserted into cell strainer on top of 50 mL tube) look like and how chambers are properly placed. Four chambers can be centrifuged at the same time with our setting

5. Dissect a female mouse sacrificed by cervical dislocation in accordance with local regulations for animal experimentation, using scissors and tweezers and collect both ovaries.

Transfer ovaries without fat tissue into the 800 μL drop under the binoculars.

6. Maintain each ovary with tweezers and scratch and poke it using a needle mounted on a 1 mL syringe until the ovary has been completely fragmented and no whole tissue bits are left (put empty tissue debris aside).
7. Under the binoculars, identify fully grown prophase I-arrested oocytes, and collect them from the mixture using the glass pipette mounted on the aspiration tube, using mouth pipetting to manipulate the oocytes. Prophase I-arrested oocytes are recognized due to the presence of a germinal vesicle (GV).
8. Transfer all GV-stage oocytes into a drop of M2-BSA + dbcAMP in the prewarmed petri dish covered with oil using mouth pipetting.
9. Clean oocytes by removing ovarian debris and by denuding oocytes from the surrounding follicular cells by repeatedly aspirating them in and out of the pulled glass pipette before transferring them to clean drops (M2-BSA + dbcAMP). Seek to select only oocytes that are fully grown with a smooth appearance and a centrally located GV.
10. Pull another glass filter pipette using an alcohol burner and connect to the aspiration tube.
11. Transfer oocytes into droplets of dbcAMP-free M2-BSA. In order to ensure that dbcAMP is washed out completely, transfer oocytes sequentially through five to six droplets of M2-BSA medium with as little medium carry over as possible. Place the oocytes in the incubator at 38 °C.
12. After 60 min, oocytes should start undergoing germinal vesicle breakdown (GVBD). GVBD marks meiosis I resumption, it is followed by extrusion of the first polar body occurring approximately 7 h in this strain background. About 16 h after GVBD, oocytes are arrested in metaphase II. The next steps of the protocol should be executed at the correct time after GVBD, corresponding to the cell cycle stage that is going to be analyzed.

3.3 Zona Pellucida Treatment

1. Prewarm one 500 μL aliquot of M2-BSA medium and one 500 μL aliquot of tyrodic acid solution in the 38 °C water bath about 10 min before starting. Tyrodic acid is used to dissolve zona pellucida (ZP) from oocytes.
2. In a petri dish, prepare the first row of three abundant drops of tyrodic acid solution and the second row of three drops of M2-BSA medium without oil (*see* **Note 8**).
3. By mouth pipetting, transfer oocytes in the first drop of tyrodic acid solution and gently aspirate them in and out three times

before transferring them in the second and the third drop, until the ZP disappears.

4. As soon as the ZP vanishes, transfer oocytes in the first drop of M2-BSA. In order to ensure that the tyrodic acid is washed out completely, transfer oocytes sequentially through three droplets of M2-BSA medium with as little medium carry over as possible (*see* **Note 9**).
5. Transfer ZP-free oocytes back into a drop of M2-BSA medium in the petri dish overlaid with oil by mouth pipetting, and maintain oocytes in the incubator at 38 °C.

3.4 Cold Treatment and Fixation

Until fixation, oocytes must stay at 38 °C. Before oocytes reach the cell cycle stage for fixation, start the following preparative steps: Prewarm one 500 μL aliquot of M2-PVP medium and one 500 μL aliquot of concavaline A in a 38 °C water bath about 10 min before starting.

1. Place the chambers on 38 °C heating plate (*see* **Note 10**).
2. Heat up the centrifuge at 38 °C.
3. Wash chambers with 1× PBS three times. Add enough liquid in such a way that the chamber is completely filled. With absorbent paper wrapped around the tip of tweezers, remove the PBS from the side of the chamber to limit the risk of touching the coverslip (*see* **Note 11**).
4. Incubate chambers with concavaline A for 10 min.
5. Wash chambers with M2-PVP.
6. Take out oocytes from incubator. Add M2-PVP to the chamber till the level exceeds by a little bit the edge of the chamber. Immediately, quickly transfer oocytes in a line at the center of the chamber, on the glued coverslip. Close the chamber by gently placing a coverslip on top of the chamber with the help of tweezers, without introducing air bubbles. This step must be done as fast as possible to prevent evaporation of the media (*see* **Note 12**).
7. The following steps are the most critical, as oocytes may easily be lost. Hence, it is useful to check at each step if oocytes were lost to improve handling of oocytes. Carefully place chambers in the centrifuge and cyto-spin the samples at 1200 g and 38 °C, by placing the chambers on top of 15 mL caps that fit onto cell strainers, which are placed onto 50 mL tubes without cap (*see* Fig. 1b). The settings on the centrifuge should be set to "slow acceleration" and "no break to stop." If there is no cold treatment, skip to **step 9**.
8. Prepare cold treatment: Add 10 mL of 1× cold treatment solution into a petri dish placed on top of an ice-water bath, as flat as possible.

9. Prepare fixation solution in a 15 mL tube. Heat the solution in 38 °C water bath.
10. As soon as centrifugation is finished, take out the chambers and place them on a 38 °C heating plate.
11. Slide off very gently the coverslip with the help of angled tweezers (*see* **Note 13**). If there is no cold treatment condition, skip to **step 13**.
12. Cold treatment aims to destabilize unstable microtubules, only leaving strong microtubule bundles for subsequent staining. Dip the chambers delicately into the 10 mL cold treatment solution using tweezers. For oocytes in meiosis I, soak the chambers for exactly 4 min, if oocytes are in meiosis II, leave the chambers for 6 min. During this time, pour the heated fixation solution in a petri dish.
13. Immediately dip the chambers in the preheated fixing solution and incubate for 30 min.
14. Prepare the permeabilization and blocking solution in a 15 mL tube.
15. Wash chambers three times by transferring them sequentially through three different petri dishes filled with 1× PBS.
16. Dip the chambers in the permeabilization and blocking solution and incubate overnight at 4 °C.

3.5 Immunostaining and Image Acquisition

1. After overnight permeabilization and blocking at 4 °C, wash the chambers three times by transferring them sequentially through three different glass petri dishes filled with 1× PBS. Each wash should last 10 min.
2. Prepare primary antibody solution (1:100), in 3% PBS-BSA. Calculate 100 μL of antibody solution per chamber. The antibody solution should be carefully added to the chamber without touching the glass bottom and without creating turbulence that may cause loss of oocytes. Incubate for 2 h in a humidified environment (glass plate, *see* **Note 14**).
3. Wash chambers three times for 10 min with 150 μL 1× PBS to remove unbound primary antibodies. These washes are done by carefully adding 1× PBS directly into the chamber from the edge and removing the wash solution by absorbing the solution with paper, without touching the inside of the chamber.
4. Prepare secondary antibody solution (1:200) in 3% PBS-BSA. Again, carefully add 100 μL per chamber and incubate for 1 h without light exposure.
5. Wash chambers three times for 10 min with 1× PBS, to remove unbound secondary antibodies. Maintain the chambers in the dark.

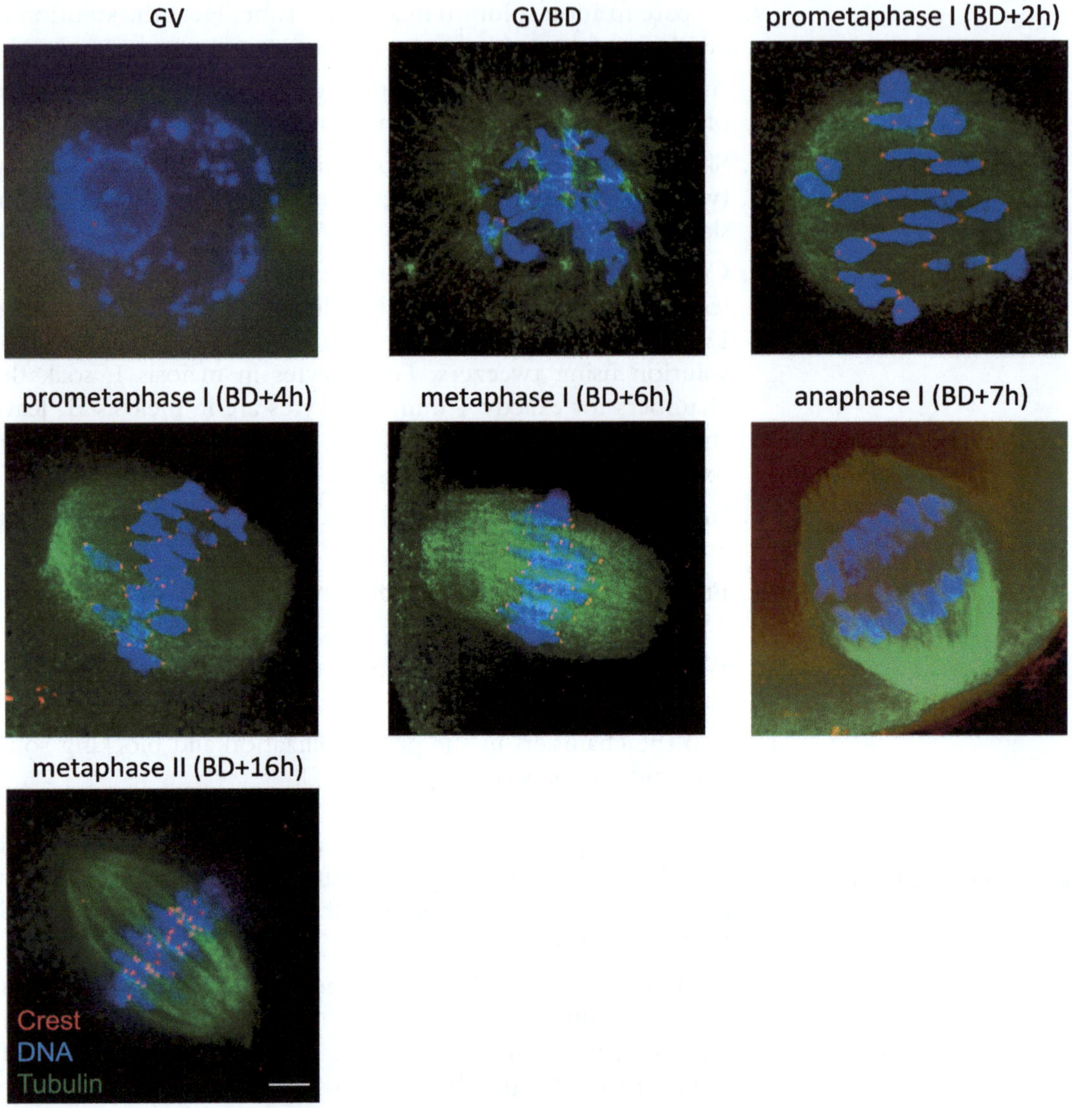

Fig. 2 Whole-mount immunofluorescence staining of oocytes fixed at the indicated cell cycle stages. Microtubules were stained with an anti-tubulin antibody (green), centromeres with anti-CREST serum (red), and DNA with DAPI (blue). As shown, no microtubule nucleation is observed in GV-stage oocytes. Nucleation begins at GVBD; a short barrel-shaped spindle starts to become visible in prometaphase I and is properly put in place at metaphase I. In metaphase II, the elongated spindle shape typical for meiosis II and distinct from metaphase I is visible. The CREST signal is not visible at GV and GVBD stage, before chromosome condensation. During prometaphase I, centromeres are visible as two dots that are very close to each other (often they cannot be distinguished and appear as one dot), whereas in anaphase I, the distance between centromeres increases and sister kinetochores appear distinctly separated – a process called kinetochore individualization. In metaphase II, microtubules bound to sister kinetochores exert pulling forces, that separate the two centromeres well apart. The scale bar indicates 5 μm

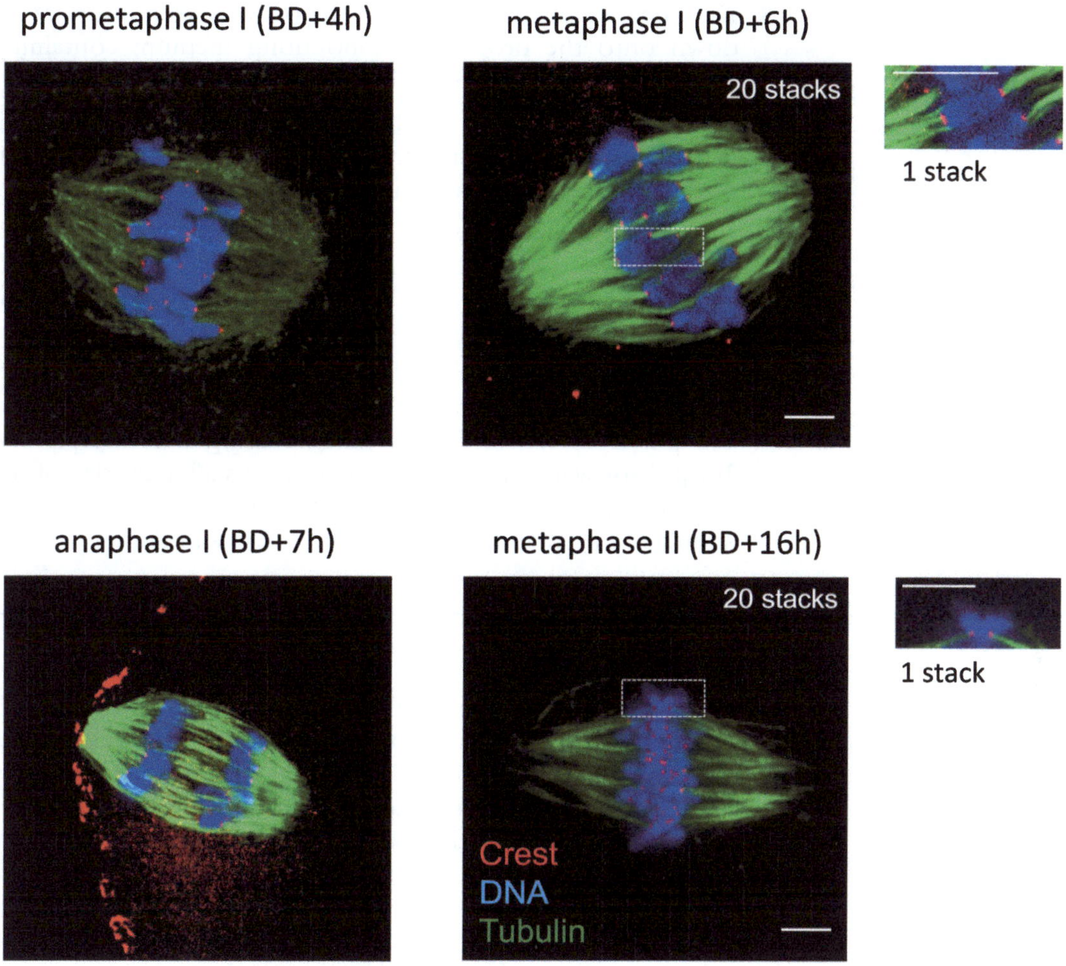

Fig. 3 Whole-mount immunofluorescence staining after cold treatment of oocytes fixed at the indicated cell cycle stages. Microtubules were stained with an anti-tubulin antibody (green), centromeres with anti-CREST serum (red), and DNA with DAPI (blue). At GVBD + 4 h, kinetochore-microtubule attachments are not yet stabilized, as stabilization occurs at GVBD + 6 h during metaphase I, when chromosomes are properly aligned on the metaphase plate. For this reason, after cold treatment microtubule staining is weak at GVBD+4H and much more intense at metaphase I and anaphase I. Also, it is important to note that visualization of microtubule-chromosome attachments is easier to observe in meiosis I than in meiosis II, as chromosomes are less packed at metaphase I stage. The scale bar indicates 5 μm

6. Prepare clean microscope slides. Pipet no more than 8 μL of Vectashield mounting solution supplemented with DAPI onto the slide. An excess of mounting medium will lead to blurry images at acquisitions.
7. Detach and peel off the coverslip from the flat steel support using a scalpel, and once unglued, hold the coverslip with tweezers (*see* **Note 15**).

8. Delicately lay the coverslip with the oocytes fixed on it, faced-down onto the drop of the mounting medium containing DAPI. Avoid creating air bubbles.
9. Seal corners first and then the edges of the coverslip with nail polish and store protected from light at 4 °C until acquisition of the immunofluorescence images.
10. Use a confocal microscope to image spindle microtubules, centromeres, and DNA. DAPI (labelling DNA), Alexa-Fluor488 (labelling microtubules), and Cy3 (labelling centromeres) are detected using 405 nm, 491 nm, and 561 nm wavelength emission lasers, respectively.
11. Acquire images of around 30 Z-stacks for each channel at 0.6 μm Z intervals. Process images using Metamorph and ImageJ software. Examples of acquisitions after overlay of the Z sections of oocytes in the absence of cold treatment are shown in Fig. 2 and of cold-treated oocytes in Fig. 3.

4 Notes

1. Medium stock must be carefully prepared, keeping the following suggestions in mind. For all stock solutions, weighing of powders has to be precise. All solutions have to be sterilized by filtration using 0.22 μm filters. Penicillin/streptomycin solution (0.625 g of penicillin + 0.5 g of streptomycin in a final volume of 10 mL water to make a 100× solution) should be prepared in advance and maintained at −20 °C, before making stock solution A. For solution E, first prepare 40 mL of stock solution, adjust pH using NaOH 10 M, and then water to obtain 50 mL. The pH will change again; hence, it has to be readjusted a second time. For solutions B and C, prepare only 10 mL stock solutions just before preparation of M2-BSA because they cannot be stored without affecting the quality of the medium. We strongly suggest preparing new B and C stock solutions each time M2 medium is made. Finally, the pH of M2-BSA and M2-PVP has to be adjusted carefully to 7.35 (±0.3) with 1 M NaOH and if possible, without using HCl.
2. Oocytes are manipulated by mouth pipetting technique. To do so, hand-pulled slender pipettes are connected to an aspiration tube (homemade) with a suction mouthpiece at the other tip. To prepare the slender pipettes, pull a glass filter pipette using an alcohol burner, snip the glass pipette in the middle, and connect to the aspiration tube (we use parafilm for sealing the pipette with the tube through an adaptor tube with a larger diameter).

3. Oocytes must be kept at 38 °C at all times. Thus, medium must be preheated and covered by mineral oil to avoid evaporation. The oil stock is kept at 38 °C protected from light to avoid oil oxidation (toxic for oocytes). Before use, the stock must be washed three times with water.
4. Protective equipment must be worn while handling formaldehyde solution. Place the fixation solution in a 38 °C water bath before adding formaldehyde, which should be added at the very last minute under a chemical hood.
5. Metaphase II is a cell cycle arrest during which the spindle is much harder to label than in meiosis I, probably due to the fact that oocytes are harder to permeabilize and the spindle is less accessible for antibody staining. To increase the tubulin signal, an anti-DMA1 antibody already coupled to FITC fluorophore can be used as a primary antibody. A secondary Alexa Fluor 488 donkey anti-mouse antibody can additionally be used to further increase the signal.
6. Alternatively to Vectashield supplemented with DAPI, Hoechst can be used. In this case, 1:200 Hoechst (10 mg/mL) is added to the secondary antibody solution, and Vectashield mounting media without DAPI is used.
7. It is crucial that the coverslip is properly sealed to prevent leakage throughout centrifugation and the various incubation steps. Once the coverslip placed, if it does not lie flat and there is even a distinguishable gap between the coverslip and the steel, fill up the "empty" side with paraffin by capillarity (*see* Fig. 1a).
8. Each time we trash our petri dishes, we save the clean lid. Hence, for washing steps like this one, we place our drops on a lid and not the dish to reduce our plastic consumption.
9. The ZP surrounds the oocyte. Persistence of ZP or any other cellular debris will alter subsequent antibody labelling. Tyrodic acid treatment leads to disintegration of the ZP so that oocytes become permeable for antibody staining. We advise to start the tyrodic acid treatment around 1 h before fixation step. This allows oocytes to have enough time to recover. The oocyte should be washed in tyrodic acid solution sufficiently long enough to dissolve the ZP but not too long to avoid the risk of compromising the oocyte's survival. Depending on the cell cycle stage of the oocyte, the ZP can be more or less easily removed. Indeed, the more a given oocyte has progressed toward meiosis II, the easier the ZP can be removed, shortening the time that the acid treatment will be applied (metaphase II oocytes will have their ZP disappear a lot quicker than GV stage oocytes; hence, they are much more sensitive to overtreatment causing lysis).

10. More than four chambers at once can be a lot to handle at the same time.
11. We usually mark a dot on the steel chamber to always absorb back the different solutions at the same position to avoid accidently removing the oocytes.
12. The amount of M2-PVP added to the chambers is crucial for the upcoming steps. Not enough media will cause the chamber to dry up and the appearance of air bubbles after placing the second coverslip on top. On the contrary, an excess amount of medium may prevent the suction effect on the upper coverslip. As the bottom coverslip, the top coverslip needs to be centered on top of the chamber with no protruding corners outside the chamber. When the coverslip has been placed on the chamber, never lift it, as this will produce a sucking force that will immediately detach all oocytes! Instead, it is possible to slide it carefully sideways with the help of tweezers. Also, when placing the upper coverslip, it is important to avoid creating air bubbles, whose movements may equally lead to the loss of oocytes or compromise their integrity.
13. The upper coverslip should be removed as soon as the chambers are taken out of the centrifuge. With the use of angled tweezers, place them on top of one corner and slide the coverslip off slowly and delicately. Remove the coverslip in the direction of the corner that is the closest to the edge or – in the event of the formation of an air bubble – the corner that allows removal of the coverslip without sliding the air bubble over the oocytes.
14. To prepare a humid plate, we humidify paper and wrap it around the edges of a 145 × 20 mm petri dish. Place the chambers in the center, and cover the dish with its lid for incubation to avoid evaporation.
15. This step is one of the most delicate ones. To detach the coverslip in the safest way, maintain two corners with your fingers in such a way that when it will start to detach, you can hold it safely between your fingers. Gently with a scalpel, detach the coverslip by sliding the scalpel under the coverslip, which should slowly detach on its own. Once removed from the steel, using the scalpel, scratch off as much of residual paraffin on the coverslip as you can. Otherwise, when mounting the slide, the paraffin will create a gap between the coverslip and the slide, leading to failure to focus and blurry acquisitions. If oocytes need to be labelled again, the coverslip can be detached from the slide and directly relabelled on a petri dish, without the need of mounting a chamber.

Acknowledgments

We thank D. Cladière for comments on the manuscript. Research in the group is supported by the Agence Nationale de la Recherche (ANR-19-CE13-0015) and la Fondation de la Recherche Médicale (Equipe FRM DEQ 202103012574).

References

1. Maro B, Verlhac MH (2002) Polar body formation: new rules for asymmetric divisions. Nat Cell Biol 4:E281–E283
2. Chaigne A, Verlhac MH, Terret ME (2012) Spindle positioning in mammalian oocytes. Exp Cell Res 318:1442–1447
3. Maro B, Howlett SK, Webb M (1985) Non-spindle microtubule organizing centers in metaphase II-arrested mouse oocytes. J Cell Biol 101:1665–1672
4. Van Blerkom J (1991) Microtubule mediation of cytoplasmic and nuclear maturation during the early stages of resumed meiosis in cultured mouse oocytes. Proc Natl Acad Sci U S A 88: 5031–5035
5. Schuh M, Ellenberg J (2007) Self-organization of MTOCs replaces centrosome function during acentrosomal spindle assembly in live mouse oocytes. Cell 130:484–498
6. Clift D, Schuh M (2015) A three-step MTOC fragmentation mechanism facilitates bipolar spindle assembly in mouse oocytes. Nat Commun 6:7217
7. Courtois A, Yoshida S, Takenouchi O et al (2021) Stable kinetochore-microtubule attachments restrict MTOC position and spindle elongation in oocytes. EMBO Rep 22:e51400
8. Kitajima TS, Ohsugi M, Ellenberg J (2011) Complete kinetochore tracking reveals error-prone homologous chromosome biorientation in mammalian oocytes. Cell 146:568–581
9. Schuh M, Ellenberg J (2008) A new model for asymmetric spindle positioning in mouse oocytes. Curr Biol 18:1986–1992
10. Verlhac MH, Lefebvre C, Guillaud P et al (2000) Asymmetric division in mouse oocytes: with or without Mos. Curr Biol 10:1303–1306
11. Gryaznova Y, Keating L, Touati SA et al (2021) Kinetochore individualization in meiosis I is required for centromeric cohesin removal in meiosis II. EMBO J 40:e106797
12. Schatten G, Simerly C, Schatten H (1985) Microtubule configurations during fertilization, mitosis, and early development in the mouse and the requirement for egg microtubule-mediated motility during mammalian fertilization. Proc Natl Acad Sci U S A 82:4152–4156
13. Maro B, Johnson MH, Pickering SJ, Flach G (1984) Changes in actin distribution during fertilization of the mouse egg. J Embryol Exp Morphol 81:211–237
14. Brinkley BR, Cartwright J (1975) Cold-labile and cold-stable microtubules in the mitotic spindle of mammalian cells. Ann N Y Acad Sci 253:428–439
15. Touati SA, Buffin E, Cladiere D et al (2015) Mouse oocytes depend on BubR1 for proper chromosome segregation but not for prophase I arrest. Nat Commun 6:6946
16. Vallot A, Leontiou I, Cladiere D et al (2018) Tension-induced error correction and not kinetochore attachment status activates the SAC in an Aurora-B/C-dependent manner in oocytes. Curr Biol 28:130–139

Chapter 14

Imaging and Analysis of Drosophila Neural Stem Cell Asymmetric Division

Anne-Marie Berisha, Gregory Eot-Houllier, and Régis Giet

Abstract

Cell division is a conserved process among eukaryotes. It is designed to segregate chromosomes into future daughter cells and involves a complex rearrangement of the cytoskeleton, including microtubules and actin filaments. An additional level of complexity is present in asymmetric dividing stem cells because cytoskeleton elements are also regulated by polarity cues. The neural stem cell system of the fruit fly represents a simple model to dissect the mechanisms that control cytoskeleton reorganization during asymmetric division. In this chapter, we propose to describe protocols that allow accurate analysis of microtubule reorganization during cell division in this model.

Key words *Drosophila*, Central nervous system, Neural stem cells, Live imaging, Indirect immunofluorescence, Microtubule, Chromosomes, Microscopy, Asymmetric cell division

1 Introduction

Cell division (mitosis) is a universal process that allows the increase of cell number and also promotes diversity within tissues. This process is characterized by a dramatic rearrangement of the cytoskeleton during mitosis. The stable interphase microtubule cytoskeleton is disassembled to reorganize into a bipolar spindle whose primary function is the segregation of chromosomes. Mitotic spindle assembly defects lead in some case to chromosome segregation errors or failure in cell division, causing a genomic instability observed in most cancer cells [1]. In addition, the mitotic spindle orientation plays an essential role during the development of tissues that relies on accurate stem cell divisions. These stem cells are generally present in restricted numbers in the tissues and ensure their development and repair (homeostasis). Drosophila larval neural stem cells (NSCs) or neuroblasts are widely used as a model system to study asymmetric stem cell division. These cells are located in the central brain of the Drosophila larvae and in the

Anna Castro and Benjamin Lacroix (eds.), *Cell Cycle Control: Methods and Protocols*, Methods in Molecular Biology, vol. 2740, https://doi.org/10.1007/978-1-0716-3557-5_14,

ventral ganglion (Fig. 1a, b). During mitotic entry, polarity cues located at the apical cortex control cell-fate protein localization at the basal cortex. Asymmetric delivery of these molecules during anaphase is possible after the mitotic spindle have aligned along the apico-basal axis during early mitosis. The two daughters will adopt different cell fates: One maintains the NSC fate identity and continues to proliferate (self-renewal), while the other ganglion mother cell (GMC) is subjected to differentiation into neurons or glial cells (Fig. 1c, d). Defects in spindle orientation lead to abnormal delivery and/or equal partition of these cell-fate determination factors. As a consequence, both daughter cells may in turn adopt stemness, leading to an amplification of stem cell-like cells and tumor formation [2–4]. Similar to those in mammalian models, these tumors can be transplanted into adult host flies, leading to the host's rapid death [5–7]. NSC division is also asymmetric in size. Polarity cues and astral microtubules contribute to the positioning of the cell division furrow toward the basal cortex [8, 9]. This results in a significant size difference between future daughter cells. Moreover, each of the centrosomes retains a specific identity associated with a different PCM composition that directly influences the nucleation capacity of the microtubules of these centrosomes [10–12]. Recently, the dynamics of nuclear envelopes associated with nucleus reformation during anaphase was also shown to be an asymmetric process in fly NSCs [13]. Therefore, the use of *Drosophila* NSCs, combined with the power of genetics and the numerous transgenic lines available to the fly community, make them a particularly versatile model system for the dissection of asymmetric cell division in different genetic backgrounds and in the presence of potential cell division inhibitors [14]. In this review, we will describe some protocols used in the laboratory to characterize some aspects of NSC division in fixed and live preparations.

2 Materials

2.1 *Drosophila* Stocks

1. Wild-type strains such as w^{1118} or Canton S.
2. The transgenic Ubi-RFP::α-tubulin line allows the expression of RFP-tagged α-tubulin, under the control of an ubiquitin promoter, to visualize the microtubule network in live samples [15]. We also use a recombinant inscuteable-GAL4 UAS-m-Cherry-α-tubulin that produces higher signal/noise ratio and is mostly expressed in the central brain region of the central nervous system [7, 9].

2.2 Buffers and Reagents

1. PBS: Phosphate buffer saline.
2. Complete Schneider media: Schneider's *Drosophila* medium, 10% fetal calf serum, 100 UI/mL penicillin/streptomycin. The complete Schneider media is stored in 2 mL aliquots at −20 °C.

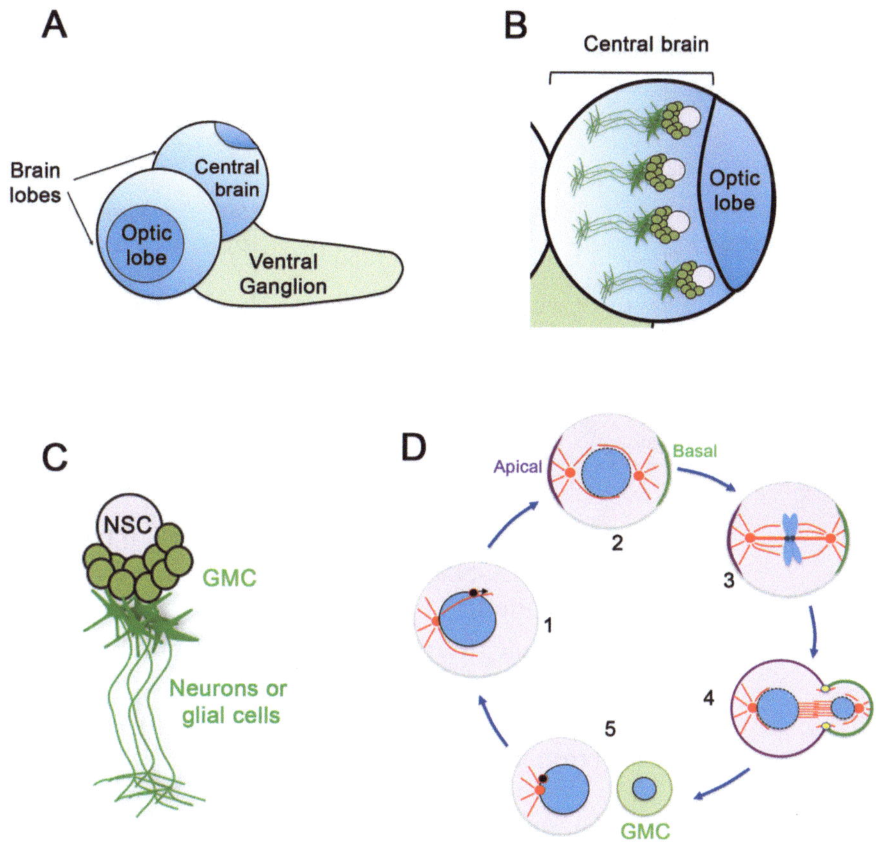

Fig. 1 Overview of the *Drosophila* central nervous system and of the type 1 NSC cell division. (**a**) Schematic lateral view of the central nervous system (CNS) of a third instar larvae. The CNS is composed of two brain lobes and a ventral ganglion. (**b**) Scheme of a brain lobe. Each brain contains an optic lobe and a central brain in which a restricted pool of ≈100 neural stem cells (NSCs) divides asymmetrically. (**c**) The most abundant type 1 NSC generates a self-renewing NSC and a ganglion mother cell (GMC) that will divide to generate differentiating cells (neurons or glial cells). (**d**) Scheme of the asymmetric cell division of an NSC. (1) During interphase the mother centrosome (black dot) is inactive and does not nucleate microtubules. The mother centrosome migrates on the MT network to reach the basal side of the NSC. (2) Full centrosome separation is completed before nuclear envelope breakdown and both centrosomes nucleate robust MT asters. The NCS mitotic entry is marked by the polarization of the NCS: the PAR complex (PAR6, Bazooka, and aPKC) and the MT force generator complex (Mud, Pins, Dynein, and Gai) are recruited at the apical cortex (purple). Concomitantly, adaptor proteins (Miranda, Pon) and their cargoes (the cell fate proteins Prospero, Numb, and Brat) are targeted to the basal cortex (green). (3) The apical MT force generator complex orients the spindle along the apico-basal axis. (4) During anaphase and telophase, an asymmetric furrow (yellow) is established thought the action of polarity cue and astral MTs. (5) After cytokinesis polarity cues and cell fate proteins are distributed into a self-renewing NSC (pink) and a GMC (green), respectively. The GMC will further divide to generate two differentiating cells

Table 1
Primary antibodies

Antibody	Host	IF dilution	Identifier	Source
α-Tubulin (DM1A)	Mouse	1/250	T-6199	Sigma
Detyr-tubulin (YL1/2)	Rat	1/200	Mab1864	Millipore
Phospho-histone H3(S10)	Mouse	1/250	05–1336	Millipore
Phospho-histone H3(S10)	Rabbit	1/250	06–570	Millipore
PKCξ (C-20)	Rabbit	1/250	SC-216	Santa Cruz
Prospero	Mouse	1/10	MR1A	DSHB
Deadpan	Rat	1/50	Ab195172	Abcam

3. Testis buffer: 10 mM Tris, 0.183 M KCl, 0.047 M NaCl, and 1 mM EDTA-NaOH pH 8.0. Adjust to pH 6.8 with a 1 M HCl stock solution.
4. Microtubule fixation buffer: Testis buffer containing 10% formaldehyde and 0.01% Triton ×-100 (*see* **Note 1**).
5. Standard fixation buffer: PBS containing 4% paraformaldehyde (electron microscopy grade) and 0.1% Triton ×-100 (*see* **Note 1**).
6. PBST: PBS containing 0.1% Triton ×-100.
7. PBSTB: PBST containing 1% BSA.
8. ProLong gold mounting media.
9. Nail polish.

2.3 Antibodies

1. Primary antibodies (see Table 1).
2. Secondary antibodies.

   Anti-rat, anti-mouse, and anti-rabbit antibodies conjugated to 488, 568, and 647 Alexa-Fluor (1:500).

2.4 Dissection and Sample Preparation

1. Petri dishes.
2. Pairs of fine forceps n°5 (Dumont).
3. Watches glass-type.
4. μ-Slide 15-well 3D (formerly μ-Slide Angiogenesis, 81506, ibidi).
5. Tungsten needles.
6. A binocular microscope (SMZ745, Nikon) equipped with Schott KL 1500 LED illumination system.

2.5 Microscopes

1. For live imaging.

   A spinning disk system consisting of a DMi8 microscope (Leica) equipped with a 63× (1.4 N.A.) oil objective, a CSU-×1 spinning disk unit (Yokogawa) and an Evolve EMCCD camera (Photometrics). The microscope is controlled by the Inscoper Imaging Suite dedicated software (Inscoper).

2. For fixed samples acquisition.

   SP5 confocal microscope (Leica) equipped with a 40× (1.25 N.A.) and a 63× (1.40 N.A.) objectives, controlled with Leica LAS software.

3 Methods

3.1 Dissection of Larval Brains

1. In a fresh vial supplemented with yeast, transfer 10 females and 10 males of the w^{1118} or RFP::α-tubulin expressing flies.
2. Every 2–3 days transfer the adults into new vials supplemented with fresh yeast.
3. About 96–120 h after egg laying, collect wandering third instar larvae with a pair of fine forceps in a drop of PBS (*see* **Note 2**).
4. Wash the larvae three times for 20 s in 100 μL drops of PBS in a Petri dish to remove fly media.
5. Transfer the larvae in a watch glass-type containing complete Schneider media for live imaging of the RFP::α-tubulin flies (*see* **Note 3**). w^{1118} larvae subjected to fixation are transferred in a drop of PBS or testis buffer depending on the subsequent fixative and antigens to be analyzed (*see* **Note 1**).
6. Under a binocular microscope hold the larva at the middle near the fifth abdominal segment with a first pair of forceps.
7. With a second pair of forceps, grasp the base of the mouth hooks at the anterior end of the larva.
8. Frankly pull the anterior end of the larva. This step allows separating the mouth hooks, the brain, the salivary gland, and disks from the main larval body.
9. For live imaging, carefully remove the mouth hooks, the fat bodies, the disks, and the salivary glands surrounding the brain and proceed to **step 3.3** (*see* **Notes 4** and **5**).
10. For fixation and immunostaining, keep the brain connected with the mouth hooks (*see* **Note 6**).
11. Transfer the dissected brain in one drop (approx. 50 μL) of dissecting medium (*see* **Note** 7).
12. Repeat until 2–5 brains are dissected and proceed to Subheading 3.2 (*see* **Note 8**).

3.2 Immunostaining of Larval Brain

1. Under a fume hood, add 50 μL of the appropriate fixation buffer to the dissected brains.
2. Gently aspirate the brains and transfer in a 1.5 mL tube containing 500 μL of the standard or microtubule fixation buffer (*see* **Note 7**). Turn the tube 5 times.
3. Incubate for 20 min at 25 °C. The brains connected by the mouth hooks sink at the bottom of the tube.
4. Remove the fixation buffer.
5. Add 500 μL PBST and let the brains sinks at the bottom of the tube.
6. Carefully remove supernatant and repeat **step 5**.
7. Wash three times in 1000 μL of PBST for 10 min on a rotating wheel. Let the brains sink between each wash.
8. Incubate the brains with 100–200 μL of the primary antibodies diluted in PBSTB overnight at 4 °C on a horizontal shaker.
9. Remove carefully the primary antibody solution without aspirating the tissues.
10. Wash two times in 500 μL PBST, similarly to **steps 5** and **6**.
11. Wash in 1000 μL of PBST three times for 10 min on a rotating wheel.
12. Incubate with 100–200 μL of the secondary antibodies solution in PBST at least 1 h at room temperature.
13. Remove the secondary antibody solution without aspirating the tissues.
14. Wash two times in 500 μL of PBST, similarly to **steps 5** and **6**.
15. Wash in 1000 μL of PBST three times for 10 min on a rotating wheel.
16. Transfer the brains in a watch glass-type containing 200 μL of PBS (*see* **Note 7**).
17. Using fine forceps, remove remaining tissue around the optic lobes including the mouth hooks.
18. Carefully transfer the brains on a drop (20 μL) of ProLong Gold at the surface of a glass slide.
19. Ensure that the ventral ganglion is positioned at the contact of the glass slide surface (Fig. 2a).
20. Gently place a 20 × 20 mm coverslip on the labelled brains.
21. Let the mounting media spread by capillarity.
22. Seal the coverslip with nail polish.
23. Mark the position of the brains on the slide side with an indelible marker (*see* **Note 9**).

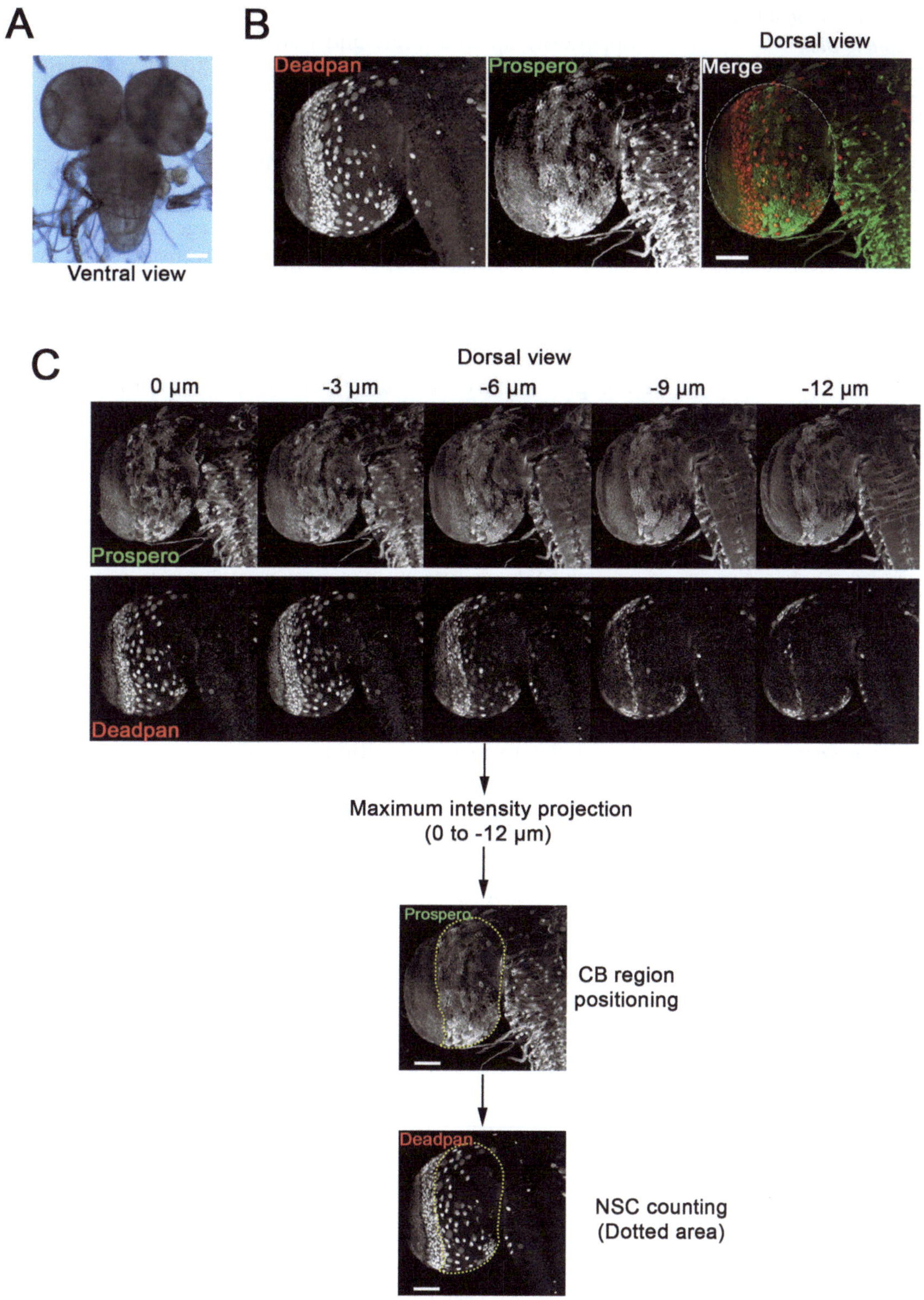

Fig. 2 Quantification of NSCs number in the central brain. (**a**) Picture of a freshly dissected third instar larva brain observed from a binocular microscope. (**b**) Representative images of larval brain lobe of a WT brain stained for the NSC marker Deadpan (red) and the neuronal marker Prospero (green). The white dotted line

3.3 Live Imaging by Spinning Disk Confocal Microscopy

1. After the dissection (Subheading 3.2, **step 9**), use the cut Eppendorf tip to aspirate and transfer the brains in one well of the angiogenesis μ-slide containing 30 μL of complete Schneider media (*see* **Note 7**). Around 1 to 4 brains can be placed in each well (*see* **Note 5**).
2. If necessary, gently turn the brains with fine forceps or tungsten needles to position the ventral side on the glass (*see* **Note 10**).
3. Place the angiogenesis μ-slide on the stage of the spinning disk confocal microscope enclosed in a thermostatic chamber at 25 °C.
4. Using a bright-field illumination, scan the well to locate each brain lobe.
5. Using the 561 nm illumination, position the objective in the central brain and target a lobe with numerous NSCs (*see* **Note 11**).
6. Record the positions of each lobe.
7. Repeat **steps 4–6** for each brain.
8. Select the duration of the experiment, the desired number of Z steps, and the appropriate exposure time for each Z acquisition. On a routine basis, for RFP-tubulin, we acquire 16 Z plane every 1 μm (15 μm depth). Each Z acquisition is performed every 30–60 s on 1–4 brains (*see* **Note 12**).

3.4 Analysis of Fixed NSCs by Confocal Microscopy

3.4.1 Quantification of NSC Number in the Central Brain

Abnormal spindle orientation can lead to a symmetric distribution of identity determinants, resulting in an increase in the number of NSC-like in the central region. To count the number of NSCs in the brain, we use the Deadpan marker (Dpn) [6, 15]. The midbrain region, which tells us where to count NSCs, is determined using the Prospero marker. Images are acquired on an SP5 confocal microscope with a 40× objective (N.A. 1.3). Images are acquired with a width of 1024 × 1024 pixels. The top and bottom planes showing detectable positive cells in the central brain region are defined using the live mode to cover the entire depth of the brain. Z-stack acquisition is performed every 3 or 4 μm. Quantifications of Deadpan-positive NSCs in the Prospero-labeled central brain region are performed using Fiji software [16, 17]. Several

Fig. 2 (continued) outlines the brain lobe. The optic lobe (OL) is localized on the left side of the brain, whereas the central brain region (CB) is on the right side of the brain. Scale bar: 50 μm. (**c**) Workflow for the quantification of NSCs in the central brain. Images corresponding to z-stacks taken every 3 μm from the ventral part of a brain stained for Prospero (top) and Deadpan (bottom). Maximum intensity z-projections for each protein are shown at the bottom. Central brain position is defined by delimitating the region that is positively labelled with the Prospero. The selected region (dotted line) is transferred to projection corresponding to Deadpan staining (bottom, right) for NSC counting. Several maximum projections are needed to quantify the NSC of the whole brain. Scale bars: 50 μm

maximum z-projections covering the entire central lobe are made, and the central brain region is positioned using the "Polygon selection" tool in Fiji using the Prospero neuronal marker (Fig. 2b). Dpn-positive NSCs are manually counted in the region for each projection using Fiji's "multi-point" tool (see example in Fig. 2c). An average of 100 NSCs are present in each brain lobe of a WT strain.

3.4.2 Analysis of Mitotic Spindle Shape and Alignment

Analysis of mitotic spindle assembly and orientation defects can be detected on fixed NSCs after immunostaining for tubulin, to reveal the microtubule network, for phospho-Histone H3 to label mitotic chromosomes, and for atypical protein kinase C (aPKC), to label the apical NCS cortex (Fig. 3a and b). The images are acquired on a SP5 confocal microscope with a 63× objective (N.A. 1.4). The images are acquired with a minimal width of 512 × 512 pixels (with a 4 or 5 × zoom). The upper and lower plans showing mitotic cells in the central brain region are defined using the live mode to cover the entire mitotic cells. Z-stack acquisition is performed every 0.5 μm. Spindle shapes are analyzed using the Fiji software and relevant mitotic figures are analyzed. Distances, such as spindle length, or cell diameter can be measured using the "line" Tool of Fiji. If the two spindle poles and the center of the apical aPKc crescent are in the same focal plane, spindle alignment along the apico-basal axis can be measured directly using the "angle" tool. If this is not the case, proceed as described in Subheading 3.5.2 to accurately measure distances and angles.

3.5 Analyses of Live NSC Division by Confocal Microscopy

The common analyses performed in our lab are described below, including mitotic timing analysis, analysis of centrosome separation, and measure of spindle length. All these analyses are performed using Fiji software.

3.5.1 Mitotic Progression

Defect in mitotic spindle assembly maintains the spindle assembly checkpoint (SAC) in an active state. Persistence of the SAC inhibits the degradation of cyclin B and Securin and delays anaphase onset. Therefore, the time between mitotic entry and exit is a direct readout of mitotic defects [16]. The analysis of mitotic duration is very informative as slightly delayed mitotic progression reveals subtle mitotic assembly defects that are not always spotted on fixed microtubule preparations. Entry into mitosis occurs when the nuclear envelope breaks down (NEBD). NEBD is defined as the mitotic entry point when the nucleus becomes permeable to cytoplasmic tubulin (Fig. 4a). The end of mitosis is marked by the anaphase onset, i.e., the moment when the sister chromatids start to migrate toward their respective poles. This corresponds to the first time point where we observe an elongation of the spindle associated with an increase of the chromosome "ghost" area in the middle of the spindle (Fig. 4a). The mitosis duration for the cell displayed in Fig. 4a is 7 min (*see* **Note 13**).

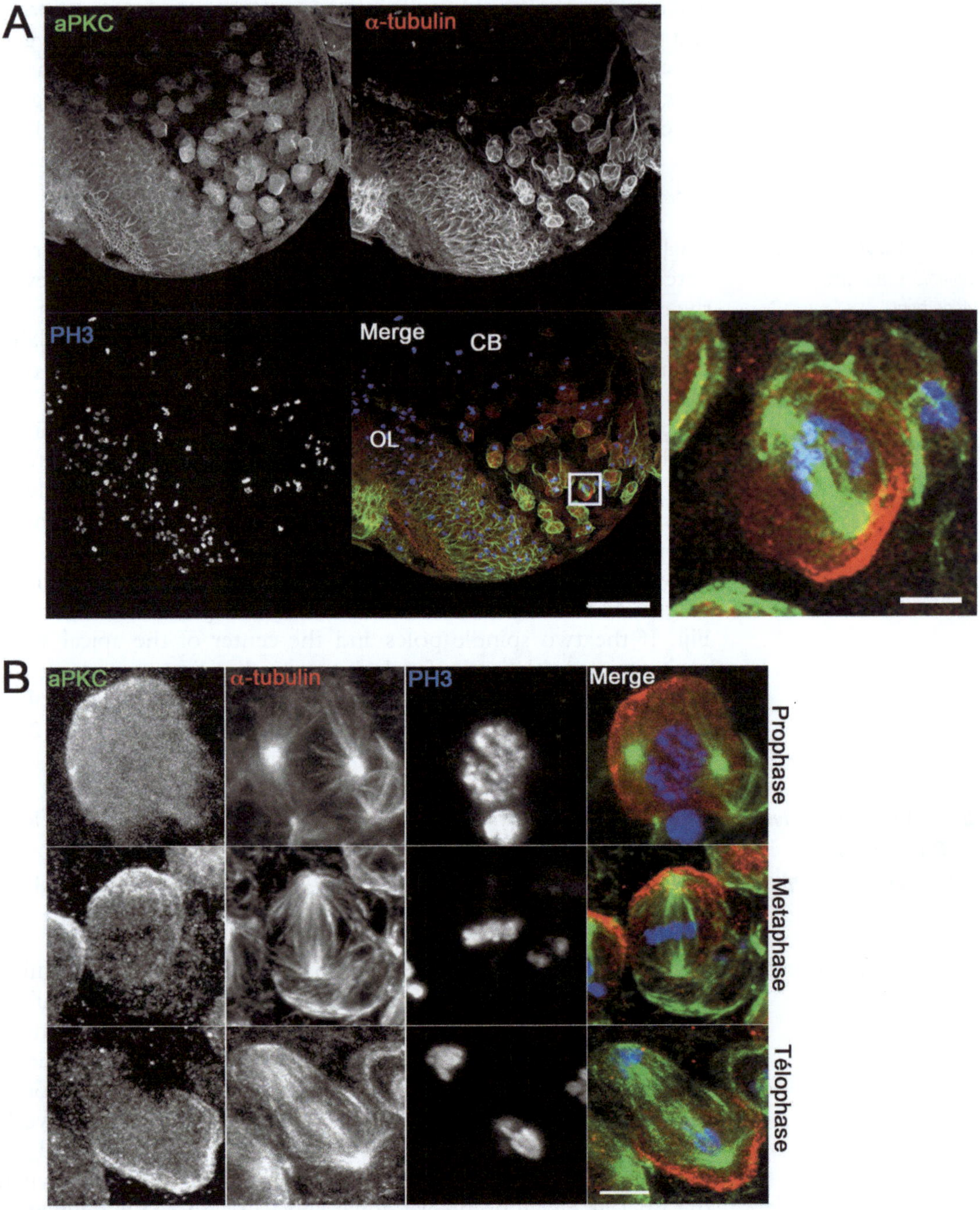

Fig. 3 Microtubules staining of NSCs for mitotic spindle analysis. (**a**) Representative images of a brain lobe stained for the cell polarity protein aPKC (red), mitotic phosphorylated histone H3Ser10 (blue), and α-tubulin (green). Single channels are also shown in monochrome taken with a 40× objective. The optic lobe (OL) is located on the left bottom side of the lobe, whereas the central brain region (CB) is on the top right side of the lobe. Scale bars: 50 μm. A higher magnification view of the metaphase NSC, taken with a 63× objective (zoom5) is shown on the bottom right of the panel. Scale bar: 5 μm. (**b**) Representative images of NSCs during prophase, metaphase, and anaphase. The cells are stained with the same markers as in (A). Scale bar: 5 μm

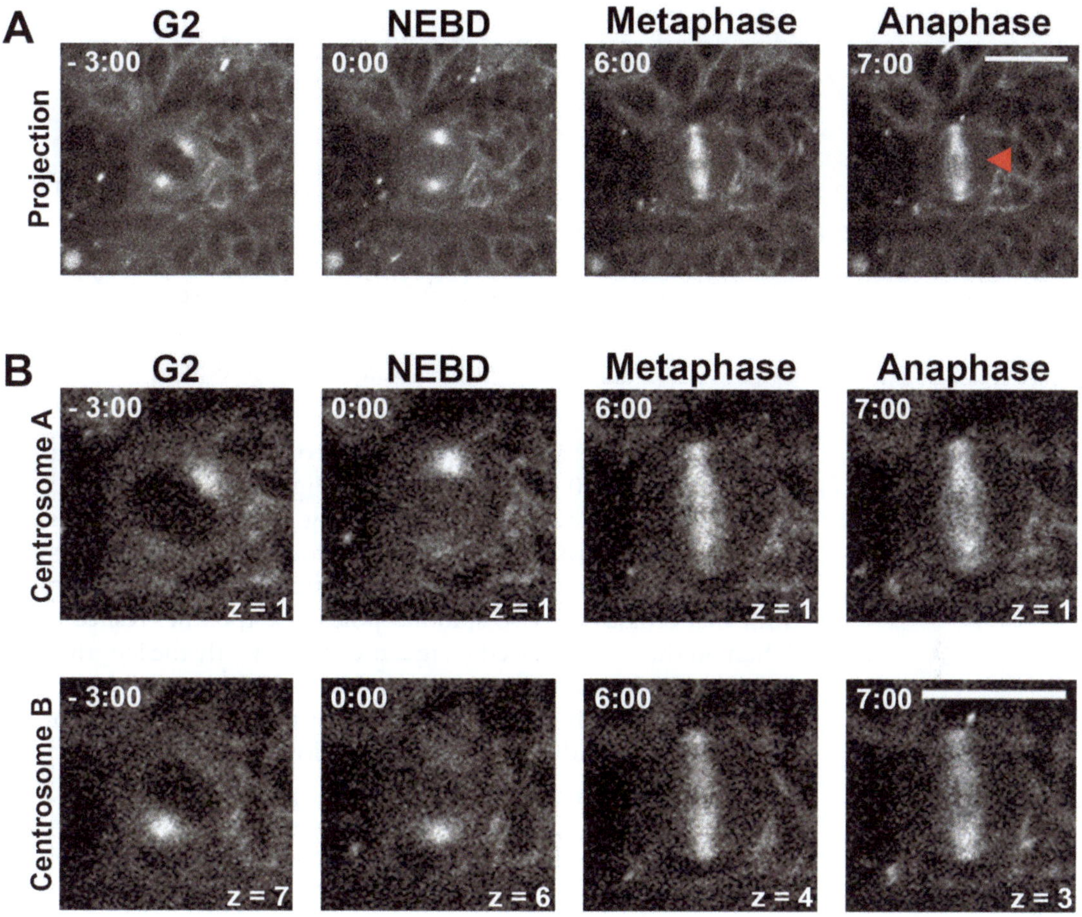

Fig. 4 Analysis of NSC division by live imaging. (**a**) All four panels show a projection of multiple stacks of a time lapse of a dividing NSC in the brain lobe from a RFP::Tubulin expressing fly. Four time points corresponding to G2 phase (−3:00), Nuclear envelope breakdown (NEBD, 0:00), metaphase (6:00), and anaphase (7:00) are shown. The red arrowhead indicates the elongation of the spindle corresponding to anaphase onset. (**b**) Selected panels corresponding to best focal planes showing the centrosomes (G2 and NEBD) or spindle poles (metaphase and anaphase), from the projections shown in A. The z position of each image (μm) is indicated at the bottom right of each panel. Scale bars: 10 μm. The time (top left) is min:sec

3.5.2 Centrosome Separation

Unlike other mammalian cell types, in *Drosophila* NSCs, centrosome separation starts immediately after cytokinesis [10, 12]. Thus, 3 min before NEBD the centrosomes are already fully separated (Figs. 1c and 4). At this time point, the α angle between the A centrosome, the nucleus center (NC), and the B centrosome is a solid read out of accurate centrosome separation [17–19].

To measure the α angle, after image calibration, the coordinates in space (*x*, *y*, *z*) of the two centrosomes (A and B) and the nucleus center (NC) are collected 3 min before NEBD using the "point" and "measure" tools in Fiji software (Fig. 4b; *see* **Note 14**). The following formula is applied to obtain the α angle.

$$\alpha\ \text{Angle} = 180 \times \frac{\arccos \frac{\vec{a}\cdot\vec{b}}{|\vec{a}| \times |\vec{b}|}}{\pi}$$

With

$$\vec{a}.\vec{b} = ((x_{NC} - x_A) \times (x_{NC} - x_B)) + ((y_{NC} - y_A) \times (y_{NC} - y_B)) + ((z_{NC} - z_A) \times (z_{NC} - z_B))$$

$$\overrightarrow{|a|} = \sqrt{(x_{NC} - x_A)^2 + (y_{NC} - y_A)^2 + (z_{NC} - z_A)^2}$$

$$\overrightarrow{|b|} = \sqrt{(x_{NC} - x_B)^2 + (y_{NC} - y_B)^2 + (z_{NC} - z_B)^2}$$

3.5.3 Spindle Length

The length of the spindle is measured from one centrosome to the other, 1 min to 30 s before anaphase (Fig. 4b).

Most of the time, when reviewing 4D acquisition files the spindles are not within the same Z confocal plane. To measure the length of the mitotic spindle, draw a straight line with the "line" tool that crosses the two spindles poles, use the Fiji "reslice" tool. Then on the novel resliced image, measure directly the length of the mitotic spindle using the "straight line" and "measure" tool.

Another method is to pick the positions (x, y, and z) of the centrosomes A and B and to use the following formula to calculate spindle length.

$$\text{Spindle length} = \overrightarrow{|ab|} = \sqrt{(x_B - x_A)^2 + (y_B - y_A)^2 + (z_B - z_A)^2}$$

4 Notes

1. All antibodies listed in the Table 1 work in the immunofluorescence protocol using the microtubule fixation buffer. Dissection in PBS of the brain lobes and their subsequent fixation in the standard fixation buffer are used when no tubulin immunostaining is required (Deadpan, Prospero, and phospho-Histone H3 labelling). However, PBS-based fixatives do not preserve spindle shape and microtubules of NSCs. For this reason, testis buffer and the microtubule fixation buffer are used for dissection and fixation, respectively, of brain samples.
2. It is recommended to use young wandering instar larvae. In younger brains, the NSCs are closer to each other, increasing the probability to spot more dividing cells by live imaging.
3. Use one aliquot of Schneider buffer per experiment. Heat to 25 °C, mix well, and centrifuge at 10,000 g to pellet debris. Use the supernatant for dissection and live imaging in the ibidi culture media.

4. Avoid direct contact of the brain lobes with the forceps. NSCs will not divide properly if the tissue is damaged.
5. For live imaging, rapidly transfer the brain in the ibidi-imaging chamber to prevent evaporation of the Schneider media in the petri dish.
6. For fixed brains, we leave the brains connected to the mouth hooks. This helps brains settlement to the bottom of the tube during the incubations with antibodies and during each washing steps. It also facilitates visualization of brain tissues during experimental manipulation due to the dark color of the mouth hooks.
7. The 200 μL pipet tip diameter is usually smaller than the size of the brain. To avoid brain deformation and stress when transferring the brain tissues by pipetting, cut the pipet tip at 5 mm from the tip end. Pipetting freshly dissected brains in a buffer without detergent or without Schneider media results in irreversible sticking of the tissues to the tip wall.
8. About 2–5 brains can be dissected and transferred in the same drop of dissecting medium.
9. Brain tissues become transparent in the mounting media. Labelling the brain position with a marker allows fast screening of the samples under the confocal microscope.
10. For live imaging, brains, lobes are oriented ventral sides down. In this position, the NSCs are stuck to the coverslip and close to the microscope objective, which increases the quality of the acquisitions.
11. NSCs are localized below the brain surface and in the central part of the brain. NSCs are the largest cells (10–15 μm of diameter) and can be easily distinguished by live microscopy using RFP-tubulin.
12. It is frequent that the brain slightly settles down during the first minutes of acquisitions. It is necessary to control the focus after few minutes.
13. Sample bleaching occurs during acquisition. The user will have to determine the compromise between the length of the acquisition, the laser power, the number, and the size of the Z series, especially if RFP-tubulin is combined with another GFP-marker. Average mitosis duration is 5–7 min in wild-type NSCs, and the cell cycle length is around 50 min [9]. It is therefore not rare during a 2 h film acquisition to see the same NCS dividing three times.
14. The nucleus center is located on the Z plane where the nucleus has the largest diameter. Note that the centrosomes and the nucleus center are rarely on the same Z plane.

References

1. Hanahan D, Weinberg RA (2011) Hallmarks of cancer: the next generation. Cell 144:646–674
2. Homem CC, Knoblich JA (2012) Drosophila neuroblasts: a model for stem cell biology. Development 139:4297–4310
3. Siller KH, Doe CQ (2009) Spindle orientation during asymmetric cell division. Nat Cell Biol 11:365–374
4. Wodarz A, Nathke I (2007) Cell polarity in development and cancer. Nat Cell Biol 9: 1016–1024
5. Caussinus E, Gonzalez C (2005) Induction of tumor growth by altered stem-cell asymmetric division in Drosophila melanogaster. Nat Genet 37:1125–1129
6. Caous R, Pascal A, Rome P et al (2015) Spindle assembly checkpoint inactivation fails to suppress neuroblast tumour formation in aurA mutant drosophila. Nat Commun 6:8879
7. Gallaud E, Richard-Parpaillon L, Aude P et al (2021) Aurora kinases promote mitotic progression and asymmetric division through activation of Polo kinase. BioXriv
8. Cabernard C, Prehoda KE, Doe CQ (2010) A spindle-independent cleavage furrow positioning pathway. Nature 467:91–94
9. Thomas A, Gallaud E, Pascal A et al (2021) Peripheral astral microtubules ensure asymmetric furrow positioning in neural stem cells. Cell Rep 37:109895
10. Rebollo E, Sampaio P, Januschke J et al (2007) Functionally unequal centrosomes drive spindle orientation in asymmetrically dividing drosophila neural stem cells. Dev Cell 12:467–474
11. Januschke J, Reina J, Llamazares S et al (2013) Centrobin controls mother-daughter centriole asymmetry in drosophila neuroblasts. Nat Cell Biol 15:241–248
12. Rusan NM, Peifer M (2007) A role for a novel centrosome cycle in asymmetric cell division. J Cell Biol 177:13–20
13. Roubinet C, White IJ, Baum B (2020) Asymmetric nuclear division of neural stem cells contributes to the formation of sibling nuclei with different identities. bioXriv
14. Pirani V, Métivier M, Gallaud E et al (2020) A novel benzodiazepine derivative that suppresses microtubule dynamics and impairs mitotic progression. J Cell Sci 133:jcs239244
15. Basto R, Lau J, Vinogradova T et al (2006) Flies without centrioles. Cell 125:1375–1386
16. Métivier M, Gallaud E, Thomas A et al (2020) Drosophila dTBCE recruits tubulin around chromatin to promote mitotic spindle assembly. Curr Biol 31:684–695.e6
17. Gallaud E, Caous R, Pascal A et al (2014) Ensconsin/Map7 promotes microtubule growth and centrosome separation in drosophila neural stem cells. J Cell Biol 204:1111–1121
18. Metivier M, Monroy BY, Gallaud E et al (2019) Dual control of Kinesin-1 recruitment to microtubules by Ensconsin in drosophila neuroblasts and oocytes. Development 146: dev171579
19. Hannaford MR, Liu R, Billington N et al (2022) Pericentrin interacts with Kinesin-1 to drive centriole motility. J Cell Biol 221. https://doi.org/10.1083/jcb.202112097

Chapter 15

Cell Cycle Mapping Using Multiplexed Immunofluorescence

Katarzyna M. Kedziora and Wayne Stallaert

Abstract

The development of technologies that allow measurement of the cell cycle at the single-cell level has revealed novel insights into the mechanisms that regulate cell cycle commitment and progression through DNA replication and cell division. These studies have also provided evidence of heterogeneity in cell cycle regulation among individual cells, even within a genetically identical population. *Cell cycle mapping* combines highly multiplexed imaging with manifold learning to visualize the diversity of "paths" that cells can take through the proliferative cell cycle or into various states of cell cycle arrest. In this chapter, we describe a general protocol of the experimental and computational components of cell cycle mapping. We also provide a comprehensive guide for the design and analysis of experiments, discussing key considerations in detail (e.g., antibody library preparation, analysis strategies, etc.) that may vary depending on the research question being addressed.

Key words Cell cycle, Single-cell, Proteomics, Multiplexed imaging, Immunofluorescence, Machine learning

1 Introduction

Decades of research of the cell cycle and its regulation have uncovered an extensive network of biochemical reactions that are coordinated in space and in time to ensure the successful replication of DNA and its segregation into two new daughter cells. This accumulated knowledge has culminated in a single, generalized model that is used to understand the cell cycle and its regulation across biological systems [1]. However, the recent development and use of technologies that allow measurements from single cells as they progress through the cell cycle [2–7] has revealed that individual cells do not always traverse the cell cycle along the same path. For example, daughter cells diverge along two distinct molecular trajectories following cell division [8, 9], and different stresses induce cell cycle exit from different phases into distinct states of cell cycle arrest [10]. Cells can also reorder key regulatory events [11] or use

Anna Castro and Benjamin Lacroix (eds.), *Cell Cycle Control: Methods and Protocols*, Methods in Molecular Biology, vol. 2740, https://doi.org/10.1007/978-1-0716-3557-5_15,

different sets of effectors [12] to trigger the molecular switches that govern progression through the cell cycle [13].

These observations suggest that cell cycle heterogeneity may play an important role in various biological and disease contexts, but studying this phenomenon presents considerable technological and conceptual challenges. Cell cycle heterogeneity among individual cells within a population necessitates single-cell measurements of cell cycle activity. While techniques such as single-cell RNA sequencing (scRNA-seq) can provide such unbiased measurements in individual cells, many key cell cycle regulatory events involve mechanisms that cannot be measured by gene expression. For example, the targeted degradation of proteins drives the cyclic dynamics that give cyclins their name [14–16], and the activity of many cell cycle effectors is governed by changes in their subcellular localization (e.g., nuclear/cytoplasmic shuttling) or posttranslational modifications such as phosphorylations, neither of which can be directly measured by gene expression. On the other hand, population-based proteomic techniques (e.g., Western blots, mass spectrometry) are insufficient if our goal is to understand regulation of the cell cycle in single cells. They also require synchronization of cell cycle states across populations which is invasive in cell cultures and impossible in tissues. Furthermore, given the extensive effector network regulating cell cycle progress, how does one choose where to zoom in to look for heterogeneity in cell cycle progression?

In this chapter, we present an alternative approach: *cell cycle mapping*. Instead of zooming in on a specific cell cycle effector or pathway, cell cycle mapping zooms out to provide a "bird's-eye view" of the paths cells can take through the cell cycle. Using multiplexed immunofluorescence (mIF), proteomic measurements of cell cycle "state" are made in individual, asynchronously cycling cells, and manifold learning is used to project these high-dimensional data into two dimensions, generating a cell cycle map (Fig. 1).

Although there currently exist several mIF approaches (e.g., 4i, CyCIF, IBEX, CODEX) [17–20], each of these techniques uses iterative immunofluorescence to acquire 10–100 s images of protein abundance and subcellular distribution in individual cells. While each technique possesses its own set of advantages and disadvantages, this chapter will describe the use of iterative indirect immunofluorescence imaging (4i) to perform mIF [17]. One major advantage of 4i is that it does not require custom conjugation of primary antibodies to detection reagents (e.g., fluorescent dyes, heavy metal isotopes, oligonucleotide barcodes). However, because 4i uses secondary antibodies conjugated to fluorescent dyes to detect primary antibody binding, experimental throughput is limited by the availability of primary antibodies of different host species.

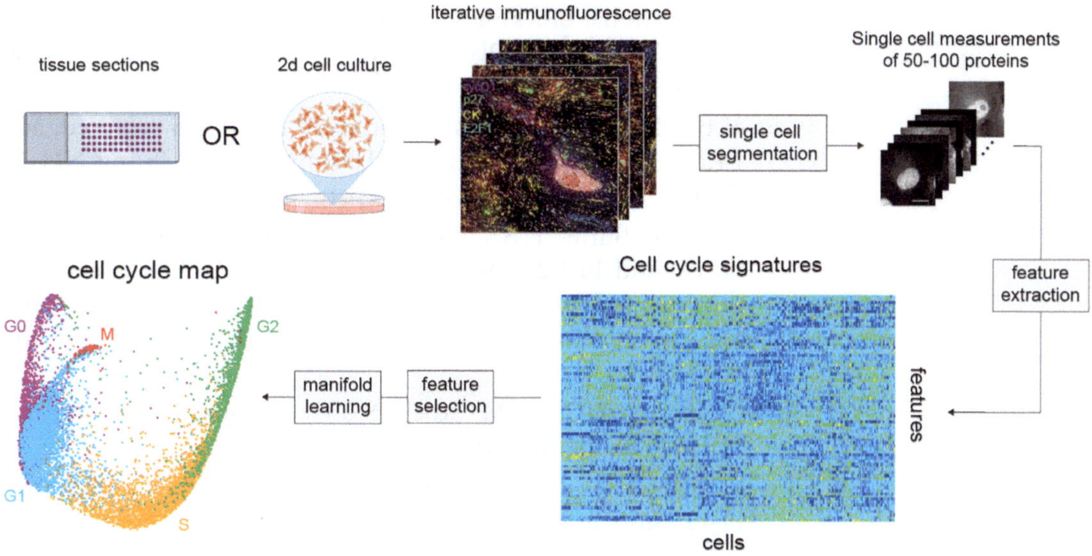

Fig. 1 Schematic of cell cycle mapping. Using tissue sections or fixed cell cultures, iterative immunofluorescence is performed to make measurements of 50–100 proteins in each cell. Following cell segmentation, 100–1000s of unique single-cell features are extracted per cell to generate high-dimensional cell cycle signatures. Feature selection is performed to enrich for cell cycle-dependent features, and manifold learning is used to find the manifold in this feature subspace that represents progression through the cell cycle and project it onto a two-dimensional embedding, or "cell cycle map." Each step is described in detail in this chapter

In this chapter, we will discuss the design and execution of mIF experiments to generate single-cell, proteomic signatures of cell cycle state, as well as image and data analysis methods to generate and interpret the resultant cell cycle maps. Our goal is to provide the reader with a complete overview of how to use cell cycle mapping to investigate heterogeneity in cell cycle regulation.

2 Materials

2.1 Reagents for 4i

1. A curated and validated antibody library (selection and validation of the antibody library is an essential component of any mIF experiment, and there are several factors to consider. See **Notes 1–5** for more details).
2. Paraformaldehyde (PFA, 8%): Dilute stock 16% PFA 1:1 (v/v) in phospho-buffered saline (PBS). Protect from light and store at 4 °C for 1 week.
3. Triton X-100. Dilute to 0.1% in PBS.
4. Hoechst 33342 ([stock] = 10 mg/mL in H_2O).

5. Conventional blocking solution (cBS): 1% bovine serum albumin (BSA) in PBS. Dissolve 0.5 g BSA in 50 ml PBS. Store at 4 °C for 1 week.
6. 4i blocking solution (sBS). For every ml of sBS required, add 14.6 mg maleimide ([final] = 100 mM) and 5.35 mg NH_4Cl ([final] = 100 mM) to cBS. Prepare immediately before use.
7. Imaging buffer (IB). 700 mM N-acetyl-cysteine (NAC) in H2O. Add 114.24 mg NAC per 1 mL of H_2O (add less H_2O than required until after pH corrected). Adjust pH with NaOH pH to 7.4. Prepare fresh immediately before each imaging round.
8. Guanidine HCl (8 M) stock: Slowly dissolve 46 g in 60 mL H_2O on magnetic stirrer. Store at 4 °C.
9. TCEP-HCl (0.5 M): Dissolve 2 g TCEP in 13.9 mL H_2O. Store at 4 °C.
10. Elution buffer (EB) premix: Combine 20 mL guanidine HCl, 20 mL urea (8 M), and 2 g L-glycine. Store at 4 °C.
11. EB working solution: Per ml of EB, combine 810 μL EB premix, 140 μL TCEP-HCl, and 50 μL HCl (to a pH of ~2.5). Prepare immediately before use.

2.2 Microscope

Please see **Notes 6–10** for key considerations when selecting/building a microscope for mIF.

3 Methods

3.1 Multiplexed Immunofluorescence (mIF)

3.1.1 Sample Preparation

Cells should be plated on a glass-bottomed dish to be compatible for imaging. Multiwell plates are useful to study multiple conditions in the same experiment and can range from 6- to 96-well format, depending on the balance between the number of cells and the number of conditions that is optimal for a given experiment. The number of cells required to make a cell cycle map will be influenced by several factors that can be difficult to assess a priori. Experimental contexts with higher degrees of cell cycle heterogeneity will require a larger number of cells to successfully resolve all trajectories. More cells will also allow rarer states to be detected. Plates can be coated with chemical reagents to increase cell adhesion or with extracellular matrix components, if the cell line requires it. Adhesion reagents should be tested for autofluorescence prior to use; however, we have successfully used poly-D-lysine and fibronectin with no major issues. Cell density at time of fixation is critical. If cells are too dense, contact inhibition will influence cell cycle progression and induce cell cycle arrest. If cells are too sparse, paracrine signaling will be reduced, which could also influence cell cycle progression in some systems. Sparse cells are also more likely to be damaged or lost during 4i. Optimal plating density is typically ~80% and should be optimized for each condition prior to the 4i experiment.

3.1.2 4i Protocol

Prior to fixation, samples should be examined for appropriate cell density, general health of the cells, and any other parameters specific to the experiment that can be examined visually (e.g., a morphological change associated with differentiation). Perform all washes and incubation steps on a rocker. Orbital shakers or 3d rockers risk swirling small volumes around edges of wells and drying the center of the well. Volumes are for 96-well plates and should be adjusted for other plate sizes.

1. Fix samples by adding 8% PFA directly to samples (1:1 v/v with media, final concentration: 4% PFA), and incubate for 30 min at room temperature (RT).
2. Rinse samples 3× with PBS (200 μl/well at RT for all PBS rinses).
3. Permeabilize cells by incubating with 0.1% Triton X-100 (50 μL/well) for 15 min at RT.
4. Rinse 1× with PBS.
5. Incubate with Hoechst (50 μL/well, 1:2000 in PBS) for 15 min at RT.
6. Wash 1× with PBS for 5 min.
7. Add 100 μL/well PBS and image Hoechst staining on microscope (*see* Subheading 3.3) to confirm sample integrity and that appropriate cell density was obtained in all wells.
8. Rinse samples 3× with H_2O (200 μL/well).
9. Wash 3× with EB (50 μL/well) for 10 min at RT on shaker (*see* **Note 11**).
10. Rinse 1× with PBS.

    Optional step: Obtain unstained images prior to each round (*see* Subheading 3.3.1, **Note 12**).

    (a) Incubate with Hoechst (50 μL/well, 1:2000 in PBS) for 15 min at RT.

    (b) Rinse 1× with PBS.

    (c) Add 100 μL/well IB and image residual signal and cellular autofluorescence.

    (d) Rinse samples 3× with H_2O (200 μL/well).
11. Incubate with sBS for 1 h at RT (50 μL/well).
12. Rinse 1× with PBS. Thoroughly remove remaining PBS from each well.

13. Incubate with primary antibodies (50 μL/well) in cBS overnight at 4 °C.
14. Wash 3× with PBS for 5 min.
15. Incubate with secondary antibodies (50 μL/well; 1:500 in cBS) and Hoechst (1:2000) for 1 h at RT (*See* **Note 13**).
16. Wash 3× with PBS for 5 min.
17. Add 100 μL/well IB and image on microscope (*see* Subheading 3.3).
18. Repeat **steps 8–17** for all remaining antibodies.

3.2 Imaging

3.2.1 Imaging Session

All good practices of collecting microscopy images for quantitative analysis are also applicable to multiplexed experiments (reviewed in [21–23]). However, we would like to emphasize here some settings that are particularly important for a successful imaging of a multiplexed experiment:

A single biomarker has to be imaged with the same settings across all the samples/conditions. Pixel intensities across an image need to be captured with the highest dynamic range but without introducing intensity values that saturate the camera/sensor. Therefore, it is important to carefully consider the entire set of samples before committing to the imaging settings (e.g., excitation intensity, exposure time, etc.). This initial time investment will help mitigate potential quality control issues later.

The order in which fluorescence channels are imaged is important. It is a standard practice to image the channels with a descending order according to the excitation wavelength to minimize the possibility of bleaching remaining fluorophores. However, if a single channel requires significant light exposure due to the low signal, it should be scanned last.

If tiling is used to collect signals from extensive areas, we recommend imaging every field of view with a set of spectral settings first before moving to other fields of view. In this way, stitching will be performed only once for each imaging round.

3.2.2 Quality Control

Unlike many microscopy experiments, the iterative nature of mIF experiments allows for the possibility of re-staining and reimaging the sample at a later stage. However, early detection of imaging problems can significantly reduce the total time of the experiment and improve the quality of the data. Therefore, it is important to perform an imaging quality check after each round of imaging, before re-staining the sample. This includes checking the general quality of the images (e.g., focus, stitching failures), as well as looking for any artifacts such as missing images or image aberrations. It is also important to check for oversaturated or undersaturated cells as these necessitate imaging with corrected microscope settings. Additionally, it is worth checking for the presence of larger artifacts that may obscure a significant number of cells (e.g., dust particles, fluorescent aggregates, etc.). These artifacts may not cover large areas, but they can add up over different channels and rounds of imaging and cause a significant loss of cells to be analyzed. In many cases, an additional set of washes can help to solve these problems.

3.3 Image Analysis

The goal of the image analysis step in a mIF experiment is to generate a high-dimensional "signature" of the cell cycle state for each cell. This high-dimensional "signature" is a set of features (e.g., mean nuclear intensity of protein X, number of foci formed by protein Y) extracted from a corrected and aligned set of images. However, the output of the imaging step (the input data for the analysis step) can vary significantly, depending on the software used by the microscope. Some software packages may provide more advanced and fully automated image preprocessing tools, while others may be more basic. As a result, the number of steps that researchers have to perform themselves in order to generate cell cycle "signatures" from mIF experiments can also vary widely.

The analysis pipeline can be implemented using different software packages, and it may be convenient to perform different steps of the analysis with the help of different platforms. One can choose from freeware, commercial software, or even custom software built in-house. Factors to consider when making this choice include the level of flexibility required, the convenience of use, whether a graphical interface is preferred, the user's skills and coding experience, and their willingness to develop their own code. Ultimately, the best choice will depend on the individual's specific requirements and preferences. However, it is worth emphasizing that all the processing steps can be performed with freely available open software tools, which are often the first to offer implementations of the new state-of-the-art-analysis algorithms. At the time of writing the popular open-source software packages include:

- Python image analysis tools (Scikit [24], Napari [25]).
- Fiji/ImageJ [26]
- MCMICRO [27]
- CellProfiler [28]
- Cellpose [29]
- QuPath [30]
- Squidpy [31]
- DeepCell [32]

Below, we provide a brief description of all the necessary steps required for the preprocessing of raw imaging data. We particularly focus on the goal of each analysis step, recommend state-of-the-art algorithms, and suggest their implementations. In each of the image analysis sections below, we will indicate the images/data that are the *input* and *output* of each step.

3.3.1 Background and Shading Correction

Input—set of raw images (tiles).

Output—set of corrected images (tiles).

Background and shading corrections are necessary to obtain quantitative data from 4i experiments. Shading correction, also known as flat field correction, is a process that is used to remove variations in the intensity of an image (usually presented as a loss of the signal at the edges—"vignetting") that are caused by the optics of the microscope. In a stitched image, problems with the flat field correction usually manifest as a characteristic grid-like pattern. Background correction, on the other hand, is used to remove unwanted signals (e.g., camera offset). The image collected by the microscope (Im_{col}) can be therefore represented as:

$$Im_{col} = Im_{signal}{}^{*}\ V + B$$

where Im_{signal} is the image of the signal of interest, V is the illumination pattern, and B is the additive background.

To calculate the corrected image (Im_{signal}), the V (illumination pattern) and B (background) components need to be defined. As mIF experiments provide many individual images acquired with the same microscope settings, they are a great candidate for a retrospective shading and background correction. In this approach, the data themselves are the source of calculating V and B components.

One of the robust algorithms for retrospective background correction is BaSiC [33]. BaSiC automatically finds V and B components from the collection of images and usually does not require any parameter tuning. It can be found implemented as a Fiji plugin [34], a Python package [35], or within the MCMICRO suite [27].

After the correction step, quality control inspection is necessary. Check the images for any artifacts and unexpected variation in fluorescence intensity. If you still observe uneven background intensity, it may indicate a problem detecting autofluorescence. This problem is common when imaging tissue slices and rare while imaging cell cultures. In such a situation, additional imaging of an unstained sample is recommended (ideally before every imaging round) to obtain images for direct subtraction (*see Optional Steps 10a-d* in Subheading 3.1.2). As this step requires precise matching of images acquired in two separate imaging sessions (before and after staining), it is recommended to be performed on stitched and aligned images.

3.3.2 Stitching

Input—set of corrected images (tiles).

Output—mosaic images for every channel.

Stitching is a process of arranging individual fields of view (tiles) into a single image. Individual tiles are acquired with an overlap (see **Note 9**) so that each neighboring pair of tiles shares the cells at the edge. This overlap is necessary as the simple arrangement of tiles based on their expected position is often not sufficient due to the finite precision of the movement of the microscope stage. Therefore, the optimization algorithm utilizes the information from the overlapping region to provide an artifact free stitched image.

Some example advanced stitching algorithms are MIST (Microscopy Image Stitching Tool) [36] and ASHLAR (Alignment by Simultaneous Harmonization of Layer/Adjacency Registration) [37]. MIST can be found implemented in Fiji and Matlab [38], and there is also an unofficial implementation in Python available [39]. ASHLAR was specifically developed for processing of the multiplexed images and is available in Python [40] or incorporated in the MCMICRO suite [27]. Additionally, ASHLAR provides registration capacity to align data from different imaging rounds (*see* Subheading 3.3.3).

After the stitching step, the quality control of the entire mosaic images is necessary. Stitching problems (e.g., visible seams, duplication of cells) may appear in the sparse regions, where there are not enough cells in the overlapping regions of the tiles to successfully calculate their relative position. In such a situation, simple stitching of the failed regions (based on the requested position of the tiles) would be preferable as it will most probably introduce smaller artifacts.

3.3.3 Image Alignment

Input—mosaic images for every channel.

Output—mosaic images in a common coordinate system.

Since samples are removed from the microscope after each round of imaging (to elute and re-stain with the next set of antibodies), it is nearly impossible to ensure that the positioning of the sample is precisely the same from round-to-round. Performing image alignment (also known as registration) step ensures that all the images from the iterative rounds are correctly positioned relative to each other.

Alignment algorithms can be divided into two categories: rigid alignment and nonrigid (elastic) alignment. Rigid alignment allows for translation or rotation of one image relatively to the other. Elastic transformation, on the other hand, is capable of correcting shifts and rotations of individual parts of the stitched image. It is often required when imaging tissue slices, which can deform during the re-staining process.

Finding the correct transformation between images from different rounds is an optimization process implemented differently by different software packages. If a constant reference image is present in every round (e.g., nuclear staining by DAPI/Hoechst), the alignment task becomes simpler. A good starting point is the StackReg algorithm [41], implemented as Fiji plugin [42] and Python library [43]. A much more powerful and flexible tool is Insight Toolkit (ITK) [44] and its Python implementation SimpleITK [45]. As it was initially developed to align different modalities of medical images, it is capable of aligning images containing patterns that are corresponding but not identical. It may be necessary when reference images are not present for all rounds. Finally, ASHLAR (*see* Subheading 3.3.2) provides alignment capabilities as well.

After performing this step, inspect the results to ensure perfect alignment of individual cells across all imaging rounds. In case of problems, make sure that the errors in the alignment are not the consequence of the errors in stitching. If the images are not stitched correctly, the correct rigid alignment simply does not exist. Finding the optimal alignment is also sensitive to the choice of the reference image (the one that will not be transformed itself, usually from the first imaging round). Alignment issues may be solved by changing the reference image itself. Finally, image alignment is also sensitive to the presence of background fluorescence and imaging artifacts. If removal of these artifacts is not possible, calculation of the alignment transformations may be easier using segmented masks instead (*see* Subheading 3.3.4). These transformations can then be applied to the fluorescence images themselves. Alignment of masks should be more successful, since the segmentation process gives the opportunity to detect objects based on size or shape, which naturally removes many of the imaging artifacts.

3.3.4 Segmentation of Objects

Input—selected aligned mosaic images.

Output—a mask of objects (nuclei, cells, or both).

In the context of a mIF experiment, segmentation is the process of identifying groups of pixels that correspond to individual objects (usually cells or nuclei in these experiments). The input to the segmentation step is a selected image (from a preprocessed and aligned stack) that provides the best contrast for the detection of objects of interest, and the output is a binary mask of the image that distinguishes individual objects (e.g., nuclei) from non-object pixels.

Usually, a single representative image can be chosen to perform the segmentation, and the resultant mask can be applied to all other fluorescence channels, since the objects of interest should be in identical positions in properly aligned images. However, if there are significant changes in the morphology of the cells between the imaging rounds (due to the sample processing), it may be necessary to segment a representative image from each round to obtain more accurate measurements of features.

As degradation of the sample is possible through the consecutive rounds of stripping and re-staining, images from early rounds often provide better contrast for segmentation. However, if cells are lost during the experiment, it is advisable to use an image from the final round for segmentation. In this situation, only cells that remained attached until the end of the imaging protocol (and therefore have measurements of all the desired markers) are included in the downstream analysis.

Segmentation can vary significantly, depending on the sample being analyzed. However, the development of deep learning algorithms (reviewed in [46–48]) has significantly improved the quality of segmentation results. It is noteworthy (and convenient!) that these improvements are often achieved with off-the-shelf pre-trained networks that do not require parameterization or retraining.

At the time of writing, state-of-the-art segmentation packages include the following: StarDist [49], Cellpose [29] and Mesmer [32], but new pretrained networks are published regularly. StarDist is currently available as a Python library [50], Fiji [51], or QuPath plugins [52]. Cellpose can be used as a Python library or a self-standing software [53], as an extension of QuPath [54], or with the help of a wrapper [55] from Fiji. Finally, Mesmer is available as a part of DeepCell Python package [56] and a cloud implementation [57].

3.3.5 Feature Extraction

Input—mosaic images in a common coordinate system and mask (s) of objects (nuclei, cells, or both).

Output—table with features of individual cells.

Once all images are successfully aligned and segmented, single-cell "features" can be extracted from the images. Measurement of these features is straightforward, relatively fast, and available in almost any image analysis software (e.g., Fiji, Scikit-Image, CellProfiler, MCMICRO).

The commonly extracted features can be divided into following categories:

- Object shape: Calculated based solely on the mask of an object (e.g. nuclear area or eccentricity).
- Fluorescence intensity: Calculated as the mean, median (less sensitive to outliers), or integrated pixel intensity in an object (e.g., nuclei, cytoplasm, etc.).
- Fluorescence distribution: The spatial distribution of fluorescence intensities in an object (e.g., the mean signal at the

periphery of a cell representing its plasma membrane, the accumulation of the signal in foci, quantification of the spatial intensity distribution by Zernike parameters).

- Signal colocalization: The co-occurrence of two signals in a given compartment (e.g., foci).
- Local neighborhood: A measure of the environment around a given cell—from a simple measurement of the local density of other cells to more advanced parameters, like the presence of specific cell types in a defined radius around the cell of interest.

As you can see from the above list, the number of primary and secondary features that can be extracted from a mIF experiment is much greater than the number of antibodies measured, often reaching hundreds or thousands of unique single-cell measurements.

3.4 Feature Selection

Feature extraction (*see* Subheading 3.3.5) will generate a high-dimensional molecular signature for each cell consisting of 100–1000s of unique single-cell features. This molecular signature provides a detailed quantitative description of cellular state. However, in order to resolve a map that purely represents progression through the cell cycle, it is imperative to enrich this molecular signature for cell cycle-dependent features by removing features that vary among cells in a cell cycle-independent manner. Features that vary due to an orthogonal biological process will convolute the manifold of interest and can distort the resultant lower-dimensional embedding. There are a growing number of unsupervised and semi-supervised feature selection methods for single-cell data [58]. In our previous work, we performed feature selection by training random forest models to predict cell cycle phase and age of individual cells using their full molecular signatures [7]. Feature importance scores were calculated, and the top 40 features that best predicted cell cycle phase and age were selected and used to generate cell cycle maps. This feature list may be generalizable to both guide selection of an antibody library and feature selection in future experiments. However, the number of features required to generate a map that provides sufficient resolution of the cell cycle may vary by research question and may also influence the size of the antibody library that is used. Empirical inspection of the maps generated with different feature sets is often helpful. In Fig. 2, we show an example of how feature set size influences the resultant cell cycle map. Using just the top 5 features (ranked using the random forest model described above), a map is generated that clearly shows progression through cell cycle phases (Fig. 2b). However, as the feature set size is increased, better separation of G0, G1, and M phases is obtained.

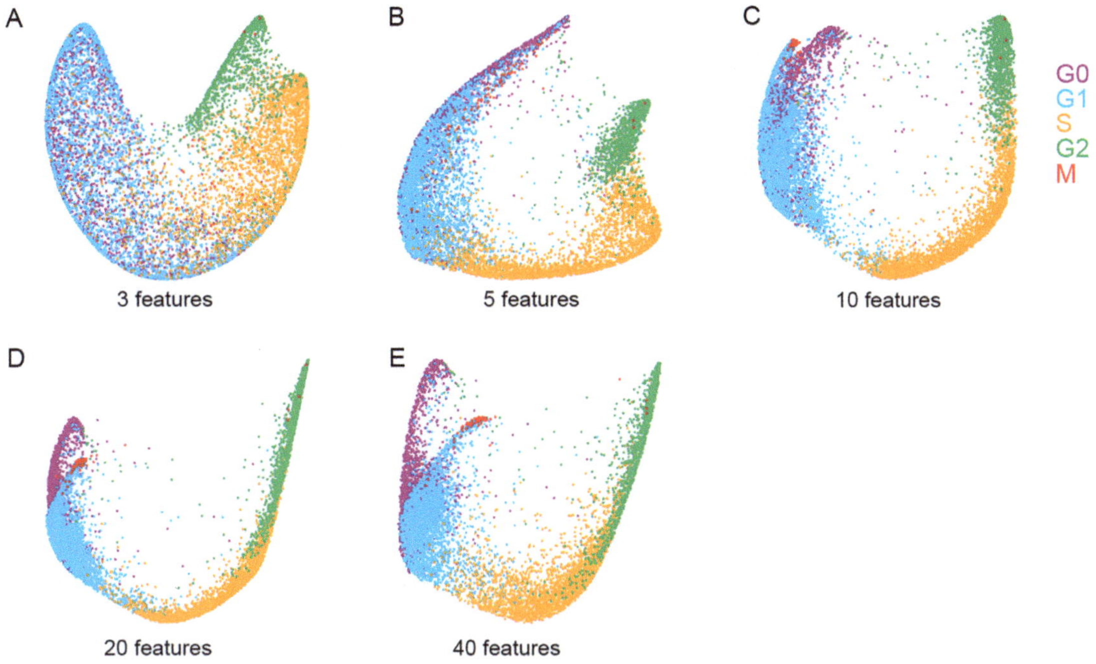

Fig. 2 Optimizing feature selection. (**a–e**) Cell cycle maps are generated using the top (**a**) 3, (**b**) 5, (**c**) 10, (**d**) 20, and (**e**) 40 features predicted by random forest-based feature selection. Individual cells are colored by cell cycle phase, as indicated

3.5 Manifold Learning

The next step is to identify the manifold that represents the cell cycle in this high-dimensional feature subspace and project it onto a lower dimension embedding for visualization. A nonlinear dimensionality reduction method that preserves both local and global similarities is ideal. One of the algorithms successfully used to generate cell cycle maps is PHATE (Potential of Heat-diffusion for Affinity-based Trajectory Embedding), due to its capacity to visualize continuous and branched manifolds [59]. The embeddings or "maps" generated by PHATE are highly dependent on several variable parameters in the algorithm (e.g. number of k-nearest neighbors, etc.), and we recommend reading the PHATE tutorial available on GitHub [60] to gain insight into these parameters and their effect on the embeddings. Optimal parameter settings must be determined empirically, as described in Subheading 3.6.

3.6 Map Interpretation

Map interpretation and manifold learning (*see* Subheading 3.5) are often performed in an iterative manner, since the quality of the map that is generated can only be assessed by a visual interrogation of the results. Best practice is to perform a parameter screen to find the combination that produces the best map.

There are several strategies to assess the quality of a given map. Overlaying the values of key features onto the map is often very instructive. DNA content (i.e., the integrated fluorescence intensity of Hoechst in the nucleus) is an objective measure of progression through the proliferative cell cycle, since DNA content doubles as cells progress from G1, through S phase, and into G2. Trajectories, over which there is a clear increase in DNA content, likely represent the progression of cells through these phases. Another useful measure for interpreting maps is RB phosphorylation. The phosphorylation status of RB is often used as a molecular correlate to distinguish actively proliferating cells (high phospho-RB) from arrested cells (low phospho-RB) [61, 62]. Overlaying phospho-RB measurements on the map can help distinguish arrested G0 cells from proliferative cells in G1/S/G2/M. Previous studies report the typical dynamics of many important cell cycle effectors (e.g., cyclins) during cell cycle progression. Once a putative proliferative cell cycle trajectory is identified (using DNA content and phospho-RB overlays), cyclin dynamics can be inferred along this path. However, one should be careful when using previously reported dynamics of specific effectors as a readout of cell cycle progression, since many of these experiments will be performed to look at how the cell cycle differs in a particular condition, which may include changes from the expected dynamics of a cell cycle protein.

In the end, the best map is the map that is most interpretable. This is an important point. No map obtained through this method is an objective visualization of the cell cycle. All maps should be considered data-driven models that can be used to generate hypotheses that should be tested using orthogonal methods.

4 Notes

1. *Choosing antibody targets.* The ability to resolve a cell cycle map is highly dependent on the number and quality of antibody targets (i.e., proteins) that are measured. Naturally, a larger antibody library will provide a more precise description of cell cycle state for each cell, which may increase the resolution of cell cycle trajectories and the probability of identifying alternative routes through the cell cycle. Of course, increasing the size of the library also increases the experimental cost and time. Previous cell cycle maps were constructed using ~50 antibodies targeting key cell cycle and signaling proteins [7, 10]. However, reducing the library to a smaller set of antibodies targeting core cell cycle effectors may produce cell cycle maps that are sufficiently resolved for certain research questions (*see* Subheading 3.4). There are also several tools available to guide primary antibody selection. For example, BenchSci [63] searches

research articles for antibodies that were successfully used for specific applications (e.g., immunofluorescence), biological samples (e.g., cells vs. formalin-fixed paraffin-embedded tissues, etc.), and species (e.g., human, mouse, rat tissue, etc.) and provides example images to support their use.

2. *Antibody validation.* As a starting point, one can select antibodies that have been successfully used in previous 4i studies [3, 7, 10, 17, 64] or validated for immunofluorescence by the manufacturers. However, we recommend validating all antibodies in-house prior to use. Antibodies can be validated by knockout/knockdown; however, knockouts of cell cycle proteins can be challenging due to the deleterious effects they often have on cells. As an alternative, antibodies can be validated by targeted perturbations known to change the abundance or spatial distributions of the proteins (e.g., mitogen stimulation, serum starvation, induction of DNA damage, etc.). After antibody validation, we recommend testing a range of antibody concentrations to determine the minimal concentration that provides the largest dynamic range.

3. *Compatibility with 4i.* We recommend that antibody validation experiments be performed using the 4i protocol (*see* Subheading 3.1.2), which differs from conventional IF protocols in several components (i.e., blocking buffer and imaging buffer). Antibodies must also be compatible with the elution protocol (*see* Subheading 3.1.2, **step 9**), which strips the antibody from the sample to allow additional rounds of staining. We have observed some variability in the stability of antigens over successive rounds of the 4i protocol. Such antigens, when identified, should be imaged in earlier rounds of the protocol. Furthermore, there may be differences in sample autofluorescence across imaging channels (e.g., in the green-orange region of the spectrum for tissue imaging). If this is the case, antibodies with the strongest intensity signal should be used in channels with high autofluorescence.

4. *Host species of primary antibodies.* Since secondary antibodies are used to detect binding of primary antibodies to their targets, the number of antibodies that can be imaged in each round will be limited by the number of primary antibodies from different host species (most commonly rabbit, mouse, goat) that are available. In order to maximize throughput, it is advisable to build an antibody library that contains similar numbers of primary antibodies from different host species.

5. *Secondary antibody selection.* Secondary antibodies conjugated to fluorescent dyes will be used to detect primary antibody binding. Secondary antibodies should be chosen for spectral compatibility with the microscope that will be used for imaging

(e.g., Alexa Fluor-488, Alexa Fluor-555, Alexa Fluor-647 is a typical set, see **Notes 6** ***and*** **10**) and coverage for the repertoire of primary antibody host species.

6. *Choosing a microscope.* The microscope should be repeatedly available for consecutive rounds of imaging that, depending on the number of conditions, areas of the samples and the number of fluorescence channels may require several hours of imaging each round.
7. *Imaging mode.* A wide-field fluorescence microscope equipped with an automatic stage is often the best option. As the sample is a flat monolayer of cells, we assume that a wide-field microscope illuminates and detects light from the entire depth of the sample at once. It provides sufficient quantitative information for most of the cell cycle regulators that are evenly dispersed in the nucleus or the cytoplasm. Moreover, the wide-field imaging is in general cheaper, faster, and less sensitive to focus drift than more advanced imaging setups. However, some research questions may require confocal microscopy. The key difference between the two techniques is that a confocal microscope is capable of selectively illuminating and detecting signals from a specific depth within the sample. Therefore, a confocal microscope offers a sectioning capacity and can provide 3D volumetric data. When such imaging data are available, small subcellular structures can be visualized with superior quality. Consequently, additional features may be measurable that are not adequately resolved by widefield imaging. However, these additional data come at a cost. Confocal imaging takes significantly longer, and 3D data will require additional data storage and processing resources. In summary, the choice between widefield and confocal imaging depends on the features of interest that are relevant to the specific question being investigated (*see* Subheadings 3.3.5 and 3.4).
8. *Magnification.* When choosing an objective for imaging, it is important to consider the trade-off between the size of the field of view (FOV) and resolution of the resulting image. A lower magnification objective will often provide a larger FOV, reducing imaging time by decreasing the number of images that need to be acquired and stitched together. However, it will also lead to a lower resolution of the images that may lack important details. In general, a 10× or 20× objective will be sufficient to distinguish signals in the cytoplasm or nuclei. If smaller subcellular structures are of interest, it may be necessary to consider an objective with a stronger magnification and higher resolution. Different biomarkers can also be imaged with different resolutions, but this approach will require taking into account differences in image sizes in the downstream analysis.

9. *Motorized stage*. A motorized stage of a microscope is necessary in order to consistently return to the same positions within the sample throughout all the imaging rounds. Furthermore, collecting data from a sufficient number of cells often requires imaging large areas of the sample that exceed a single field of view of a microscope even with a low magnification objective. Therefore, it is necessary to divide the area of interest into overlapping tiles in order to capture it completely. The overlap between these tiles is usually around 10–20%, but sparse samples may require a larger overlap for the stitching algorithm to work properly. Many microscopes can automate tiling and stitching of images. While it is crucial for the microscope to maintain and locate identical positions within the sample, the automated tiling and stitching capacity is not obligatory and can be integrated into a subsequent analysis pipeline (*see* Subheading 3.3).
10. *Spectral detection*. The chosen microscope has to be equipped with spectral filters to optimally acquire fluorescence signals of the selected fluorophores (see **Note 5**). Sequential fluorophore detection is commonly recommended to prevent cross talk and bleedthrough between the fluorescence channels. However, the time required to change spectral settings for sequential detection may limit the speed of image acquisition. As a rule of thumb, a filter wheel setup on a wide-field microscope will provide faster imaging of multiple fluorescent channels than a filter cube setup, if the option is available.
11. We have observed that the first round of antibody elution induces the most cell detachment and largest changes in morphology. To circumvent this, pretreat samples with EB prior to the first round of antibody incubation. Do not leave EB on the sample longer than the specified duration. Longer incubation times risk damaging the sample.
12. These images can be used to verify successful antibody elution after each round and to correct images of subsequent rounds for any residual staining.
13. EB will also deplete the Hoechst from the sample. Additional Hoechst must be added each round. Furthermore, EB can also strip nucleotides from the sample, reducing binding sites for Hoechst as the experiment progresses. Higher concentrations of Hoechst (1:1000–1:500) can be used if a reduction in fluorescence is observed.

References

1. Nurse P (2000) A long twentieth century of the cell cycle and beyond. Cell 100:71–78. https://doi.org/10.1016/s0092-8674(00)81684-0
2. Sakaue-Sawano A, Kurokawa H, Morimura T et al (2008) Visualizing spatiotemporal dynamics of multicellular cell-cycle progression. Cell 132:487–498. https://doi.org/10.1016/j.cell.2007.12.033
3. Gut G, Tadmor MD, Pe'er D et al (2015) Trajectories of cell-cycle progression from fixed cell populations. Nat Methods 12:951–954. https://doi.org/10.1038/nmeth.3545
4. Gookin S, Min M, Phadke H et al (2017) A map of protein dynamics during cell-cycle progression and cell-cycle exit. PLoS Biol 15:e2003268. https://doi.org/10.1371/journal.pbio.2003268
5. Zerjatke T, Gak IA, Kirova D et al (2017) Quantitative cell cycle analysis based on an endogenous all-in-one reporter for cell tracking and classification. Cell Rep 19:1953–1966. https://doi.org/10.1016/j.celrep.2017.05.022
6. Rappez L, Rakhlin A, Rigopoulos A et al (2020) DeepCycle reconstructs a cyclic cell cycle trajectory from unsegmented cell images using convolutional neural networks. Mol Syst Biol 16:e9474. https://doi.org/10.15252/msb.20209474
7. Stallaert W, Kedziora KM, Taylor CD et al (2022) The structure of the human cell cycle. Cell Syst 13:230–240.e3. https://doi.org/10.1016/j.cels.2021.10.007
8. Spencer SL, Cappell SD, Tsai FC et al (2013) The proliferation-quiescence decision is controlled by a bifurcation in CDK2 activity at mitotic exit. Cell 155:369–383. https://doi.org/10.1016/j.cell.2013.08.062
9. Overton KW, Spencer SL, Noderer WL et al (2014) Basal p21 controls population heterogeneity in cycling and quiescent cell cycle states. Proc Natl Acad Sci U S A 111:E4386–E4393. https://doi.org/10.1073/pnas.1409797111
10. Stallaert W, Taylor SR, Kedziora KM et al (2022) The molecular architecture of cell cycle arrest. Mol Syst Biol 18:e11087. https://doi.org/10.15252/msb.202211087
11. Liu C, Konagaya Y, Chung M et al (2020) Altered G1 signaling order and commitment point in cells proliferating without CDK4/6 activity. Nat Commun 11:5305. https://doi.org/10.1038/s41467-020-18966-9
12. Knudsen ES, Kumarasamy V, Nambiar R et al (2022) CDK/cyclin dependencies define extreme cancer cell-cycle heterogeneity and collateral vulnerabilities. Cell Rep 38:110448. https://doi.org/10.1016/j.celrep.2022.110448
13. Stallaert W, Kedziora KM, Chao HX, Purvis JE (2019) Bistable switches as integrators and actuators during cell cycle progression. FEBS Lett 593:2805–2816. https://doi.org/10.1002/1873-3468.13628
14. Zou T, Lin Z (2021) The involvement of ubiquitination machinery in cell cycle regulation and cancer progression. Int J Mol Sci 22. https://doi.org/10.3390/ijms22115754
15. Maiani E, Milletti G, Nazio F et al (2021) AMBRA1 regulates cyclin D to guard S-phase entry and genomic integrity. Nature 592:799. https://doi.org/10.1038/s41586-021-03422-5
16. Simoneschi D, Rona G, Zhou N et al (2021) CRL4AMBRA1 is a master regulator of D-type cyclins. Nature 592:789. https://doi.org/10.1038/s41586-021-03445-y
17. Gut G, Herrmann MD, Pelkmans L (2018) Multiplexed protein maps link subcellular organization to cellular states. Science 361. https://doi.org/10.1126/science.aar7042
18. Radtke AJ, Chu CJ, Yaniv Z et al (2022) IBEX: an iterative immunolabeling and chemical bleaching method for high-content imaging of diverse tissues. Nat Protoc 1–26:378. https://doi.org/10.1038/s41596-021-00644-9
19. Black S, Phillips D, Hickey JW et al (2021) CODEX multiplexed tissue imaging with DNA-conjugated antibodies. Nat Protoc 16:3802. https://doi.org/10.1038/s41596-021-00556-8
20. Lin J-R, Izar B, Wang S et al (2018) Highly multiplexed immunofluorescence imaging of human tissues and tumors using t-CyCIF and conventional optical microscopes. elife 7. https://doi.org/10.7554/eLife.31657
21. Lee J-Y, Kitaoka M (2018) A beginner's guide to rigor and reproducibility in fluorescence imaging experiments. Mol Biol Cell 29:1519–1525. https://doi.org/10.1091/mbc.E17-05-0276
22. Ogama T (2020) A beginner's guide to improving image acquisition in fluorescence microscopy. Biochemist 42:22–27. https://doi.org/10.1042/BIO20200075

23. Montero Llopis P, Senft RA, Ross-Elliott TJ et al (2021) Best practices and tools for reporting reproducible fluorescence microscopy methods. Nat Methods 18:1463–1476. https://doi.org/10.1038/s41592-021-01156-w
24. van der Walt S, Schönberger JL, Nunez-Iglesias J et al (2014) Scikit-image: image processing in Python. PeerJ 2:e453. https://doi.org/10.7717/peerj.453
25. Chiu C-L, Clack N, The Napari Community (2022) Napari: a python multi-dimensional image viewer platform for the research community. Microsc Microanal 28:1576–1577. https://doi.org/10.1017/S1431927622006328
26. Schindelin J, Arganda-Carreras I, Frise E et al (2012) Fiji: an open-source platform for biological-image analysis. Nat Methods 9: 676–682. https://doi.org/10.1038/nmeth.2019
27. Schapiro D, Sokolov A, Yapp C et al (2021) MCMICRO: a scalable, modular image-processing pipeline for multiplexed tissue imaging. Nat Methods 19:311–315. https://doi.org/10.1038/s41592-021-01308-y
28. McQuin C, Goodman A, Chernyshev V et al (2018) CellProfiler 3.0: next-generation image processing for biology. PLoS Biol 16: e2005970. https://doi.org/10.1371/journal.pbio.2005970
29. Stringer C, Wang T, Michaelos M, Pachitariu M (2021) Cellpose: a generalist algorithm for cellular segmentation. Nat Methods 18:100–106. https://doi.org/10.1038/s41592-020-01018-x
30. Bankhead P, Loughrey MB, Fernández JA et al (2017) QuPath: open source software for digital pathology image analysis. Sci Rep 7:16878. https://doi.org/10.1038/s41598-017-17204-5
31. Palla G, Spitzer H, Klein M et al (2022) Squidpy: a scalable framework for spatial omics analysis. Nat Methods 19:171–178. https://doi.org/10.1038/s41592-021-01358-2
32. Bannon D, Moen E, Schwartz M et al (2021) DeepCell Kiosk: scaling deep learning–enabled cellular image analysis with Kubernetes. Nat Methods 18:43–45. https://doi.org/10.1038/s41592-020-01023-0
33. Peng T, Thorn K, Schroeder T et al (2017) A BaSiC tool for background and shading correction of optical microscopy images. Nat Commun 8:14836. https://doi.org/10.1038/ncomms14836
34. BaSiC: a BaSiC tool for background and shading correction of optical microscopy images. Github. Available from: https://github.com/marrlab/BaSiC
35. BaSiCPy. Github. Available from: https://github.com/peng-lab/BaSiCPy
36. Chalfoun J, Majurski M, Blattner T et al (2017) MIST: accurate and scalable microscopy image stitching tool with stage modeling and error minimization. Sci Rep 7:1–10. https://doi.org/10.1038/s41598-017-04567-y
37. Muhlich JL, Chen Y-A, Yapp C et al (2022) Stitching and registering highly multiplexed whole-slide images of tissues and tumors using ASHLAR. Bioinformatics 38:4613–4621. https://doi.org/10.1093/bioinformatics/btac544
38. MIST: microscopy image stitching tool. Github. Available from: https://github.com/usnistgov/MIST
39. Fukai YT m2stitch: MIST-inspired microscope image stitching package. Github. Available from: https://github.com/yfukai/m2stitch
40. ashlar: ASHLAR: alignment by simultaneous harmonization of layer/adjacency registration. Github. Available from: https://github.com/labsyspharm/ashlar
41. Thévenaz P, Ruttimann UE, Unser M (1998) A pyramid approach to subpixel registration based on intensity. IEEE Trans Image Process 7:27–41. https://doi.org/10.1109/83.650848
42. StackReg. http://bigwww.epfl.ch/thevenaz/stackreg/. Accessed 2 May 2023
43. Lichtner G pystackreg: a python extension for the automatic alignment of a source image or a stack (movie) to a target image/reference frame. Github. Available from: https://github.com/glichtner/pystackreg
44. Johnson HJ, McCormick MM, Ibanez L (2015) The ITK software guide Book 2: design and functionality, 4th edn. Kitware, Incorporated
45. Yaniv Z, Lowekamp BC, Johnson HJ, Beare R (2018) SimpleITK image-analysis notebooks: a collaborative environment for education and reproducible research. J Digit Imaging 31: 290–303. https://doi.org/10.1007/s10278-017-0037-8
46. Moen E, Bannon D, Kudo T et al (2019) Deep learning for cellular image analysis. Nat Methods 16:1233–1246. https://doi.org/10.1038/s41592-019-0403-1

47. Bilodeau A, Bouchard C, Lavoie-Cardinal F (2022) Automated microscopy image segmentation and analysis with machine learning. Methods Mol Biol 2440:349–365. https://doi.org/10.1007/978-1-0716-2051-9_20
48. Ravindran S (2022) Five ways deep learning has transformed image analysis. Nature 609: 864–866. https://doi.org/10.1038/d41586-022-02964-6
49. Schmidt U, Weigert M, Broaddus C, Myers G (2018) Cell detection with star-convex polygons. In: Medical image computing and computer assisted intervention—MICCAI 2018. Springer International Publishing, pp 265–273
50. stardist: StarDist—object detection with star-convex shapes. Github. Available from: https://github.com/stardist/stardist
51. Schmidt U, Weigert M StarDist. In: ImageJ Wiki. https://imagej.net/plugins/stardist. Accessed 2 May 2023. Available from: https://imagej.net/plugins/stardist
52. qupath-extension-stardist: QuPath extension to run StarDist nucleus detection. Github. Available from: https://github.com/qupath/qupath-extension-stardist
53. cellpose: a generalist algorithm for cellular segmentation with human-in-the-loop capabilities. Github. Available from: https://github.com/MouseLand/cellpose
54. qupath-extension-cellpose: an extension that wraps a Cellpose environment such that WSI can be analyzed using Cellpose through QuPath. Github. Available from: https://github.com/BIOP/qupath-extension-cellpose
55. ijl-utilities-wrappers: wrappers for external software calls (Cellpose, Elastix, Ilastix...) and java utilities (object conversions and display). Github. Available from: https://github.com/BIOP/ijl-utilities-wrappers
56. deepcell-tf: deep learning library for single cell analysis. Github. Available from: https://github.com/vanvalenlab/deepcell-tf
57. kiosk-console: DeepCell kiosk distribution for kubernetes on GKE and AWS. Github. Available from: https://github.com/vanvalenlab/kiosk-console
58. Yang P, Huang H, Liu C (2021) Feature selection revisited in the single-cell era. Genome Biol 22:321. https://doi.org/10.1186/s13059-021-02544-3
59. Moon KR, van Dijk D, Wang Z et al (2019) Visualizing structure and transitions in high-dimensional biological data. Nat Biotechnol 37:1482–1492. https://doi.org/10.1038/s41587-019-0336-3
60. PHATE: PHATE (Potential of Heat-diffusion for Affinity-based Transition Embedding) is a tool for visualizing high dimensional data. Github. Available from: https://github.com/KrishnaswamyLab/PHATE
61. Pardee AB (1974) A restriction point for control of normal animal cell proliferation. Proc Natl Acad Sci U S A 71:1286–1290. https://doi.org/10.1073/pnas.71.4.1286
62. Zetterberg A, Larsson O (1985) Kinetic analysis of regulatory events in G1 leading to proliferation or quiescence of Swiss 3T3 cells. Proc Natl Acad Sci U S A 82:5365–5369. https://doi.org/10.1073/pnas.82.16.5365
63. BenchSci BenchSci: reimagining research. https://www.benchsci.com/. Accessed 8 May 2023
64. Kramer BA, Sarabia Del Castillo J, Pelkmans L (2022) Multimodal perception links cellular state to decision-making in single cells. Science 377:642–648. https://doi.org/10.1126/science.abf4062

Chapter 16

Investigating Heterogeneous Cell-Cycle Progression Using Single-Cell Imaging Approaches

Hee Won Yang

Abstract

Investigating cell-cycle progression has been challenging due to the complex interconnectivity of regulatory processes and inherent cell-to-cell heterogeneity, which often require synchronization procedures. However, recent advancements in cell-cycle sensors and single-cell imaging techniques have turned this heterogeneity into an advantage for investigating the molecular mechanisms underlying diverse responses. This has led to significant progress in our understanding of cell-cycle regulation. In this paper, we present a comprehensive live single-cell imaging workflow that leverages cutting-edge live-cell sensors. These advanced single-cell imaging procedures provide promising opportunities for elucidating the molecular mechanisms underpinnings of heterogeneous responses in cell-cycle progression.

Key words Single-cell imaging, Cell-cycle sensors, Cell-to-cell heterogeneity

1 Introduction

The proper regulation of the cell cycle is a critical process in multicellular organisms. Dysregulation of this process has been associated with various diseases, including cancer, developmental disorders, and degenerative diseases [1]. Consequently, understanding the mechanisms controlling the cell cycle is vital for developing treatments for these diseases. The initiation of cell-cycle entry is regulated by cyclin-dependent kinases 4 and 6 (CDK4/6) and the retinoblastoma protein (Rb) (Fig. 1) [2, 3]. During quiescence, nonphosphorylated Rb directly binds and suppresses E2F transcriptional factors that regulate cell-cycle genes, including the CDK2 activator cyclin E/A. To initiate the cell cycle, mitogenic signaling upregulates cyclin D, which binds and activates CDK4/6. In turn, active CDK4/6 phosphorylates and inactivates Rb, inducing E2F activation and subsequent CDK2 activation and genes required for cell-cycle entry. Active CDK2 also phosphorylates Rb, creating a positive feedback loop that reinforces the activation

Anna Castro and Benjamin Lacroix (eds.), *Cell Cycle Control: Methods and Protocols*, Methods in Molecular Biology, vol. 2740,
https://doi.org/10.1007/978-1-0716-3557-5_16,

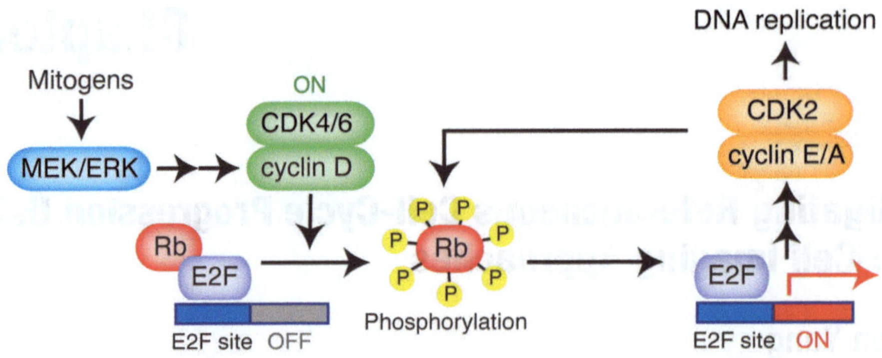

Fig. 1 Schematic illustrating the canonical model of cell-cycle entry. CDK4/6 activation initiate Rb phosphorylation and inactivation, leading to E2F activation and subsequent CDK2 activation

of E2F [4–6]. This positive feedback loop generates a bistability in cell-cycle entry, ensuring that once cells cross this point (the restriction point), it progresses through to completion.

Mammalian cells rely on a combination of intrinsic and extrinsic factors to determine their proliferation-quiescence decision [7, 8]. Due to heterogeneity in intrinsic factors such as history of mitogenic and DNA damage signaling [7, 9, 10] and contact inhibition [11], the proliferation-quiescence decision and cell-cycle progression are variable even under identical environmental conditions. To reduce cell-to-cell heterogeneity, traditional cell-cycle studies have employed various synchronization methods like serum starvation and double thymidine block [12]. However, even synchronized cells exhibit variable cell-cycle progression. For instance, when monitoring ERK and CDK2 activities simultaneously in non-transformed MCF-10A cells synchronized by serum starvation, we observed heterogenous ERK activity and subsequent CDK2 activity following mitogen stimulation in 10 randomly selected single-cell traces (Fig. 2a). Subsets of cells did not activate CDK2 and remained quiescent throughout the experimental time window (Fig. 2b). Bulk, averaged analysis has the potential to overlook small subsets of cells. Moreover, cancer cells are difficult to synchronize since they often harbor activating mutations in mitogenic signaling that can induce cell proliferation without mitogens. To address these challenges, the use of live-cell reporters in combination with single-cell techniques is proving to be highly effective. This approach enables the classification of heterogeneous cell populations while also generating quantitative data on key protein activities in real-time, specifically for cell-cycle entry within individual live cells. Single-cell imaging allows for the investigation of cellular processes at the individual cell levels, providing unprecedented insights into cellular heterogeneity and function.

In this paper, we provide an overview of available cell-cycle sensors and present a detailed protocol for studying heterogeneous

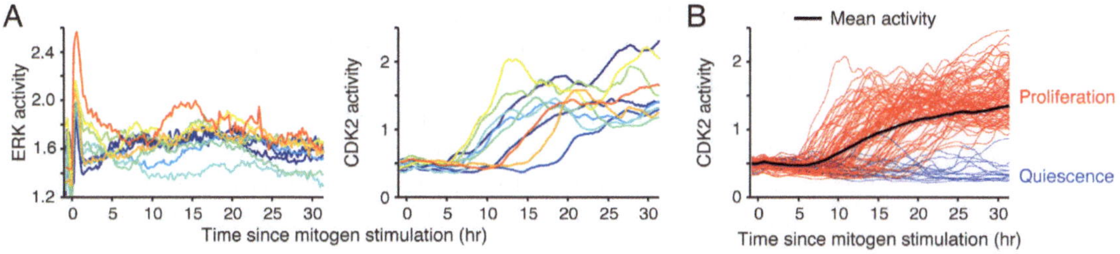

Fig. 2 Variability in mitogenic signaling and cell proliferation among synchronized cells. (**a**) Single-cell traces of ERK activity (left) and CDK2 activity (right). (**b**) Single-cell traces of CDK2 activity with classification of cells based on their CDK2 activity. (**a** and **b**) Non-transformed MCF-10A cells were serum-starved for 48 h before mitogen stimulation. FRET-based ERKAR and KTR-based CDK2 sensors were used to monitor ERK and CDK2 activities, respectively

behaviors in cell-cycle progression at the single-cell level. Our methodology employs single-cell imaging, which involves several key steps: expression of live-cell sensors, microscope setting based on utilized fluorescent proteins, segmentation of individual cells using a nucleus marker, measurements of nuclear or cytoplasmic intensity, tracking of individual cells over time, and data analysis. By implementing this protocol, researchers can acquire information about the complex processes involved in cell-cycle progression at the single-cell level. The capacity to observe individual cells in real time enables the detection of subtle variations in cell-cycle behavior that may not be apparent in bulk measurements. This can offer valuable insights into the underlying mechanisms driving cell proliferation as well as potential targets for therapeutic intervention.

2 Materials

2.1 Biological Samples

Cell lines or primary cells.

2.2 Live-Cell Sensors

Various live-cell sensors are available for investigating cell-cycle progression and associated signaling pathways.

2.2.1 Kinase Translocation Reporter (KTR) [13]

KTR is a type of genetically encoded biosensor designed to monitor kinase activity in living cells. KTR works by taking advantage of the subcellular localization changes that occur in response to substrate phosphorylation. Each KTR consists of a fluorescent protein, a nuclear localization signal (NLS), and a nuclear export signal (NES), as well as a kinase-specific substrate sequence (a docking site for kinase), which determines the specificity of KTR. When the relevant kinase phosphorylates the KTR substrate, the phosphorylation substrate undergoes a conformation change that affects its subcellular localization, shuttling between the nucleus and the cytoplasm. In the absence of phosphorylation, the NLS dominates, accumulating the KTR in the nucleus. However, upon phosphorylation, the NES becomes dominant, leading to the export of the KTR from the nucleus to the cytoplasm. This translocation between the nucleus and cytoplasm can be monitored using fluorescence microscopy, allowing researchers to monitor changes in kinase activity within individual cells. Kinase activity reported by KTR can be quantified by measuring the fluorescence intensity in both the nucleus and cytoplasm, with the cytoplasmic-to-nuclear fluorescence intensity ratio commonly used to define kinase activity. KTR has several advantages, including compatibility with live-cell imaging, capability to detect dynamic changes in kinase activity, and the requirement of only one fluorescent protein tag. Nevertheless, KTRs contain a degenerate CDK substrate motif from the CDK2 sensor [14], which leads to KTRs also detecting CDK activities measured by the CDK2 sensor during S and G2 phases [15–17]. These nonspecific signals have been found to correlate linearly with those detected by the CDK2 sensor [16, 18–20]. Using specific inhibitors, these nonspecific signals can be corrected based on the CDK2 sensor via linear regression. However, these correction factors may vary depending on the cell lines used and necessitate simultaneous measurement with the CDK2 sensor to correct the nonspecific signals (*see* **Note 1**).

Various KTR sensors are available, including the following: (1) CDK2 and CDK4/6 KTR sensors [16] in a single construct (Addgene, #126679), suitable for transient transfection or lentiviral infection. The CDK2 sensor primarily monitors CDK2 activity during G1 and S phases [21] and in the presence of CDK1 activity [22] but generally measures the collective activity of cyclin E/A-CDK1/2 complexes [23]. The CDK4/6 sensor is conjugated with mCherry. Of note, previous attempts to conjugate the CDK4/6 sensor with other fluorescent proteins, such as mVenus and mTurquoise, were unsuccessful, as the sensor primarily localized only in the nucleus. (2) The AKT and ERK sensors [24] in a single construct (Addgene, #129632 or #129631). 3. Individual KTR sensors for ERK [13] (Addgene, #59138), JNK [13] (Addgene, #59139), p38 [13] (Addgene, #59142), and AKT [25] (Addgene, #125131).

2.2.2 Förster Resonance Energy Transfer (FRET)

FRET sensors are a powerful tool for live-cell imaging, allowing to monitor molecular interactions, conformational changes, and signaling events in real time within live cells. These genetically encoded FRET sensors can be introduced into cells, where they can be used to monitor various cellular processes. FRET sensors have several advantages, including high sensitivity, noninvasiveness, and multiplexing capabilities. Unlike KTR, FRET sensors can also be targeted to specific subcellular compartments, such as the plasma membrane, allowing to investigate interactions of proteins and signaling within distinct cellular regions. FRET sensors have some limitations, including potential photobleaching, the need for careful calibrations and two fluorescent proteins (donor and acceptor), and variabilities in dynamic ranges.

Various FRET sensors are available, including the following: (1) ERK FRET sensors [17, 26] without cross-activity to CDK1 (Addgene, #173855, #167821, #167822, #167828, and #167829). (2) CDK1 FRET sensor [27] (Addgene, #26064). (3) Ras and Rho small GTPase sensors [28]. (4) cAMP FRET sensor [29]. (5) Src kinase FRET sensor [30].

2.2.3 Fluorescent Ubiquitination-Based Cell Cycle Indicators (FUCCI)

FUCCI is a powerful tool to monitor and visualize cell-cycle progression in real time. The FUCCI system uses genetically engineered fluorescent proteins, each of which is fused to a specific protein associated with cell-cycle regulation, providing visual cues about cell-cycle progression. These regulators are modified by ubiquitin, inducing its degradation by the proteasome.

Various FUCCI sensors are available, including the following: (1) Geminin degron is a substrate of the anaphase-promoting complex (APC), an E3 ubiquitin ligase that targets proteins for degradation. The APC is inactivated near the G1/S transition, allowing Geminin to accumulate and serve as a biomarker for this transition [31]. (2) Cdt1 degron is recognized by the E3 ubiquitin ligases and targeted for degradation during the S phase. This makes Cdt1 an effective biomarker for both the G1/S and S/G2 transitions, with previous studies demonstrating its utility in monitoring these cell-cycle transitions [32, 33]. (3). Stem-loop binding protein (SLBP) is degraded following S phase and thus can be used as a S/G2 transition reporter [34].

2.2.4 Histone-Based Biosensors

These sensors are based on fluorescently labeled histone proteins, such as histone 2B. This can be used to visualize chromatin condensation and chromosome segregation during cell division.

2.2.5 Proliferating Cell Nuclear Antigen (PCNA)

PCNA is a protein that plays an essential role in DNA replication. Expression levels of PCNA are increased and high during S phase. Furthermore, PCNA makes puncta formation during S phase. These features make PCNA a useful marker for visualizing cells in S phase [35].

2.3 Nucleus Marker

To segment, isolate, and automatically track individual cells over time, a fluorescently conjugated nucleus maker, such as histone 2B and nucleus localization signal, is needed.

2.4 Live-Cell Imaging-Compatible Culture Plate or Chamber

Multi-well plates (glass-bottom plate for high-resolution and short-distance objectives or plastic-bottom plate for low-resolution and long-distance objectives).

2.5 Inhibitors for Calibrating Nonspecific Signals in KTR

To calculate nonspecific signals in KTR during S and G2 phases, a specific kinase inhibitor and additional monitoring the CDK2 sensor are needed. By selecting cells activating the CDK2 sensor in the presence of a specific kinase inhibitor, the contribution of the CDK2 sensor can be calculated using linear regression [16, 18].

2.6 Inverted Fluorescence Microscope Equipped with a Live-Cell Imaging System

For live-cell imaging, a microscope with an environmental chamber that can control temperature, CO_2, and humidity is needed.

2.7 Standard Cell Culture Equipment and Supplies

The presence of phenol red has been shown to quench fluorescence in both the green and red emission channels, leading to a decrease in signal-to-noise ratio. Therefore, the use of phenol-red free growth media can increase signal-to-noise and improve the accuracy of fluorescence measurements.

2.8 Imaging Acquisition Software

2.9 Imaging Analysis Software

3 Methods

3.1 Establishing Cells Expressing Sensors

Optimal methods for DNA introduction vary depending on the efficacy and toxicity of transfection reagents. When considering DNA transfection, it may be necessary to choose between transient introduction using transfection reagents or establishing stable cell lines using virus infection. In cases where expression levels are heterogeneous, fluorescence-activated cell sorting (FACS) can be utilized to isolate cells expressing relatively homogeneous levels of the fluorescent protein (*see* **Note 2**).

3.2 Calibrating Nonspecific Signals in KTR Sensors

To determine cross-reactivity to CDK2 activity during S and G2 phases, specific inhibitors to relevant kinases can be used. Furthermore, this correction method is based on the activity measured by the CDK2 sensor, which requires co-expression of the CDK2 sensor for the correction. To assess cross-reactivity, a live-cell experiment should be performed in the presence of a relevant kinase inhibitor. After aligning single-cell traces by mitosis, a non-specific signaling that linearly correlates with the CDK2 sensor activity can be calculated using linear regression. For example, to account for the CDK2 contribution in CDK4/6 reporter measurements in MCF-10A cells, a previous study has calculated the correction factor in the presence of the CDK4/6 inhibitor palbociclib [16] (corrected CDK4/6 activity = CDK4/6 reporter activity – 0.35 × CDK2 reporter activity).

3.3 Imaging Cell-Cycle Sensors

3.3.1 Cell Preparation

1. Prior to plating, ensure the cells are heathy, within their passage limit, and growing in a suitable medium.
2. Prepare a glass- or plastic-bottomed multi-well plate or an appropriate imaging chamber.
3. Seed the cells onto the imaging plate or chamber, adjusting the number of cells based on the experiment duration. Maintain cell density between 30% and 80% to avoid contact inhibition.
4. Allow 5–10 min for cells to settle down to the bottom of the imaging plate or chamber. Then, transfer to a cell incubator and allow another 12–24 h for cells to adhere and reach the desired confluency. For suspension cells, create a thin layer on an imaging plate or chamber with a coverslip, using live-cell imaging-compatible mounting medium.

3.3.2 Transient Introduction of DNA Constructs Encoding Cell-Cycle Sensors

1. If necessary, DNA constructs of cell-cycle sensors can be introduced using suitable transfection reagents, following the manufacturer's instructions (*see* **Note 3**).
2. After DNA introduction, allow adequate time for fluorescent protein expression and maturation (usually 12–24 h).

3.3.3 Microscope Setup and Live-Cell Imaging

1. Prior to imaging, ensure the microscope and environmental chamber are properly calibrated to maintain temperature, CO_2 levels, and humidity. Preheat these components to stabilize the environment.
2. Verify sufficient data storage space on the microscope computer.
3. Turn on the microscope, camera, and any required microscope components.
4. Position the prepared live-cell imaging plate or chamber on the microscope stage.
5. Select regions of interests. Alternatively, set a number of positions per each well position.
6. Adjust the focus using bright-field or fluorescent protein tagged with a nucleus marker for a clear image acquisition.

7. Select appropriate excitation and emission filters for the florescent proteins used in the experiment.
8. Choose suitable objectives and exposure time for each channel, considering signal-to-noise ratios of fluorescent proteins. Aim for protein signals at least twofold higher than background signals. Optimize the imaging setup to minimize phototoxicity and photobleaching while maintaining adequate signal-to-noise ratios. Keep total exposure time under 600 milliseconds (*see* **Note 4**).
9. Set the time interval and total imaging duration based on the desired temporal resolution and experiment length. For adherent cells, 10–12 min time interval is typically sufficient to track cells over time.
10. Initiate imaging acquisition using the image acquisition software, ensuring one time-point imaging is completed within the set interval.
11. Monitor cell health during imaging and check for signs of phototoxicity or photobleaching. Adjust imaging parameters if necessary.

3.4 Image Analysis

To analyze the acquired images, use specialized software, such as Image J, Imaris, and CellProfiler, or customized computational scripts using Matlab, R, or Python.

Here are some example pipelines:

1. CellProfiler [36] (https://cellprofiler.org/): CellProfiler is a powerful open-source image analysis software to process and quantify various biological features from images. It is well suited for processing live single-cell imaging data by segmenting cells, identifying them, and extracting relevant features such as cell size, shape, and intensity.
2. EllipTrack [37] (https://github.com/tianchengzhe/EllipTrack): EllipTrack is an effective tool for tracking single cells and extract diverse biological features. It implements a global track-linking algorithm and utilizes training datasets.
3. A Matlab-based cell tracking script from Dr. Tobias Meyer's lab (https://github.com/MeyerLab/image-analysis-ratnayeke-2022 or https://github.com/scappell/Cell_tracking): These provide cell tracking scripts along with example scripts for data plotting.
4. A Python-based cell tracking script from Dr. Markus Covert's lab (https://github.com/braysia/CellTK). Additionally, KTR measurement protocols [38] introduce detailed protocols for measuring KTR and offers a python-based tracking imaging analysis.

4 Notes

1. Calibration factors for KTR sensors are often different depending on cell lines and KTR sensors [16, 18, 21].
2. To minimize the heterogeneity of a nuclear marker, it is recommended to perform FACS on cells with high nuclear marker expression. Nucleus segmentation depends on the intensity of the nuclear marker; therefore, implementing this strategy will enhance the accuracy of nucleus segmentation.
3. Be aware of the potential effects of fluorescent markers on cellular physiology; choose markers with minimal impact on cell-cycle progression.
4. Ensure that the imaging conditions are optimized to minimize phototoxicity and photobleaching. Maintain sterile conditions during sample preparation and live-cell imaging to prevent contamination. Verify localization and signal-to-noise ratio of the fluorescent markers prior to live-cell imaging experiments.

References

1. Hanahan D (2022) Hallmarks of cancer: new dimensions. Cancer Discov 12(1):31–46. https://doi.org/10.1158/2159-8290.CD-21-1059
2. Choi YJ, Anders L (2013) Signaling through cyclin D-dependent kinases. Oncogene 33: 1890–1903. https://doi.org/10.1038/onc.2013.137
3. Malumbres M, Barbacid M (2009) Cell cycle, CDKs and cancer: a changing paradigm. Nat Rev Cancer 9:153–166. https://doi.org/10.1038/nrc2602
4. Bartek J, Bartkova J, Lukas J (1996) The retinoblastoma protein pathway and the restriction point. Curr Opin Cell Biol 8(6):805–814. https://doi.org/10.1016/s0955-0674(96)80081-0
5. Fisher RP (2016) Getting to S: CDK functions and targets on the path to cell-cycle commitment. F1000Research 5:2374. https://doi.org/10.12688/f1000research.9463.1
6. Planas-Silva MD, Weinberg RA (1997) The restriction point and control of cell proliferation. Curr Opin Cell Biol 9:768–772. https://doi.org/10.1016/S0955-0674(97)80076-2
7. Yang HW, Chung M, Kudo T et al (2017) Competing memories of mitogen and p53 signalling control cell-cycle entry. Nature 549: 404–408. https://doi.org/10.1038/nature23880
8. Heldt FS, Barr AR, Cooper S et al (2018) A comprehensive model for the proliferation–quiescence decision in response to endogenous DNA damage in human cells. Proc Natl Acad Sci 115:2532–2537. https://doi.org/10.1073/pnas.1715345115
9. Barr AR, Cooper S, Heldt FS et al (2017) DNA damage during S-phase mediates the proliferation-quiescence decision in the subsequent G1 via p21 expression. Nat Commun 8(1):14728. https://doi.org/10.1038/ncomms14728
10. Arora M, Moser J, Phadke H et al (2017) Endogenous replication stress in mother cells leads to quiescence of daughter cells. Cell Rep 19(7):1351–1364. https://doi.org/10.1016/j.celrep.2017.04.055
11. Fan Y, Meyer T (2021) Molecular control of cell density-mediated exit to quiescence. Cell Rep 36(4):109436. https://doi.org/10.1016/j.celrep.2021.109436
12. Ligasova A, Koberna K (2021) Strengths and weaknesses of cell synchronization protocols based on inhibition of DNA synthesis. Int J Mol Sci 22(19). https://doi.org/10.3390/ijms221910759
13. Regot S, Hughey JJ, Bajar BT, Carrasco S, Covert MW, Albeck JG et al (2014) High-sensitivity measurements of multiple kinase activities in live single cells. Cell 157:1724–1734. https://doi.org/10.1016/j.cell.2014.04.039

14. Spencer SL, Cappell SD, Tsai FC et al (2013) The proliferation-quiescence decision is controlled by a bifurcation in CDK2 activity at mitotic exit. Cell 155:369–383. https://doi.org/10.1016/j.cell.2013.08.062
15. Gerosa L, Chidley C, Frohlich F et al (2020) Receptor-driven ERK pulses reconfigure MAPK signaling and enable persistence of drug-adapted BRAF-mutant melanoma cells. Cell Syst 11(5):478–94e9. https://doi.org/10.1016/j.cels.2020.10.002
16. Yang HW, Cappell SD, Jaimovich A et al (2020) Stress-mediated exit to quiescence restricted by increasing persistence in CDK4/6 activation. elife 9. https://doi.org/10.7554/eLife.44571
17. Ponsioen B, Post JB, Buissant des Amorie JR et al (2021) Quantifying single-cell ERK dynamics in colorectal cancer organoids reveals EGFR as an amplifier of oncogenic MAPK pathway signalling. Nat Cell Biol 23(4): 377–390. https://doi.org/10.1038/s41556-021-00654-5
18. Kim S, Carvajal R, Kim M, Yang HW (2023) Kinetics of RTK activation determine ERK reactivation and resistance to dual BRAF/MEK inhibition in melanoma. Cell Rep 42(6):112570. https://doi.org/10.1016/j.celrep.2023.112570
19. Kim S, Armand J, Safonov A et al (2023) Sequential activation of E2F via Rb degradation and c-Myc drives resistance to CDK4/6 inhibitors in breast cancer. Cell Rep 42 (11):113198. https://doi.org/10.1016/j.celrep.2023.113198
20. Zhang M, Kim S, Yang HW (2023) Non-canonical pathway for Rb inactivation and external signaling coordinate cell-cycle entry without CDK4/6 activity. Nat Commun 14 (1):7847. https://doi.org/10.1038/s41467-023-43716-y
21. Kim S, Leong A, Kim M, Yang HW (2022) CDK4/6 initiates Rb inactivation and CDK2 activity coordinates cell-cycle commitment and G1/S transition. Sci Rep 12(1):16810. https://doi.org/10.1038/s41598-022-20769-5
22. Arora M, Moser J, Hoffman TE et al (2023) Rapid adaptation to CDK2 inhibition exposes intrinsic cell-cycle plasticity. Cell 186(12): 2628–43 e21. https://doi.org/10.1016/j.cell.2023.05.013
23. Schwarz C, Johnson A, Kõivomägi M et al (2018) A precise Cdk activity threshold determines passage through the restriction point. Mol Cell 69:253–64.e5. https://doi.org/10.1016/j.molcel.2017.12.017
24. Chavez-Abiega S, Gronloh MLB, Gadella TWJ et al (2022) Single-cell imaging of ERK and Akt activation dynamics and heterogeneity induced by G-protein-coupled receptors. J Cell Sci 135(6). https://doi.org/10.1242/jcs.259685
25. Maryu G, Matsuda M, Aoki K (2016) Multiplexed fluorescence imaging of ERK and Akt activities and cell-cycle progression. Cell Struct Funct 41(2):81–92. https://doi.org/10.1247/csf.16007
26. Lin S, Hirayama D, Maryu G et al (2022) Redundant roles of EGFR ligands in the ERK activation waves during collective cell migration. Life Sci Alliance 5(1):e202101206. https://doi.org/10.26508/lsa.202101206
27. Gavet O, Pines J (2010) Progressive activation of CyclinB1-Cdk1 coordinates entry to mitosis. Dev Cell 18(4):533–543. https://doi.org/10.1016/j.devcel.2010.02.013
28. Komatsu N, Aoki K, Yamada M et al (2011) Development of an optimized backbone of FRET biosensors for kinases and GTPases. Mol Biol Cell 22:4647–4656. https://doi.org/10.1091/mbc.E11-01-0072
29. Nikolaev VO, Bunemann M, Hein L et al (2004) Novel single chain cAMP sensors for receptor-induced signal propagation. J Biol Chem 279(36):37215–37218. https://doi.org/10.1074/jbc.C400302200
30. Ting AY, Kain KH, Klemke RL et al (2001) Genetically encoded fluorescent reporters of protein tyrosine kinase activities in living cells. Proc Natl Acad Sci U S A 98(26): 15003–15008. https://doi.org/10.1073/pnas.211564598
31. Sakaue-Sawano A, Kurokawa H, Morimura T et al (2008) Visualizing spatiotemporal dynamics of multicellular cell-cycle progression. Cell 132:487–498. https://doi.org/10.1016/j.cell.2007.12.033
32. Sakaue-Sawano A, Yo M, Komatsu N et al (2017) Genetically encoded tools for optical dissection of the mammalian cell cycle. Mol Cell 68(3):626–40 e5. https://doi.org/10.1016/j.molcel.2017.10.001
33. Grant GD, Kedziora KM, Limas JC et al (2018) Accurate delineation of cell cycle phase transitions in living cells with PIP-FUCCI. Cell Cycle 17:2496–2516. https://doi.org/10.1080/15384101.2018.1547001
34. Bajar BT, Lam AJ, Badiee RK et al (2016) Fluorescent indicators for simultaneous reporting of all four cell cycle phases. Nat Methods 13(12):993–996. https://doi.org/10.1038/nmeth.4045

35. Zerjatke T, Gak IA, Kirova D et al (2017) Quantitative cell cycle analysis based on an endogenous all-in-one reporter for cell tracking and classification. Cell Rep 19(9): 1953–1966. https://doi.org/10.1016/j.celrep.2017.05.022

36. Stirling DR, Swain-Bowden MJ, Lucas AM et al (2021) CellProfiler 4: improvements in speed, utility and usability. BMC Bioinform 22(1):433. https://doi.org/10.1186/s12859-021-04344-9

37. Tian C, Yang C, Spencer SL (2020) EllipTrack: a global-local cell-tracking pipeline for 2D fluorescence time-lapse microscopy. Cell Rep 32(5):107984. https://doi.org/10.1016/j.celrep.2020.107984

38. Kudo T, Jeknić S, Macklin DN et al (2018) Live-cell measurements of kinase activity in single cells using translocation reporters. Nat Protoc 13(1):155–169. https://doi.org/10.1038/nprot.2017.128

Chapter 17

MAARS Software for Automatic and Quantitative Analysis of Mitotic Progression

Tong Li, Yannick Gachet, and Sylvie Tournier

Abstract

In this chapter, we describe a software called MAARS (Mitotic Analysis And Recording System) that enables automatic and quantitative analysis of mitotic progression on an open-source platform. This computer-assisted analysis of cell division allows the unbiased acquisition of multiple parameters such as cell shape or size, metaphase or anaphase delays, as well as various mitotic abnormalities. This chapter describes the power of such an expert system to highlight the complexity of the mechanisms required to prevent mitotic chromosome segregation errors, leading to aneuploidy.

Key words High-content analysis, Mitosis, Fission yeast, Neuploidy

1 Introduction

High-content screening (HCS) combines automated fluorescence microscopy with quantitative image analysis to ensure unbiased interpretation of acquired data. This approach allows unbiased multiparametric data extraction from single cells (for review, see [1]), first, to faithfully capture cell phenotype metrics and, second, to identify affected phenotypes in cells undergoing aberrant treatment/condition. One of the challenges of microscopy-based screening is the lack of tailored computational analysis for the biological question. A fully automated HCS image analysis pipeline consists of a combinatorial approach of image processing, image analysis, statistical analysis, and machine learning-based decision-making. Automation at this level not only enables quantitative, highly reproducible analysis but also leads to the identification of subtle phenotypic abnormalities, thanks to the unbiased data processing at all stages. In brief, an HCS image analysis pipeline includes the following: (i) image acquisition. This is usually done with an epifluorescence imaging system. (ii) Object segmentation or instance segmentation. This step aims to find the object of

Anna Castro and Benjamin Lacroix (eds.), *Cell Cycle Control: Methods and Protocols*, Methods in Molecular Biology, vol. 2740, https://doi.org/10.1007/978-1-0716-3557-5_17,

interest. This can be either a cell or a region of tissue, depending on the biological question. (iii) Feature extraction/selection. This step extracts phenotypic features describing each object found in the previous step and keeps those that are relevant for the study. Finally, (iv) statistical analysis. One would accept or reject the hypothesis of the study by examining the p-value of statistical tests. These four steps convert imaging data into a quantifiable description of the study and, presumably, its biological significance. Various tools exist to perform such HCS studies in different species or for different biological questions, such as the Columbus software in the high-content imaging system Operetta or open-source tools such as Micropilot [2]. However, these software do not always meet the needs of all biological research, where requirements are constantly evolving. This results in a lack of flexibility.

Here, we describe a high-content analysis (HCA) approach, called MAARS, which combines automated fluorescence microscopy with quantitative image analysis of mitosis [3]. Based on the concept of image processing and basic machine learning method (i.e., decision tree), MAARS provides automatic, multiparametric, and quantitative single-cell analysis of mitotic progression in the fission yeast *Schizosaccharomyces pombe* [3], a renowned model for the study of chromosome segregation [4–7]. *S. pombe* is a rod shape, symmetrically dividing eukaryote that possesses three chromosomes [8], which can be tracked by live imaging during mitosis with the centromeres fluorescently tagged [8]. As in metazoans, fission yeast mitosis begins in M-phase and is divided into three phases [9, 10]. In phase 1 (prophase), a short spindle is formed, and kinetochores, a large protein complex assembled at the centromeres of each chromosome, begin to align at the spindle center [11]. In phase 2 (prometaphase/metaphase/anaphase A), the spindle maintains a constant length of approximately 2 μm in wild-type cells and in anaphase A, and thanks to the degradation of the cohesin complex, sister chromatids separate allowing their movement toward the spindle poles [12]. The abrupt change in the rate of spindle elongation signals the beginning of anaphase B (phase 3) [9].

The MAARS analysis pipeline uses the optical refraction of cell membrane to find cell contours in fission yeast. It was previously used to segment bacterial populations [13]. This optical refraction of cell membrane can easily be recorded using the transmitted light acquisition on a standard microscope by moving through the Z dimension of cells. The ability to segment cells using transmitted light is extremely valuable as it not only has low phototoxicity but also saves wavelengths for other fluorescently labelled biological components. Specifically, bright-field microscopy produces a coherent intensity pattern similar to the inverse derivative of a Gaussian distribution. This unique property is used to segment the cell edges and can be used to obtain a region of interest (ROI). Several algorithms have exploited these properties for segmentation. For example, PombeX identifies this pattern using k-means clustering [14], while Pomegranate algorithm relies on pattern projection and adaptive thresholding to accurately represent 3D cells from 2D segmentation [15]. Other methods use traditional segmentation on bright-field images by edge detection [16] or adaptive thresholding [17]. Today, deep learning is the state of the art in cell segmentation [18–21]. With continuous development of the trained model, these tools will become the first choice for most cell segmentation tasks in image analysis.

Following non-invasive segmentation, MAARS algorithm faithfully extracts dynamic single-cell metrics from time-lapse images of mitotic progression. These features include (a) cell shape/size, (b) spindle positioning, (c) spindle elongation from prophase to telophase, and (d) semi-automated kinetochore lagging signal and finally provide a graphical user interface to visualize kinetochore MT attachment error correction.

2 Materials

2.1 Cell Culture

1. Yeast extract liquid media: 0.5% YE. Adjust pH at 5.5 and sterilize. Before use, add 3% glucose (from a 50% stock solution). It is preferable to use double distilled water when preparing media.
2. Yeast extract plates: 0.5% YE. Adjust pH at 5.5 and add 2% agar before sterilization. Before use, add 3% glucose (from a 50% stock solution).
3. Edinburgh minimal medium composition: 14.7 mM potassium hydrogen phthalate; 15.5 mM Na2HPO4; 93.5 mM NH4Cl; 2% glucose; 20 mL/l salt stocks; 1 mL/l vitamin stocks; and 0.1 mL/l mineral stocks.
4. 50× Salt stocks (filter-sterilize and store at 4 °C): 0.26 M MgCl2.6H20; 4.99 mM CaCl2.2H20; 0.67 M KCl; 14.1 mM Na2SO4.
5. 1000× Vitamin stocks (filter-sterilize and store at 4 °C): 4.20 mM pantothenic acid; 8 1.2 mM nicotinic acid; 55.5 mM inositol; and 40.8 μM biotin.
6. 10,000× Mineral stocks (filter-sterilize and store at 4 °C): 80.9 mM boric acid; 23.7 mM MnSO4; 13.9 mM ZnSO4.7H2O; 7.40 mM FeCl2.6H2O; 2.47 mM molybdic acid; 6.02 mM KI; 1.60 mM CuSO4.5H2O; and 47.6 mM citric acid.

2.2 Microscopy

1. CoverWell imaging chambers (Grace Bio-Labs 635,021; 1 × 20 mm Dia × 1,7 mm depth; Fig. 1).
2. Agarose D5 (DNA grade, Euromedex).

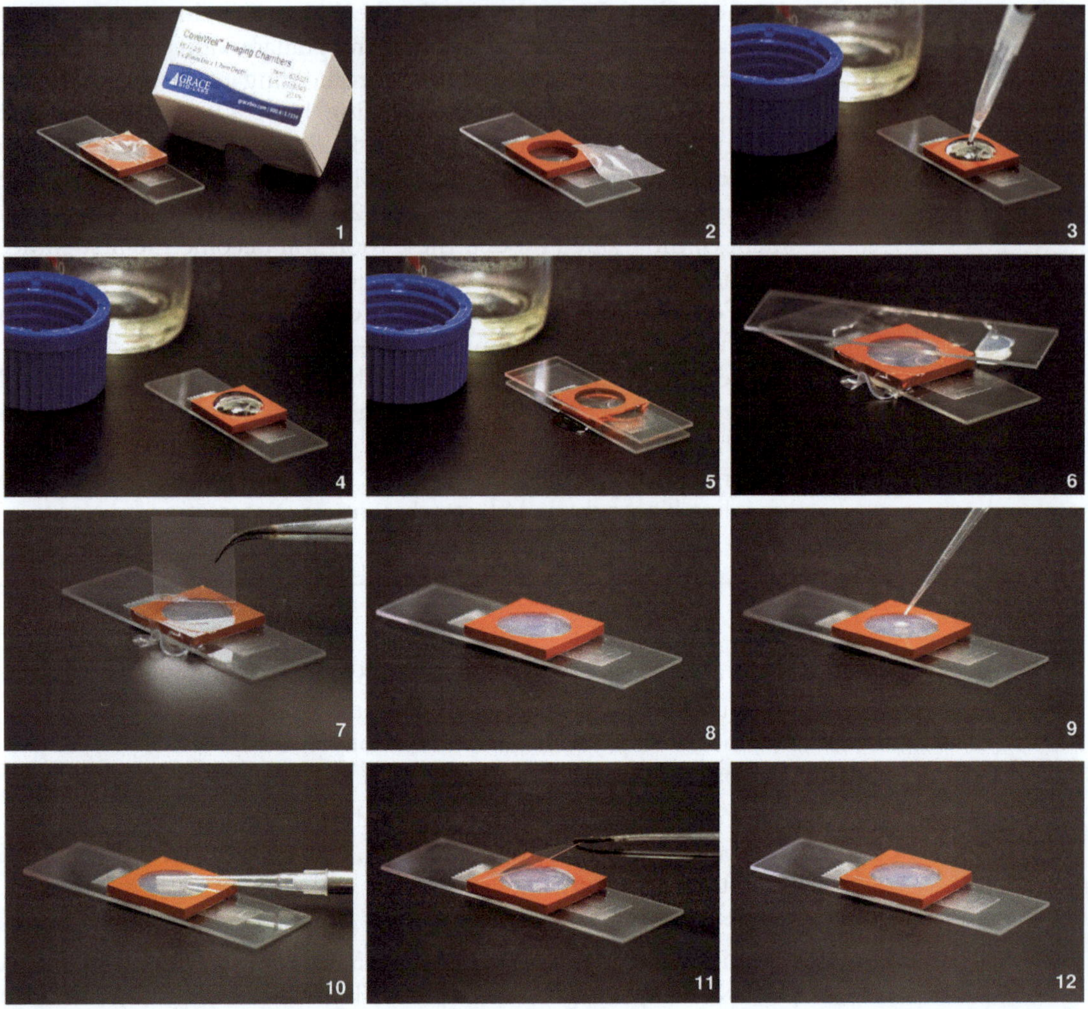

Fig. 1 Preparation of microscopy chambers for live cell imaging. Photos illustrating the various steps in the preparation of agarose chambers for live cell microscopy (*see* text for details)

MAARS is installed on an inverted wide-field microscope (Nikon Eclipse TI) equipped with a neo-scientific complementary metal-oxide-semiconductor camera (2560 × 2160 pixel resolution; Andor Technology), a LED light source (Lumencor Spectra), and a 60× objective with a numerical aperture (NA) of 1.40.

Standard genetic and PCR-based gene targeting methods were used to construct *S. pombe* strain used to analyze mitosis in fission yeast with MAARS. The fluorescent tagged strain *cdc11-cfp ndc80-gfp* [22] (available on demand) was produced by homologous recombination using modified pFA6a series of plasmids, carrying kanamycin resistance gene [23].

3 Methods

3.1 Cell Culture Preparation

After thawing cells on yeast extract plates, strains are grown in YE liquid media for 2 days at 25 °C before harvesting and plating on agarose plugs for imaging (*see* **Note 1**). All cell lines are fluorescently tagged with GFP and CFP on Ndc80 (kinetochores) and Cdc11 (spindle poles), respectively, to reflect mitotic progression with minimal aberration to cells [22]. All fluorescently labelled strains are integrated at their native chromosomal loci and under the control of their endogenous promoters. Minimal Media (*see* Subheading 2 for composition) is preferred for imaging to avoid autofluorescence. Time-lapse acquisition takes approximately 30–45 min to ensure coverage of at least one mitotic chromosome segregation if it starts immediately after the first snapshot.

3.2 Mounting of Cells on Microscopy Slides

1. Attach a CoverWell imaging chamber to a microscope slide using double-sided adhesive tape (Fig. 1, photo 1 and 2). These chambers can be reused many times after washing, provided they are properly cleaned with distilled water after each use.
2. Place 0.3 g of agarose and 15 mL of EMM medium supplemented with vitamins and supplements (leucine, urea, histidine, adenine) in a 50 mL sterile plastic tube. Heat the tube in a microwave until the agarose is completely dissolved, minimizing evaporation to maintain a constant supplement concentration after heating. Wait until the tube is lukewarm (*see* **Note 2**) and then pour 1.5 mL of the mixture into the imaging chamber so that a dome appears above the chamber (Fig. 1, photo 3 and 4).
3. Place a microscope slide at an angle of approximately 45° to flatten the agarose mixture, expel air, and prevent bubble formation. This step must be performed as soon as possible. Otherwise with agar solidified, the entire imaging area will be tilted and won't be analyzable in the downstream step. Hold the slide until the agarose pellet solidifies (Fig. 1, photo 5).
4. Wait approximately 30 min before removing the upper slide from the imaging chamber (Fig. 1, photo 6).
5. Remove excess moisture and agarose residue from the plug by gently sliding a glass coverslip over the edges of the chamber (Fig. 1, photo 7 and 8). Take care not to press too hard to avoid damaging the agarose plug. The surface must be as smooth as possible (*see* **Note 3**).
6. Apply a monolayer of cells (maximum 2 μL of concentrated cells) to the center of the agarose plug (Fig. 1, photo 9 and 10) and spread the cells gently using either the side of a tip or a

coverslip. Allow the agarose to absorb the excess medium for a few minutes (*see* **Note 4**).

7. Place a coverslip on the plug and gently press on the coverslip to remove any air (Fig. 1, photo 11 and 12).
8. Leave the slide in the incubator for 30 min at 25 °C before processing (*see* **Note 5**). This is to reduce the stress on the cells after they have been mounted.

3.3 Installation of MAARS Software

1. Description of MAARS software.

   MAARS is a Micro-Manager plugin designed to perform either on-the-fly image analysis in combination of real-time microscope control or post-acquisition analysis of previously acquired data. MAARS is written in Java and uses numerous image processing libraries including Fiji [24], Trackmate [25], LOCI bio-formats [26], and Micro-Manager [27]. MAARS is licensed under open-source license BSD (Berkeley Software Distribution License). Documentation and source code are available in a public repository at (https://github.com/bnoi/MAARS/releases/tag/v1.0.0). The plugin as packaged in Jar files can be used in ImageJ. Detailed tutorial and information can be found on Github (https://github.com/bnoi/MAARS/tree/master/doc/demos).
2. MAARS installation.
3. Install Java > = 1.8 and install it to the default location. Then, make sure your Fiji/Micro-Manager program is running with it. On the download page, you will find 4.jar files:

- maars_lib_[version]-SNAPSHOT.jar (Core),
- maars_bfSeg_[version]-SNAPSHOT.jar (single cell segmentation with GUI, optional),
- maars_otf_[version]-SNAPSHOT.jar (On-the-fly analysis and lagging detection with Micro-Manager, optional),
- maars_post_[version]-SNAPSHOT.jar (Post-acquisition analysis with Fiji, optional).

Copy maars_lib_XXX-SNAPSHOT.jar (the core package) along with any optional packages that is needed (otf, post, bfSeg) into your *ImageJ/Fiji* plugins folder.

Set up memory allocation in Fiji in Edit/Options/Memory & Threads. The amount of memory required varies depending on the acquisition setting if one need to perform on-the-fly image analysis a minimum of 16 Gb of RAM is recommended. It is recommended to also read the additional instructions that can be found at (https://github.com/bnoi/MAARS/blob/master/doc/install_MAARS.md).

3.4 Automated Image Acquisition and on-the-Fly Image Analysis with MAARS

MAARS was designed to be the first step toward the concept of an intelligent microscope. It is thus able to react to certain predefined conditions as they are detected during acquisition. The entire workflow starts with parameter setting to ensure optimal acquisition. Once all parameters are appropriately set, the program starts to acquire image from the sample and trigger analysis once those images are stored in RAM.

MAARS consists of three Java modules, coordinated by a *beanshell* script that runs with Micro-Manager. The first module is used to segment the image (*see* Subheading 3.4.3). The second module is designed to extract phenotypic features from segmented objects and to detect the mitotic state of cells (*see* Subheading 3.4.4). The third module collects and displays mitotic parameters via a graphical interface. It also coordinates image acquisition with analysis during execution.

For each user-defined field of view, cell segmentation is performed to check for the presence of a cell. If it is present, the mitotic detection module is triggered and tries to find out if it is a mitotic cell. The rule is simply determined by whether or not there are cells with two spindle poles. If the answer is yes, MAARS will start a 30–45 min acquisition (set by the user). If not, MAARS will not waste time acquiring a field of view with no mitotic cells and will move on to the next one. Once the stage has moved to the next field of view, the above process is repeated.

3.4.1 Multi-Position Sample Exploration

Multi-position acquisition is usually used to increase the statistical power of the experiment with a minimum amount of sample preparation. A programmable microscope stage is usually required to ensure precise and agile sample movement, allowing multi-position acquisition and in some cases Z movement for focus finding purposes. The Nikon microscope on which we've installed MAARS can do both if the user provides the necessary inputs (area to be explored, Z-focus range, etc.) before starting the acquisition. The exploration of the specimen is thus performed in the order defined by the user (*see* Fig. 2, Exploration).

3.4.2 Image Acquisition of Living Cells

When acquiring time-lapse fluorescence images of living organisms, the fine-tuning of imaging parameters is key to proper downstream analysis. Illumination must be minimized to prevent phototoxicity. This is particularly true with wide-field microscopy, which is less sensitive to weak signals. For fission yeast imaging, the LED light intensity has been preset to a minimum power. The time gap is also set to 30 or 35 s, (i.e., not less than 30 s). If the gap is longer than 35 s, we may miss an important biological event, such as a mitotic delay, and it may also cause unnecessary problems with subsequent object tracking.

With its primitive intelligence, MAARS can react to certain conditions during acquisition. Firstly, the images must be in

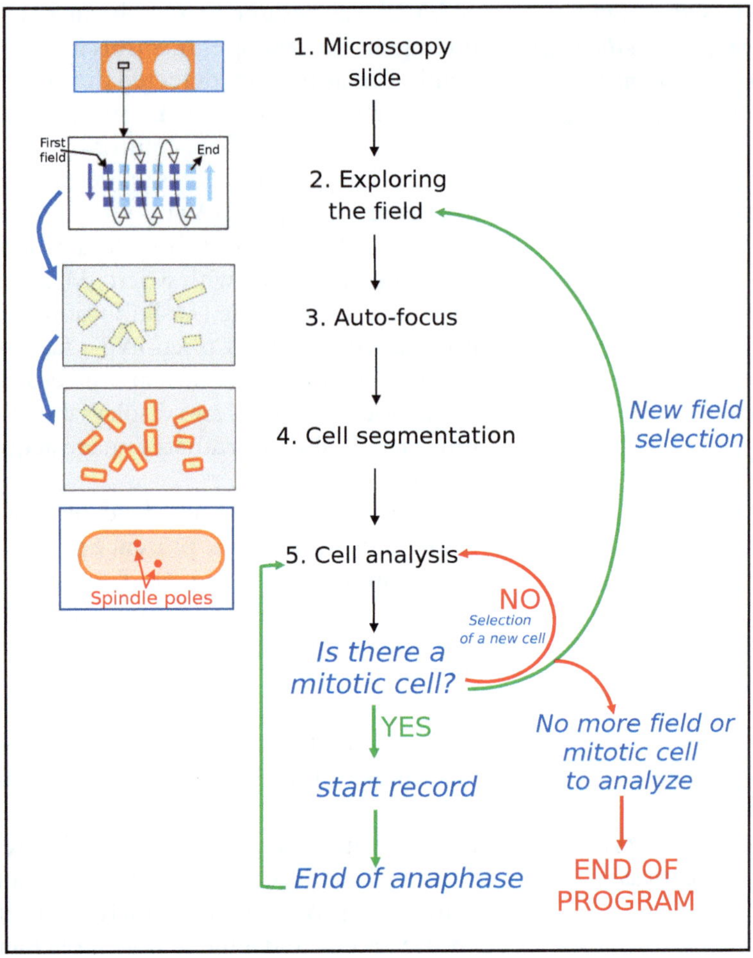

Fig. 2 Diagram of the main steps of MAARS software. The system conditions are set by the user. The system changes fields when there are no more mitotic cells to analyze and stops when there are no more fields to analyze (*see* text for details)

focus. To ensure that the images can be analyzed, an autofocus step is performed before acquisition of fluorescence channels. With these fluorescence channel images, MAARS will find not only the cells of interest but also the subcellular components within the cells (*see* Subheading 3.4.4).

The parameters that can be analyzed with MAARS are summarized in Fig. 3. Pole-to-pole distance and kinetochore localization are among the features that can be extracted from each time point (*see* Subheading 3.4.4) and are used to detect abnormalities such as chromosome lagging. If an abnormal condition has been previously defined by the user (in this case the presence of a lagging chromosome). MAARS also provides a way of visualizing the cell before the next time point is acquired. However, it should be noted that the

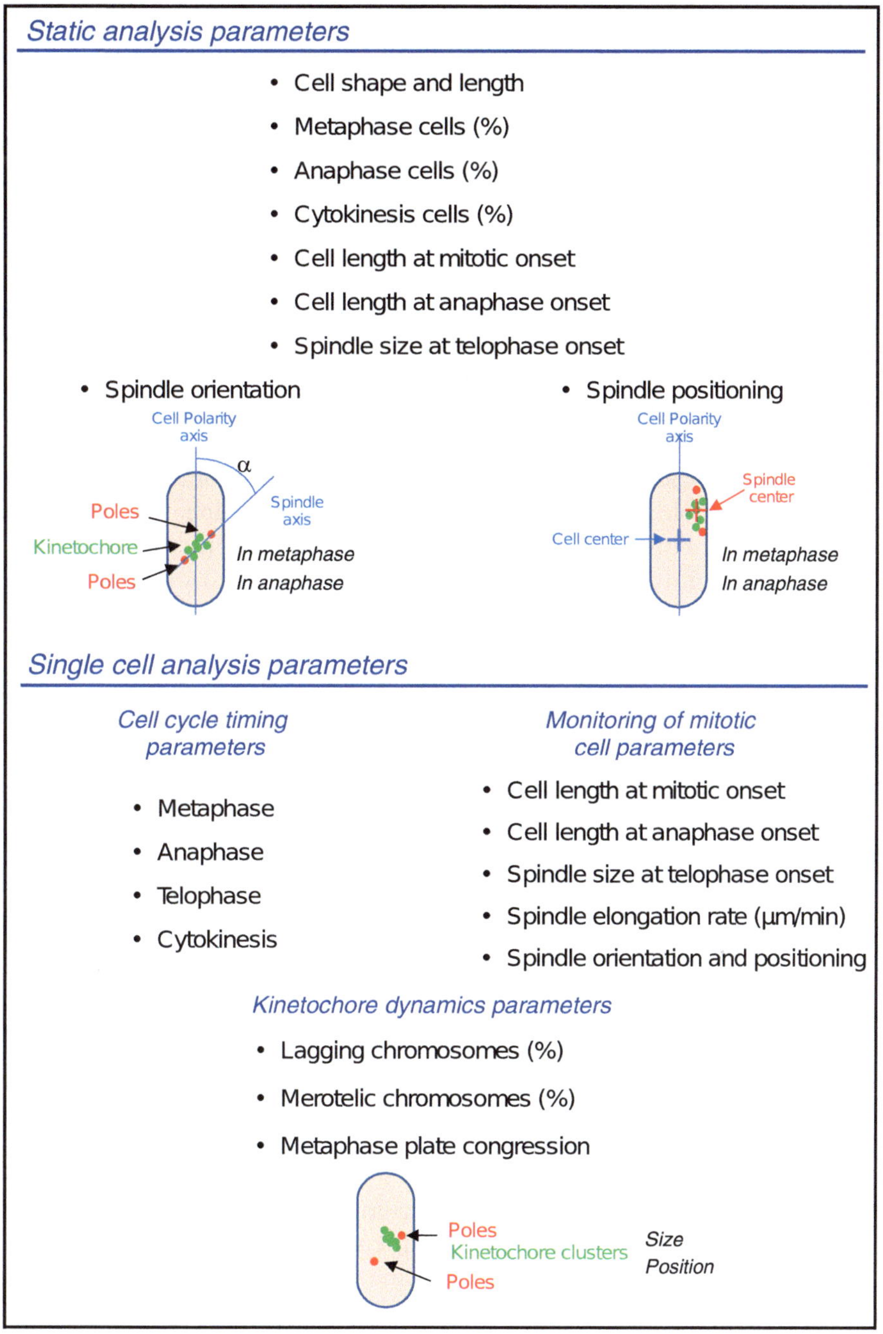

Fig. 3 List of parameters that can be analyzed by the MAARS software. The parameters are analyzed either in static or dynamic mode

ability to display these parameters on-the-fly along with the acquisition requires a certain amount of computing resources. In this case, a workstation with 32Gb of RAM and a Xeon CPU was used to ensure an optimized user experience.

3.4.3 Object Segmentation

The segmentation method used to define the outlines of fission yeast cells is adapted from a method for segmenting *P. aeruginosa* cells [13], which was also based on transmitted light segmentation without the use of fluorescent labels. The principle of this correlation imaging technique is that under bright-field illumination, pixels located at the cell wall of *S. pombe* changes from dark to bright as the sample moves away from the objective. This dark-to-bright intensity variation follows the inverse derivative of a Gaussian (Fig. 4a). This variation is reflected in 34 bright-field images (17 images below and 17 images above the longitudinal axis of the cell) taken with a Z-step of 0.3 μm. Therefore, for any XY position, if the pixel intensity variation follows this curve, it is highly likely that the corresponding position is at the cell wall. This similarity can be quantified as a correlation score between the theoretical curve and the observed one. With a simple thresholding of this score, all cells are delineated, thus, segmented. Finally, Fiji's Particle Analyzer plugin returns polygonal regions of interest (ROIs) based on the segmented objects, and these ROIs are considered as potential "cells" (Fig. 4a).

Cell segmentation may be disturbed in regions with more than one cell layer, as cell wall delineation is unable to separate cells that cross each other. The solidity parameter, which describes the irregularity of each ROI, is chosen to filter out ROIs that may not be accurately segmented. The value of 0.85 was chosen based on our empirical experience, but the user can change this threshold if irregular cell shapes are indeed expected. For example, object (a) in Fig. 4b, which corresponds to multiple cells, has a solidity of 0.6 (not very compact) and object (b), which corresponds to a single cell, has a solidity of 0.9 (compact). Thus, the solidity for compact objects is close to 1. Figure 4c illustrates the adaptability of the method to different morphologies of fission yeast mutants with aberrant cell shape (*see* **Note 8**). Figure 4d shows the performance of MAARS in segmenting an entire fission yeast microscope field.

3.4.4 Feature Extraction/Selection

1. Spot detection.

   MAARS applies a maximum Z-projection of the images prior to spot detection to maximize the chance of capturing fluorescent spots in focus. Thus, all analyzes on fluorescence are performed in 2D. Kinetochores and spindle poles, tagged with GFP/CFP, respectively, display "blob"-shaped single dots under a microscope. It is crucial to faithfully detect their relative position for each time point in order to extract accurate biological description of the cell. TrackMate, a Fiji plugin module called SpotDetector, was used for this spot segmentation task. More importantly, it allows fast and interactive fine-tuning of parameters to optimize spot detection and avoid false positives. Of all the features it calculates, these two are the most

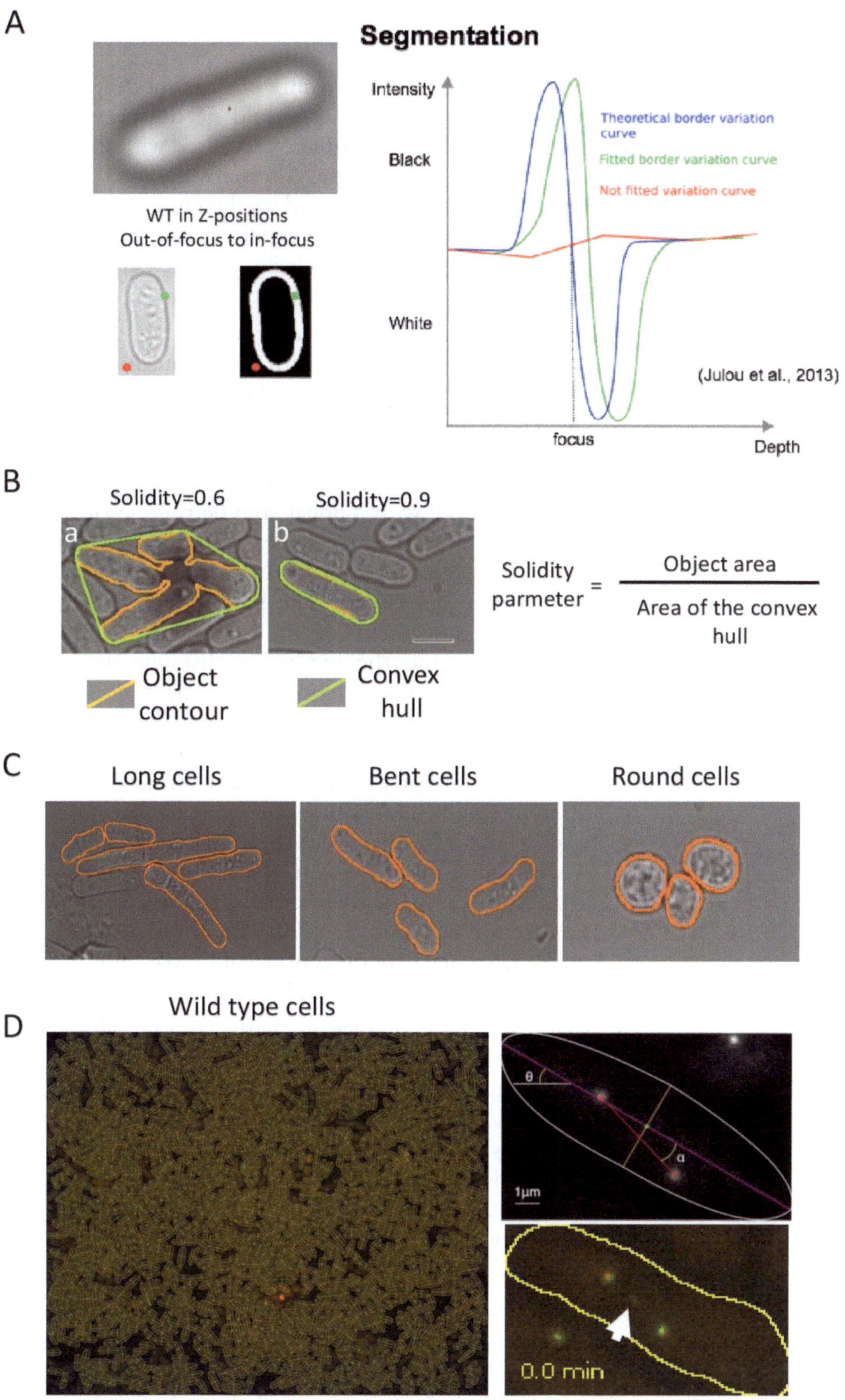

Fig. 4 Illustration of the steps of the segmentation method. (**a**) The cell contour changes in intensity as the Z changes. The green dot indicates the position of a pixel belonging to the cell contour, while the red dot indicates the position of a pixel not belonging to the cell contour (background). The graph on the right represents a relative pixel intensity (intensity measured for one pixel divided by the average of the intensities of all measured pixels) as a function of the focal plane. The black dotted vertical line marks the focal plane

critical: spot radius and quality of the spot detection. To select these parameters, the user would need a previously acquired sample of similar image quality, acquired with the same imaging setting on the microscope. Alternatively, the "Test" button in the MAARS GUI can be used on the same sample to be imaged just before the real run. It is expected that these parameters will need to be adjusted for each fluorescence channel and experiment to ensure consistent spot detection.

2. Mitotic cells and phases detection.

   In fission yeast, the length of the mitotic spindle is determined by the rectangular area between the two spindle poles. The relative position of the two poles and the distance between them can be easily calculated using the spot detection described in the last section. In this study, a cell undergoing mitosis is defined as a cell with a linear spindle pole separation. In practice, the first step is to detect cells that show at least two fluorescent signals within a ROI for at least 3 min (to avoid false-positive detection). The second step is to measure the directional movement of the spindle poles for the entire image series, and the increasing distance between the two SPBs is called spindle elongation. The trajectory of spindle elongation is then divided into phases between the minimum and maximum distance, namely, prophase, prometaphase, metaphase, and anaphase. Different mutant strains will show different spindle elongation curves. It is, therefore, difficult to have a predefined model for detecting mitosis. Instead, MAARS applies a linear fit to the elongation curve and returns its p-value as a metric to reflect its linearity, which is usually not the case for non-mitotic cells. This thresholding is applied to reduce false-negative detections and favor false-positive detections that can be eliminated later.

3. Anaphase onset detection with spindle elongation slope change.

   Changes in the slope of a spindle elongation curve can be used to detect the onset of prophase, the onset of anaphase B, and the event of merotelic attachment correction. A 3 min time window is shifted across the spindle elongation curve. Any abrupt increase in slope reflects a sudden pole-to-pole elongation. Any values greater than a predefined threshold may

Fig. 4 (continued) (the one for which the cell image is sharp). (**b**) Explanation of the solidity parameter. Solidity is close to 1 for compact objects (*see* text for details). (**c**) Examples of cells of different shape and size segmented by the implemented method. (**d**) Left. Example of a whole field of cells segmented with the implemented method. Right. Different mitotic parameters are then analyzed as shown in the images, including the presence of a lagging chromosome (white arrow) indicate an event listed above. The onset of anaphase B, as the critical time point for subsequent analysis, usually corresponds to the time of maximum slope change with 3 min windows. Longer or shorter windows may result in noisy detection.

4. Lagging chromosomes detection after anaphase B onset.

   In a wild-type cell, all kinetochores should co-localize with the SPBs during anaphase. Otherwise, the cell is considered to have a lagging chromosome. MAARS searches for cells with any kinetochore signal (Ndc80-GFP) located between the two spindle poles during anaphase B progression. Cells with a detected lagging chromosome are displayed in a pop-up window in the MAARS GUI at the end of the analysis.

5. Manual data cleaning.

   The entire automated recognition process is not error-free. Therefore, some semiautomated manual work is required. The first type of error is the presence of telophase cells in the analyzed population. In fact, the two spindle poles of cells in telophase may move away from each other by chance or due to inaccurate spot detection. In either case, it is recommended to go back to the raw image to confirm/reject the detection. Since MAARS stores all cropped images of all cells, this step can be as simple as opening a file in Fiji and doing a visual inspection.

3.5 Examples of Quantitative Mitosis Analysis with MAARS

3.5.1 Static Analysis of Mitotic Phases in Wild-Type and Mutant Cells

In fission yeast, kinetochores are anchored to SPBs during G2 (Fig. 5a). At the onset of mitosis, they leave the spindle poles (Fig. 5b, black circle) to align at the spindle midzone in metaphase [11] (Fig. 5b, red circle). In anaphase A and B, following cohesin degradation, kinetochores move back to the SPBs (Fig. 5b, pink circles). Quantitative analysis of kinetochore fluorescence intensity at the SPBs as a function of spindle length using MAARS allows visualization of these different steps in wild-type and mutant cells (Fig. 5b, c). In agreement with previous works [28, 29], in the absence of Klp6 (fission yeast Kinesin-8), a highly processive motor known to regulate microtubule dynamics [30, 31], we observe a population of cells whose kinetochores have not fully moved to the spindle poles (Fig. 5c, red circle), representing a prolonged metaphase and the presence of lagging kinetochores.

3.5.2 Static and Single Live Cell Analysis of Spindle Orientation in Wild-Type and Mutant Cells

MAARS is able to quantify spindle orientation during mitotic progression (Figs. 6 and 7). Spindle orientation is defined as the minimum angle between spindle and cell major axis, which may reflect issues of spindle positioning (Fig. 6a).

As shown in Fig. 6b, a static analysis of spindle orientation reveals a correlation between the angle of the mitotic spindle (angle to the longitudinal axis of the cell) and its size. The angle values obtained in vivo as a function of spindle size appear to be well below the theoretical values of angles based on the contour of the

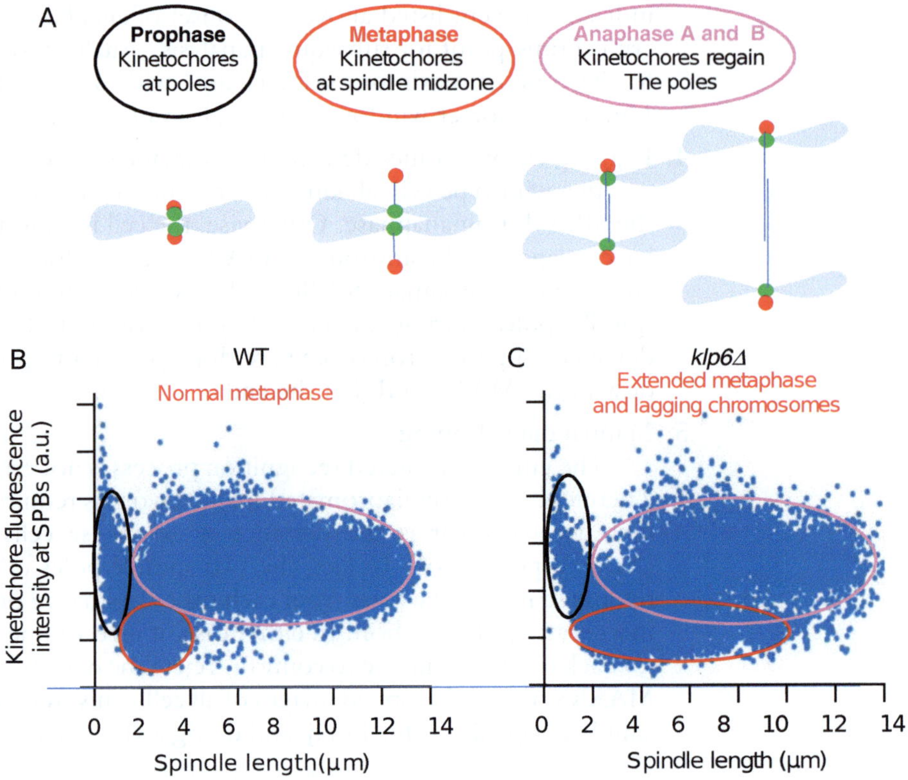

Fig. 5 Analysis of SPB kinetochore positioning during mitosis using MAARS. (**a**) Schematic illustration of the position of kinetochores relative to SPBs during mitotic progression (*see* text for details). (**b**). Analysis of the relative intensity of kinetochores at spindle poles as a function of spindle size in wild-type cells. (**c**) Analysis of the relative intensity of kinetochores at spindle poles as a function of spindle size in the Kinesin-8 mutant, *klp6Δ*. Different cell populations appear corresponding to the stages of mitosis schematized in (A). Each dot corresponds to a mitotic cell. Note the appearance of a population of mitotic cells with either extended metaphase or lagging kinetochores (i.e., kinetochores not at spindle poles in long spindles)

cells (red and gray lines), confirming that an active mechanism is involved in spindle orientation in fission yeast [32]. However, dynamic analysis of these processes with MAARS is more informative because it simultaneously reveals the dynamics of orientation and mitotic spindle elongation in single cells. As shown in Fig. 4c, the spindle angle is relatively stable throughout metaphase in wild-type cells, whereas in the absence of astral microtubules (*mod20* mutant, blue) [33], the orientation of the spindle is disturbed during metaphase. In both cases, wild type and mutant, the spindle orients at the metaphase to anaphase transition (red line).

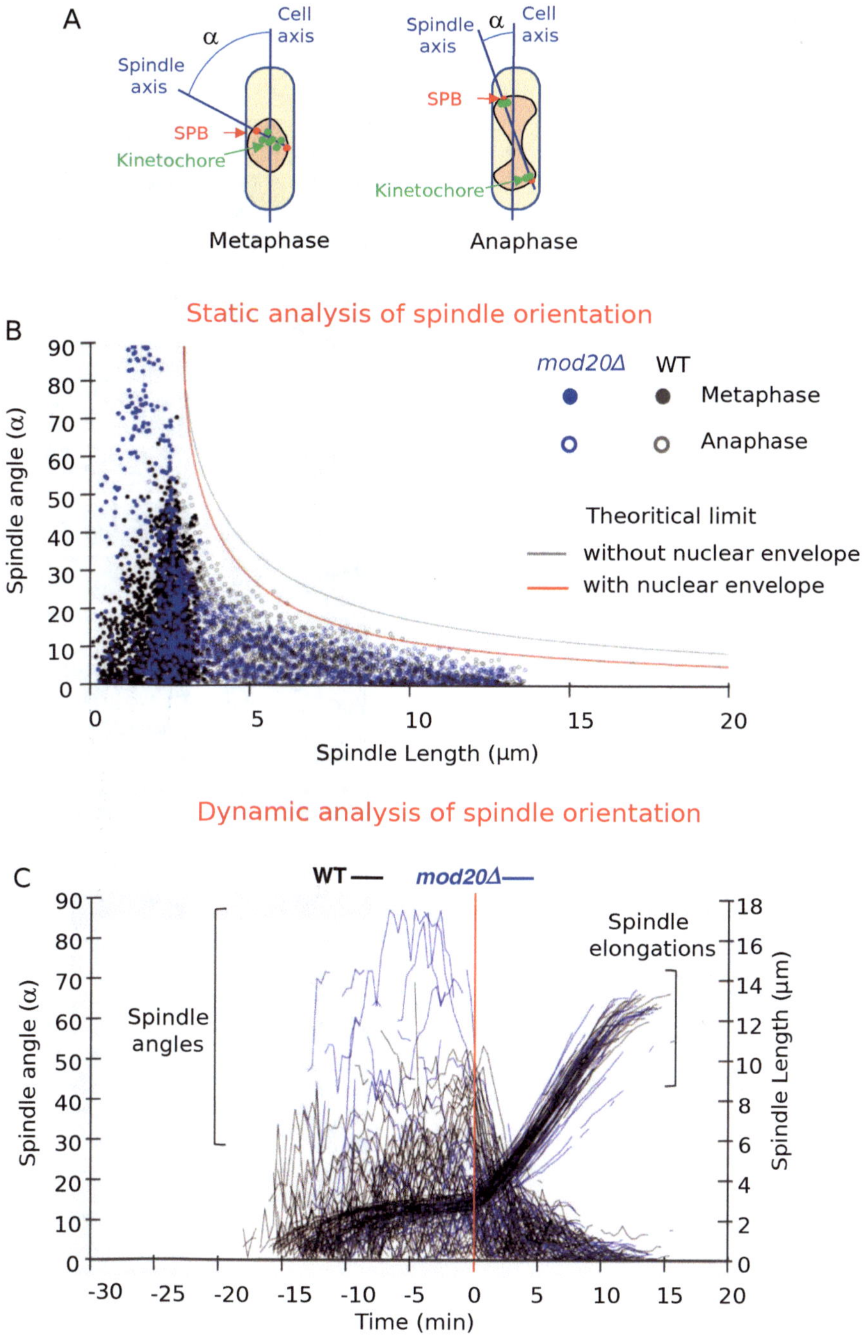

Fig. 6 Static and dynamic analysis of spindle orientation during mitosis using MAARS. (**a**) Schematic of spindle orientation measurement during mitotic progression. The nuclear envelope is shown in black, while SPBs are shown in red and kinetochores in green. MAARS determines the angle between the spindle axis and the cell polarity axis. (**b**) Static analysis of spindle orientation as a function of spindle length in wild-type (black) and mutant (*mod20Δ*, blue) cells. Each point corresponds to a mitotic cell. The gray curve represents the theoretical physical limit of spindle orientation as a function of spindle length when constrained by the cell contours. The red curve represents the theoretical physical limit of spindle orientation as a function of spindle length when constrained by the presence of the nuclear envelope within the cell (closed mitosis). (**c**) Dynamic analysis of spindle orientation as a function of spindle length in wild-type (black) and mutant (*mod20Δ*, blue) cells. Each line corresponds to a mitotic cell followed from metaphase to anaphase. The transition from metaphase to anaphase is indicated by the red line (time 0)

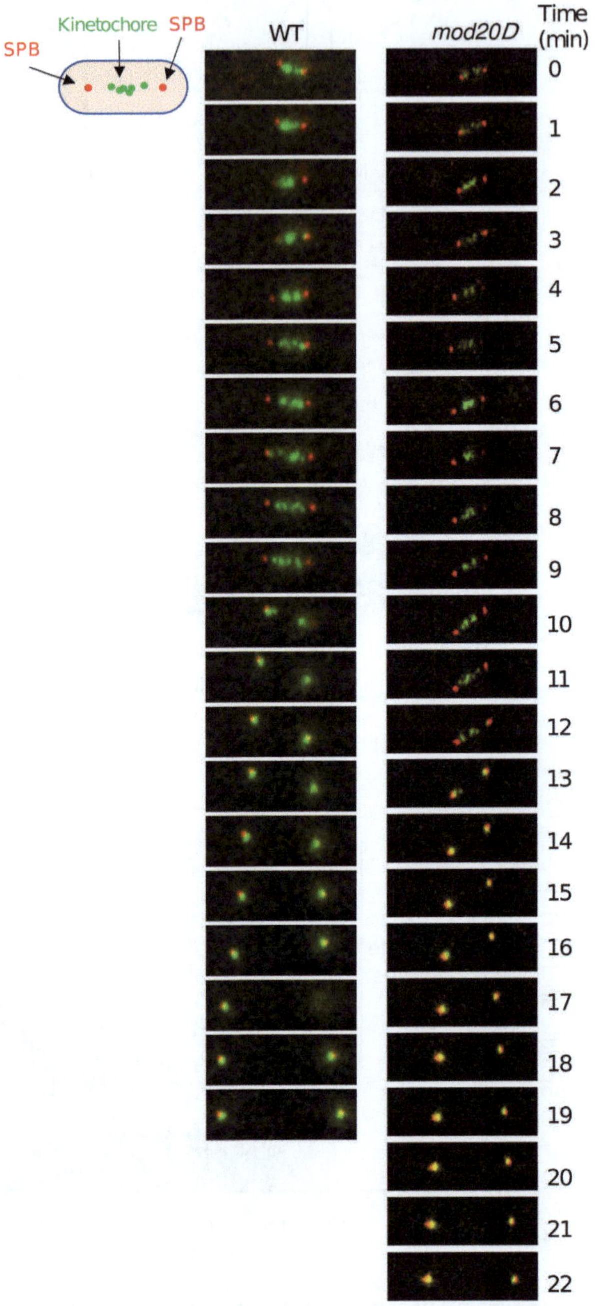

Fig. 7 Example of a wild-type and *mod20Δ* cell imaged during mitotic progression using MAARS. SPBs are shown in red and kinetochores in green

4 Notes

1. It is essential to use freshly thawed cells and fresh YE dishes as the fluorescent signal is highly dependent on growth conditions.
2. If the agarose is poured too hot, the cells may move on the slide during image acquisition.
3. It is important to remove excess moisture or agarose on the imaging chamber to prevent cells from moving during image acquisition. However, be careful not to press too hard with the coverslip on the plug so that the surface remains as smooth as possible.
4. The agarose needs to dry, but not too much, to avoid air bubbles in the final mounting. When your slide is ready, remember to check under the microscope for the presence of a single layer of cells.
5. This step allows the cells to adapt to the minimal medium with agarose and to stabilize them on the slide.
6. Theoretically, it should be possible to take bright-field images at each time point for cell tracking. However, due to the time limit of 35 s, this is difficult. Thus, only one segmentation is taken for each acquisition, and it is therefore important that the mounting of the cells as described in Subheading 3.2 is optimized so that they remain immobile during acquisition.
7. The intensity of the lamp should be minimized to avoid photobleaching and phototoxicity.
8. The adaptability of the segmentation method offers the possibility to analyze fission yeast mutants with altered shape or even other yeast species such as budding yeast, as shown in [3]. However, the decision tree used to implement the data filtering is built from *S. pombe* images. It is therefore preferable to construct species-specific decision trees for cells whose morphology is very different from that of fission yeast. Alternatively, you can use a deep learning-based solution such as Omnipose [21].

Acknowledgements

We would like to thank the *S. pombe* community for supplying strains. Tong Li was funded by la Fondation ARC. This work was funded by the ANR-blanc120601"Chromocatch" and the plan Cancer 2009–2013 "Systems Biology." Microscopy equipment was funded by the CNRS, l'ANR-blan120601 and l'Association de la Recherche sur le Cancer (ARC). The authors declare no competing interests.

References

1. Mattiazzi Usaj M, Styles EB, Verster AJ et al (2016) High-content screening for quantitative cell biology. Trends Cell Biol 26:598–611. https://doi.org/10.1016/j.tcb.2016.03.008
2. Conrad C, Wunsche A, Tan TH et al (2011) Micropilot: automation of fluorescence microscopy-based imaging for systems biology. Nat Methods 8:246–249. https://doi.org/10.1038/nmeth.1558
3. Li T, Mary H, Grosjean M et al (2017) MAARS: a novel high-content acquisition software for the analysis of mitotic defects in fission yeast. Mol Biol Cell 28:1601–1611. https://doi.org/10.1091/mbc.E16-10-0723
4. Nurse P, Thuriaux P (1980) Regulatory genes controlling mitosis in the fission yeast Schizosaccharomyces pombe. Genetics 96:627–637
5. Nurse P (1975) Genetic control of cell size at cell division in yeast. Nature 256:547–551
6. Hirano T, Funahashi S, Uemura T, Yanagida M (1986) Isolation and characterization of Schizosaccharomyces pombe cutmutants that block nuclear division but not cytokinesis. EMBO J 5:2973–2979
7. Yanagida M (1998) Fission yeast cut mutations revisited: control of anaphase. Trends Cell Biol 8:144–149
8. Kohli J, Hottinger H, Munz P et al (1977) Genetic mapping in SCHIZOSACCHAROMYCES POMBE by mitotic and meiotic analysis and induced Haploidization. Genetics 87: 471–489
9. Nabeshima K, Nakagawa T, Straight AF et al (1998) Dynamics of centromeres during metaphase-anaphase transition in fission yeast: Dis1 is implicated in force balance in metaphase bipolar spindle. Mol Biol Cell 9:3211–3225
10. Tatebe H, Goshima G, Takeda K et al (2001) Fission yeast living mitosis visualized by GFP-tagged gene products. Micron 32:67–74
11. Mary H, Fouchard J, Gay G et al (2015) Fission yeast kinesin-8 controls chromosome congression independently of oscillations. J Cell Sci 128:3720–3730. https://doi.org/10.1242/jcs.160465
12. Yanagida M (2000) Cell cycle mechanisms of sister chromatid separation; roles of Cut1/separin and Cut2/securin. Genes Cells 5:1–8
13. Julou T, Mora T, Guillon L et al (2013) Cell-cell contacts confine public goods diffusion inside Pseudomonas aeruginosa clonal microcolonies. Proc Natl Acad Sci U S A 110: 12577–12582. https://doi.org/10.1073/pnas.1301428110
14. Peng JY, Chen YJ, Green MD et al (2013) PombeX: robust cell segmentation for fission yeast transillumination images. PLoS One 8: e81434. https://doi.org/10.1371/journal.pone.0081434
15. Baybay EK, Esposito E, Hauf S (2020) Pomegranate: 2D segmentation and 3D reconstruction for fission yeast and other radially symmetric cells. Sci Rep 10:16580. https://doi.org/10.1038/s41598-020-73597-w
16. Wang L, Li S, Sun Z et al (2018) Segmentation of yeast cell's bright-field image with an edge-tracing algorithm. J Biomed Opt 23:1–7. https://doi.org/10.1117/1.JBO.23.11.116503
17. Liu G, Dong F, Fu C, Smith ZJ (2019) Automated morphometry toolbox for analysis of microscopic model organisms using simple bright-field imaging. Biol Open 8:bio037788. https://doi.org/10.1242/bio.037788
18. Stringer C, Wang T, Michaelos M, Pachitariu M (2021) Cellpose: a generalist algorithm for cellular segmentation. Nat Methods 18:100–106. https://doi.org/10.1038/s41592-020-01018-x
19. Bannon D, Moen E, Schwartz M et al (2021) DeepCell kiosk: scaling deep learning-enabled cellular image analysis with Kubernetes. Nat Methods 18:43–45. https://doi.org/10.1038/s41592-020-01023-0
20. Weigert M, Schmidt U, Haase R et al (2020) Star-convex polyhedra for 3D object detection and segmentation in microscopy. In: 2020 IEEE Winter Conference on Applications of Computer Vision (WACV). IEEE, Snowmass Village, CO, USA, pp 3655–3662
21. Cutler KJ, Stringer C, Lo TW et al (2022) Omnipose: a high-precision morphology-independent solution for bacterial cell segmentation. Nat Methods 19:1438–1448. https://doi.org/10.1038/s41592-022-01639-4
22. Tournier S, Gachet Y, Buck V et al (2004) Disruption of astral microtubule contact with the cell cortex activates a Bub1, Bub3, and Mad3-dependent checkpoint in fission yeast. Mol Biol Cell 15:3345–3356
23. Bahler J, Wu JQ, Longtine MS et al (1998) Heterologous modules for efficient and versatile PCR-based gene targeting in Schizosaccharomyces pombe. Yeast 14:943–951
24. Schindelin J, Arganda-Carreras I, Frise E et al (2012) Fiji: an open-source platform for biological-image analysis. Nat Methods 9:

676–682. https://doi.org/10.1038/nmeth.2019

25. Tinevez JY, Perry N, Schindelin J et al (2016) TrackMate: an open and extensible platform for single-particle tracking. Methods. https://doi.org/10.1016/j.ymeth.2016.09.016
26. Linkert M, Rueden CT, Allan C et al (2010) Metadata matters: access to image data in the real world. J Cell Biol 189:777–782. https://doi.org/10.1083/jcb.201004104
27. Edelstein A, Amodaj N, Hoover K et al (2010) Computer control of microscopes using micro-Manager. Curr Protoc Mol Biol Chapter 14 (Unit14):20. https://doi.org/10.1002/0471142727.mb1420s92
28. Garcia MA, Koonrugsa N, Toda T (2002) Two kinesin-like Kin I family proteins in fission yeast regulate the establishment of metaphase and the onset of anaphase A. Curr Biol 12:610–621
29. West RR, Malmstrom T, McIntosh JR (2002) Kinesins klp5(+) and klp6(+) are required for normal chromosome movement in mitosis. J Cell Sci 115:931–940
30. Gupta ML Jr, Carvalho P, Roof DM, Pellman D (2006) Plus end-specific depolymerase activity of Kip3, a kinesin-8 protein, explains its role in positioning the yeast mitotic spindle. Nat Cell Biol 8:913–923. https://doi.org/10.1038/ncb1457
31. Mayr MI, Hummer S, Bormann J et al (2007) The human kinesin Kif18A is a motile microtubule depolymerase essential for chromosome congression. Curr Biol 17:488–498. https://doi.org/10.1016/j.cub.2007.02.036
32. Gachet Y, Tournier S, Millar JB, Hyams JS (2004) Mechanism controlling perpendicular alignment of the spindle to the axis of cell division in fission yeast. EMBO J 23:1289–1300
33. Zimmerman S, Chang F (2005) Effects of gamma-tubulin complex proteins on microtubule nucleation and catastrophe in fission yeast. Mol Biol Cell 16:2719–2733. https://doi.org/10.1091/mbc.E04-08-0676

Index

Anna Castro and Benjamin Lacroix (eds.), *Cell Cycle Control: Methods and Protocols*, Methods in Molecular Biology, vol. 2740, https://doi.org/10.1007/978-1-0716-3557-5,